The growing plant cell wall: chemical and metabolic analysis

Stephen C. Fry

Department of Botany, University of Edinburgh

The growing plant cell wall: chemical and metabolic analysis

Longman
Scientific &
Technical

Copublished in the United States with
John Wiley & Sons, Inc., New York

Longman Scientific & Technical
Longman Group UK Limited
Longman House, Burnt Mill, Harlow
Essex CM20 2JE, England
and Associated Companies throughout the world.

Copublished in the United States with
John Wiley & Sons, Inc., 605 Third Avenue, New York, NY 10158

First published 1988

British Library Cataloguing in Publication Data
Fry, Stephen C. (Stephen Charles), 1953–
The growing plant cell wall.
1. Plants. Cells. Walls. Biochemistry.
I. Title
581.8′75

ISBN 0-582-01897-8

Library of Congress Cataloging-in-Publication Data
Fry, Stephen C., 1953–
    The growing plant cell wall.
    (Monographs and surveys in the biosciences)
    Bibliography: p.
    Includes index.
    1. Plant cell walls—Analysis—Laboratory manuals.
I. Title. II. Series.
QK725.F79   1988   581.87′5   88-587

ISBN 0-470-21079-6 (Wiley, USA only)

Printed and bound in Great Britain by
Biddles Ltd, Guildford and King's Lynn

Contents

List of Panels

Chapter 4 Wall polymers: chemical characterisation

Chapter 5 Wall biosynthesis

Chapter 6 Wall enzymes

Chapter 7 Wall architecture

Chapter 8 Wall turnover and sloughing

Chapter 9 Biologically-active wall oligosaccharides

Chapter 10 Wall tightening and loosening: growth

Conventions and abbreviations

CONVENTIONS

In vivo = in living cells, regardless of whether these have been grown in a field or a Petri dish.

In vitro = after extraction from cells.

Imino acids (Hyp and Pro) are included as 'amino acids'.

Cereals are included as 'grasses'.

Chromatography solvents are by volume unless otherwise stated.

% Concentrations of solutions are w/v [g solute / 100 ml of solution] if the solute is a solid at 20°C, and v/v [ml solute / 100 ml solution] if it is a liquid.

ABBREVIATIONS

A_{280}	absorbance at 280 nm
ABA	abscisic acid
AceA	aceric acid [3-C-carboxy-5-deoxy-L-xylose]
AGP	arabinogalactan protein
AIR	alcohol-insoluble residue
Ala	L-alanine
All	D-allose
AOPP	amino-oxyphenylpropionate
Api	D-apiose
Ara	L-arabinose
AraH	L-arabinitol
Arg	L-arginine
Asn	L-asparagine
Asp	L-aspartic acid
BAW	BuOH/HOAc/H_2O (12:3:5)
BSA	bovine serum albumin
BuOH	butan-1-ol
Bz	Benzene
C-	carbon-linked
CDTA	trans-1,2-diaminocyclohexane-N,N,N',N'-tetra-acetate
Ci	curie
CMC	carboxymethylcellulose
CTAB	cetyltrimethylammonium bromide
D-	optical isomer [see p 50]
DEAE	diethylaminoethyl
DMSO	dimethylsulphoxide
DP	degree of polymerisation
dpm	decays per minute
EDTA	ethylenediaminetetra-acetate
EEO	electro-endo-osmosis
EGTA	ethyleneglycol-bis-(β-aminoethyl ether)-N,N,N',N'-tetra-acetate
EryH	erythritol
EtOAc	ethyl acetate
EtOH	ethanol
-f	furanose
Fer	feruloyl
Fru	D-fructose
Fuc	L-fucose
GA_3	gibberellic acid
Gal	D-galactose
GalA	D-galacturonic acid
GalH	galactitol
GalN	D-galactosamine
GalNAc	N-acetyl-D-galactosamine
GC	gas chromatography [gas-liquid chromatography]
Glc	D-glucose
GlcA	D-glucuronic acid
GlcH	D-glucitol [D-sorbitol]
GlcN	D-glucosamine
GlcNAc	N-acetyl-D-glucosamine
Gln	L-glutamine
Glu	L-glutamic acid
GPC	gel-permeation chromatography
HEPES	N-2-hydroxyethylpiperazine-N'-2-ethanesulphonate
His	L-histidine
HOAc	acetic acid
HPLC	high-pressure liquid chromatography
Hyp	L-hydroxyproline
IAA	indol-3-ylacetic acid

Ino	myo-inositol	PIPES	piperazine-N,N'-bis(2-ethane-sulphonate)
KDO	3-deoxy-manno-octulosonic acid ['ketodeoxyoctulo-sonic acid']	PITC	phenylisothiocyanate
		pNP	p-nitrophenol (or -yl)
L-	optical isomer [see p 50]	POPOP	1,4-di-2-(5-phenyl-oxazolyl)-benzene
LCC	lignin---carbohydrate complex	PP$_i$	inorganic pyrophosphate
Leu	L-leucine	PPO	2,5-diphenyloxazole
LSC	liquid scintillation-counting	Pro	L-Proline
		PrOH	propan-1-ol
Lys	L-lysine	psi	pounds per square inch
Lyx	D-lyxose	QAE	quaternary aminoethyl
m-	meta	R$_F$	chromatographic mobility relative to solvent front
m̄	electrophoretic mobility		
M̄an	D-mannose	RG	rhamnogalacturonan
ManH	D-mannitol	Rha	L-rhamnose
Me	methyl	RI	refractive index
MeFuc	2-O-methyl-L-fucose	Rib	D-ribose
MES	2-(N-morpholino)ethanesul-phonate	RibH	ribitol
		R$_{Rha}$	chromatographic mobility relative to L-rhamnose
Met	L-methionine		
MeXyl	2-O-methyl-D-xylose	S-	sulphur-linked
MMNO	N-methylmorpholine-N-oxide	S̄A	specific (radio)activity
m̄$_{pic}$	electrophoretic mobility relative to picrate	SAM	S-adenosyl-L-methionine
		Ser	L-serine
M$_r$	'molecular weight'	SDS	sodium dodecyl (lauryl) sulphate
N-	nitrogen-linked		
N̄aOMe	sodium methoxide	t½	half-life
NDP	nucleoside 5'-diphosphate	TC̄A	trichloroacetic acid
NTP	nucleoside 5'-triphosphate	TFA	trifluoroacetic acid
O-	oxygen-linked	ThrH	threitol
ō-	ortho	Thr	L-threonine
p̄-	para	TLC	thin-layer chromatography
-p	pyranose	TMG	2,3,4,6-tetra-O-methyl-D-glucose
PĀHBAH	p-hydroxybenzoic acid hydrazide		
		Tyr	L-tyrosine
PAL	phenylalanine ammonia-lyase	UDP	uridine 5'-diphosphate
		UMP	uridine 5'-monophosphate
PAW	PhOH/HOAc/H$_2$O (2:1:1, w/v/v)	UTP	uridine 5'-triphosphate
		UV	ultra-violet
PC	paper chromatography	v	volume
PCV	packed cell volume	Val	L-valine
PE	paper electrophoresis	V$_0$	void volume [see p 79]
Phe	L-phenylalanine	V$_i$	included volume [see p 80]
PhOH	phenol	w	weight
pI	isoelectric point	Xyl	D-xylose
P$_i$	inorganic phosphate	XylH	xylitol

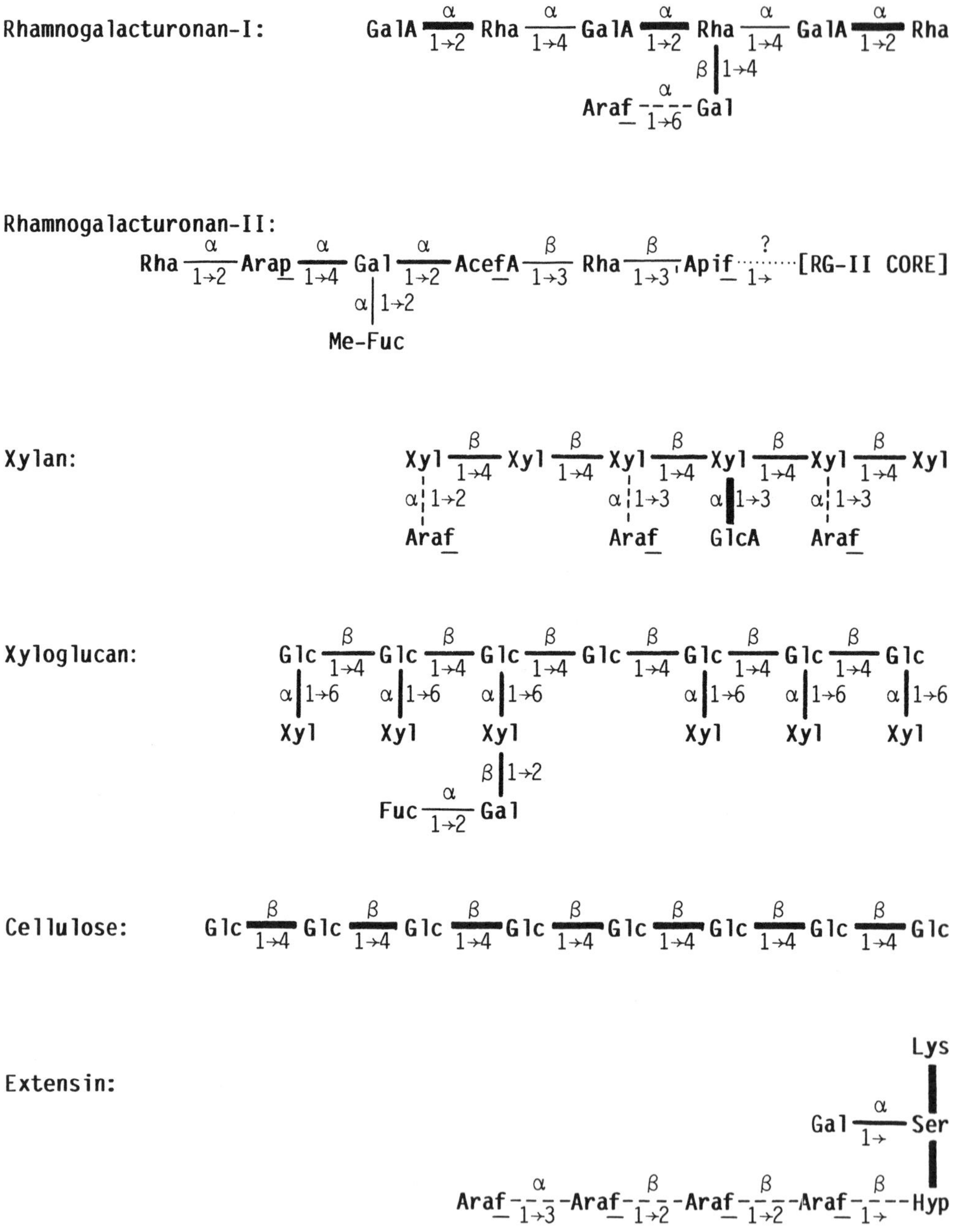

Rhamnogalacturonan-I:

GalA $\xrightarrow[1\to2]{\alpha}$ Rha $\xrightarrow[1\to4]{\alpha}$ GalA $\xrightarrow[1\to2]{\alpha}$ Rha $\xrightarrow[1\to4]{\alpha}$ GalA $\xrightarrow[1\to2]{\alpha}$ Rha

$\beta \big| 1\to4$

Araf $\underset{1\to6}{\text{---}}$ Gal (α)

Rhamnogalacturonan-II:

Rha $\xrightarrow[1\to2]{\alpha}$ Arap $\xrightarrow[1\to4]{\alpha}$ Gal $\xrightarrow[1\to2]{\alpha}$ AcefA $\xrightarrow[1\to3]{\beta}$ Rha $\xrightarrow[1\to3]{\beta}$ Apif $\overset{?}{\cdots\cdots}$ [RG-II CORE]

$\alpha \big| 1\to2$

Me-Fuc

Xylan:

Xyl $\xrightarrow[1\to4]{\beta}$ Xyl $\xrightarrow[1\to4]{\beta}$ Xyl $\xrightarrow[1\to4]{\beta}$ Xyl $\xrightarrow[1\to4]{\beta}$ Xyl $\xrightarrow[1\to4]{\beta}$ Xyl

$\alpha \vdots 1\to2$ $\alpha \vdots 1\to3$ $\alpha \big| 1\to3$ $\alpha \vdots 1\to3$

Araf   Araf   GlcA   Araf

Xyloglucan:

Glc $\xrightarrow[1\to4]{\beta}$ Glc $\xrightarrow[1\to4]{\beta}$ Glc $\xrightarrow[1\to4]{\beta}$ Glc $\xrightarrow[1\to4]{\beta}$ Glc $\xrightarrow[1\to4]{\beta}$ Glc $\xrightarrow[1\to4]{\beta}$ Glc

$\alpha \big| 1\to6$ $\alpha \big| 1\to6$ $\alpha \big| 1\to6$ $\alpha \big| 1\to6$ $\alpha \big| 1\to6$ $\alpha \big| 1\to6$

Xyl   Xyl   Xyl   Xyl   Xyl   Xyl

$\beta \big| 1\to2$

Fuc $\xrightarrow[1\to2]{\alpha}$ Gal

Cellulose:

Glc $\xrightarrow[1\to4]{\beta}$ Glc $\xrightarrow[1\to4]{\beta}$ Glc $\xrightarrow[1\to4]{\beta}$ Glc $\xrightarrow[1\to4]{\beta}$ Glc $\xrightarrow[1\to4]{\beta}$ Glc $\xrightarrow[1\to4]{\beta}$ Glc $\xrightarrow[1\to4]{\beta}$ Glc

Extensin:

Lys

|

Gal $\xrightarrow[1\to]{\alpha}$ Ser

|

Araf $\underset{1\to3}{\overset{\alpha}{\text{---}}}$ Araf $\underset{1\to2}{\overset{\beta}{\text{---}}}$ Araf $\underset{1\to2}{\overset{\beta}{\text{---}}}$ Araf $\underset{1\to}{\overset{\beta}{\text{---}}}$ Hyp

Basic structures of some important cell wall polymers. The relative acid-stability of the bonds is indicated: ▬▬, most stable; ——, ——, - - - -, progressively less stable; ·······, least stable. All sugars are in the pyranose ring-form, except those marked f (furanose).

Preface

This book is mainly a description, in bench-top detail, of techniques for the study of primary cell walls in higher plants. Deepening interest in the chemistry and metabolism of cell walls reflects the breadth of their significance in biology. Walls play crucial rôles in morphogenesis, growth, disease-resistance, recognition, signalling, digestibility, nutrition and decay. Scientists interested in questions concerning these processes need to know about methods of cell wall analysis, and the present book is written with such people in mind. The book emphasises the simplest techniques available to answer the important questions. Simple methods have the great advantage that they are less prone to misinterpretation.

Modern courses in biochemistry often pay scant attention to polymeric carbohydrates and phenolics, the emphasis being on proteins and nucleic acids. Yet the major contribution of carbohydrates and phenolics to the functioning of plant cell walls makes a knowledge of their chemistry vital: I have therefore included an introduction to this chemistry as a basis for the methods described.

The methods are presented in self-contained packages referred to as 'Panels', although any single Panel would rarely be used alone. For this reason, I have provided cross-references to useful preceding and following steps. The botanical materials recommended in the Panels are merely examples known to be suitable; many other plant materials will work. For this reason, plant names are largely excluded from the index.

A book of this nature has scope for an enormous bibliography — the relatively few papers I have selected are neither the earliest nor the latest nor the most definitive: they are generally those in which I consider the Materials and Methods section to be particularly helpful.

Stephen C. Fry
October, 1987

Acknowledgements

To Vreni in recognition of her patience, help and
encouragement throughout the preparation of the text.

I also thank Mrs J.G. Miller for excellent technical assistance and the
Agricultural and Food Research Council for financial support of my work.

1 Introduction to the growing cell wall

1.1: BACKGROUND

1.1.1: Reasons for studying primary cell walls

<u>Intrinsic interest</u>. The primary cell wall is a vital organelle with unusual physical and chemical properties. It is composed of polymers, some of whose structures (and therefore biosyntheses) are among the most complex known [345]. The wall contains enzymes (Ch 6), which are catalytically active <u>in vivo</u>, and it dramatically changes its structure during development. All these factors make the cell wall a fascinating object of study.

<u>Cell shape, size and attachment</u>. The wall governs the shape and size of the cell, and fastens it to its neighbours. This can be seen by observing the effects of enzymic removal of the wall (Section 2.7.6): an isolated protoplast is spherical (the shape adopted by an unconstrained body), its size is a simple function of the osmotic pressure of the bathing medium and it is unattached to its neighbours. The cell wall is thus where one would look to understand the control of plant growth and morphogenesis.

<u>Disease resistance</u>. The wall is a barrier to the entry of pathogens into the cytoplasm. Walls differ in their penetrability, and cells can rapidly bolster their walls in response to attempted invasion [46].

<u>Recognition</u>. Wall metabolism may form the basis of the processes whereby a plant cell can detect the approach of a pathogen. It may also govern plant/plant cell recognition during grafting and fertilisation [230,245,345].

<u>Signalling</u>. The wall contains latent regulatory molecules: certain oligosaccharides, enzymically cleaved from the polysaccharides of the wall, exert specific and potent effects when added to living plant cells [345]. Similar oligosaccharides occur <u>in vivo</u>, where they probably regulate plant growth, development and metabolism [175].

<u>Digestibility</u>. A knowledge of wall structure would aid our ability to predict and alter digestibility. Important situations where walls are digested include

<u>in vivo</u>: in cell separation during fruit ripening and abscission, in the

formation of perforation plates in xylem vessel elements, in the mobilisation of reserves during germination, and (to a small extent: see Ch 8)
wall turnover during cell expansion;

   in vitro: in the laboratory isolation of protoplasts; and

   post mortem: in the rotting of plant litter in soil, the fermentation of
waste plant materials to useful products e.g. ethanol, the nutrition of
livestock, including the preparation of silage, the modification of dietary
fibre in the human colon, and in the commercial preparation of fruit
'juice' by enzymic treatment.

1.1.2 Definitions

     A **cell wall** is a layer of structural material found external to the
protoplast. It is usually 0.1-10 μm thick and composed of polysaccharides,
with smaller amounts of glycoproteins and phenolic compounds. It contrasts
with the **plasma membrane**, which is less than 0.01 μm thick and composed of
phospholipids and glycoproteins.

     A **primary** wall is one whose polysaccharide framework was deposited
during **growth** in cell surface area. Such walls have special properties,
and are justifiably the subject of intensive study. [Growth of a surface
is defined as irreversible increase in area.]

     The primary walls of adjacent cells are usually cemented together
over part of their area. The glue responsible for this may be visible as a
separate layer, the **middle lamella.** However, it is often difficult to
define a precise boundary between the middle lamella and the wall.

     The bulk of a mature plant is **secondary** wall material, i.e. wall
whose polysaccharide framework was deposited after the cessation of surface
growth. Since new wall polysaccharide is deposited immediately outside the
plasma membrane, it follows that the secondary wall is internal to the
primary wall. In cells with very thick secondary walls it may be difficult
to discern the primary wall.

     These definitions stipulate the polysaccharide framework because this
is what gives the wall its basic shape, size and thickness; other components (e.g. lignin) may be infiltrated into an existing wall long after
the deposition of its polysaccharides.

     A polysaccharide molecule could qualify as a component of a primary
wall for either of two reasons: **(a)** because it was deposited into a wall

that was being formed de novo, creating a new boundary between two nuclei which had previously shared a common cytoplasm; and **(b)** because it was deposited into an existing wall that was expanding in area. Type (a) normally occurs immediately after mitosis, but it also occurs during the development of the endosperm, which, upon reaching a certain volume as a coenocyte, is subsequently carved up into mononucleate cells by de novo wall-formation. Type (b) occurs during the growth of a mononucleate cell and may continue after its last cell division. In many cells the two types of surface growth proceed in parallel.

Since plant cells often grow uniaxially, the side walls may continue to increase in area after the end walls have reached their final size. If new polysaccharide were to be deposited on such end walls at the same time as on the expanding side walls, then the same cell would be depositing primary wall material on its sides and secondary on its ends. It has not been established whether these two types of wall differ chemically.

A typical primary wall is about 0.1 μm thick. It is a biphasic structure, consisting of relatively rigid cellulosic **microfibrils** embedded in a gel-like **matrix** composed of non-cellulosic polysaccharides and glyco- proteins. The microfibrils lie in the plane of the cell surface. In some cells, new microfibrils are deposited with a uniform orientation (perpen- dicular to the principal axis of cell elongation); in the outer (= older) wall layers the microfibrils may seem to have been passively reorientated by cell expansion (the multinet growth hypothesis). It has been suggested that the orientation of the microfibrils governs the direction of cell elongation, since microfibrils will allow expansion in a direction that tends to move them apart (= elongation perpendicular to the microfibrils) but not in one that requires them to stretch, or slide past one another, longitudinally. However, in other cells the orientation of microfibril deposition changes rhythmically, resulting in a wall with a **helicoidal** structure [527]. It is unclear how such cells establish a principal axis of elongation.

Newly deposited microfibrils often run parallel to the microtubules that underlie the plasma membrane. It is unknown what orientates the microtubules, or how these may orientate the microfibrils.

Certain expanding cell walls are markedly different from the typical 0.1 μm thick version on which most of the biochemical work has been done.

For instance, the growing cell walls of collenchyma (e.g. the 'strings' in a celery stalk) are often about 10 μm thick; however, the scant evidence available suggests that they have an average chemical composition similar to that of thinner primary walls. In addition, the walls of the epidermis of a growing stem, although clearly expanding in area, are relatively thick and possess a cuticle on the outer surface. Thus cuticularisation does not prevent a wall being primary, as defined here.

1.1.3 Chemical composition and biological function

Primary and secondary cell walls differ in chemical composition. This is not surprising in view of the different biological rôles of the two structures. The primary wall has to resist catastrophic bursting in the face of a turgor pressure of several atmospheres, but at the same time it must be plastic enough to allow an appropriate extent of growth. Secondary walls do not grow; they may take on any of a wide variety of other important biological rôles including defence, support and storage. It is not clear whether it is the deposition of wall material with 'secondary' composition that stops growth, or whether the primary wall undergoes a change to resist further expansion and the cell responds by thereafter laying down only wall material with 'secondary' composition.

Growing primary walls are composed of ca. 65 % water, almost all of which is in the matrix. Water is an important structural component since it is necessary for the formation of gels in the matrix [381].

The polymer composition of the primary walls of grasses differs considerably from that typical of other higher plants:

Polymer		% of dry wt of unlignified primary cell wall	
		Grasses	Others
Cellulose	poly-sac-cha-rides	30	30
Pectin		5	35
Arabinoxylan		30	5
β-(1→3),(1→4)-glucan		30	0?
Xyloglucan		4	25
Extensin	glycoprotein	½	5

The figures are only approximate and are subject to variation between and within plants, but the trends are real. Since the primary walls of grasses and Dicots have similar overall properties, and respond similarly to growth-controlling treatments such as lowered pH, it seems likely that different polymers can serve the same function in different plants. [The alternative would be that certain major polymers are redundant, e.g. pectin in Dicots and arabinoxylan in grasses.]

Certain primary walls contain lignin in addition to the polymers listed above. Lignin, which is hydrophobic, partly replaces the water [381]. Lignification is most prevalent in cells with secondary walls, but it often begins in the primary walls or middle lamellae of these cells.

1.1.4 The formation and growth of primary cell walls

A new wall is usually laid down soon after mitosis, establishing a boundary between the two daughter nuclei. Membrane-bounded vesicles packed with non-cellulosic polysaccharides align themselves at the equator of the mitotic spindle, where they fuse (Fig. 1.1.4 a): their membranes become the two new plasma membranes, and their contents form the middle lamella and matrix of the two daughter cell walls. Soon afterwards, the new plasma membranes begin to extrude cellulose microfibrils, and further matrix poly-saccharides are deposited by exocytosis from Golgi-derived vesicles. It is not clear when or how extensin is first deposited.

Although de novo wall-formation usually occurs as described above, cells that stop dividing do not necessarily lose the ability to form new walls. This can be seen during endosperm development (Section 1.1.2), and also experimentally during wall regeneration by isolated protoplasts in culture (Section 2.7.6) and by protoplasts that, although not isolated, have been shrunken away from their existing cell walls by plasmolysis [65].

For the two daughter cells to gain full independence, the peripheries of the new walls must fuse with their common mother cell wall; the mother wall must then, and only then, be split between the lines of fusion. The splitting seems to involve active lysis of the mother wall (Fig. 1.1.4 b-d) a process that clearly calls for tight spatial and temporal control [265].

A new primary wall can usually grow in area. This expansion is driven by the turgor pressure exerted on the wall by the protoplast within (Ch 10). Turgor pressure is developed by the tendency of the protoplast to take up water by osmosis. The protoplast's osmotic pressure is maintained by biosynthesis of cell metabolites and by active uptake of inorganic ions

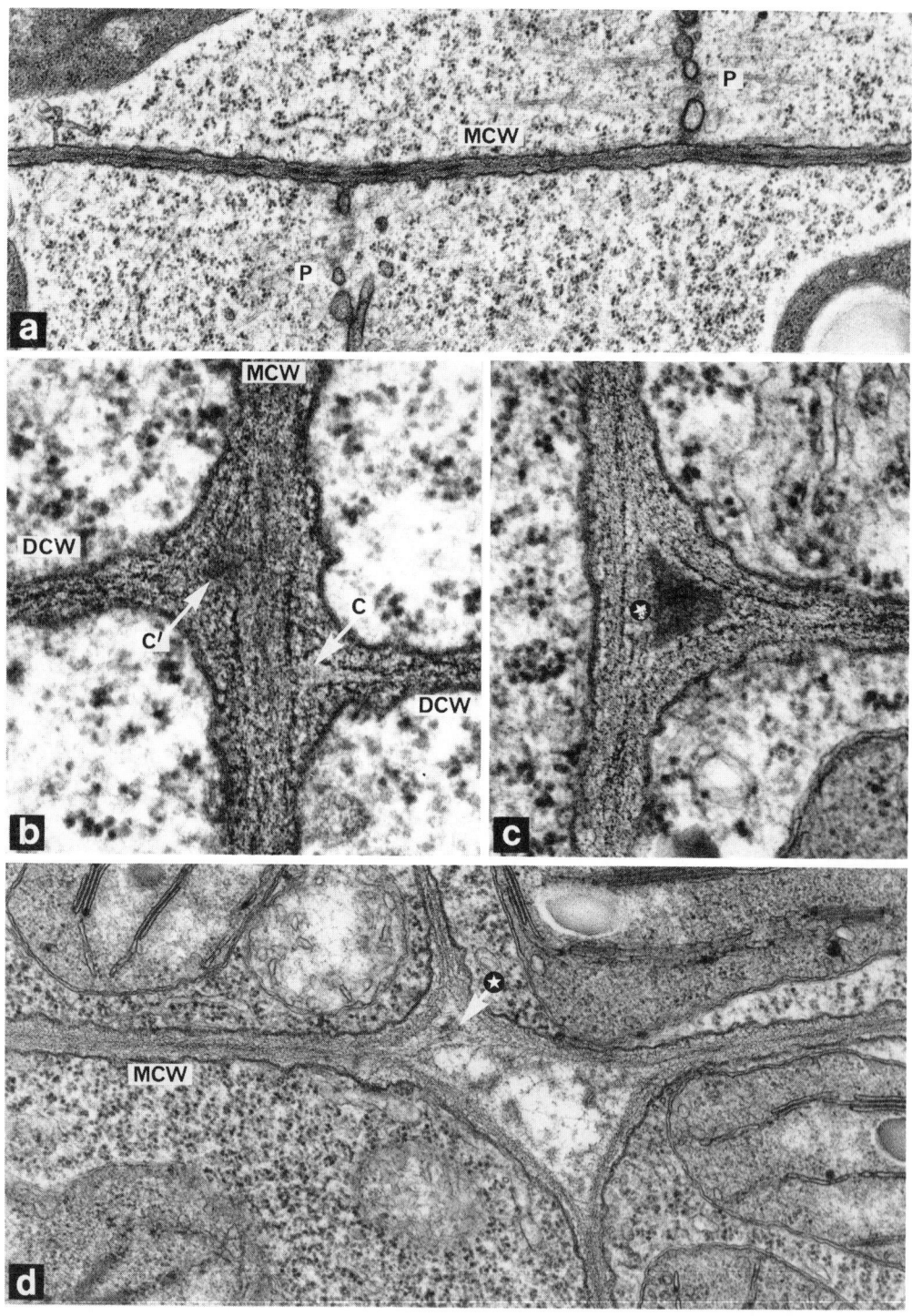

Fig. 1.1.4: Cytokinesis and formation of air spaces (legend on facing page).

e.g. K+. Therefore wall expansion is absolutely dependent on the proto-plast's metabolic activity.

In addition, an expanding wall usually maintains at least its initial thickness, i.e. wall biosynthesis usually accompanies expansion [83]. This imposes a further demand for metabolic energy during wall growth. Wall growth can be maintained for a short time in the absence of continued net deposition of building material, and some walls visibly thin during growth [410]; but, even in the most extreme examples of this, some wall synthesis does accompany growth [500]. New wall material deposited during expansion is from two sources: **(a)** the cellulose microfibrils are synthesised at the plasma membrane, and **(b)** the matrix polysaccharides are synthesised intra-cellularly and exported to the wall in membrane-bounded vesicles (Ch 5).

1.2: ISOLATION AND PURIFICATION OF PLANT CELL WALLS (see also [229,442a])

Most of the techniques described in this book require some form of wall isolation. The very simplest 'isolation' techniques should be con-sidered: valid conclusions can be drawn from work with impure walls (e.g. chemical structure of an extracted polymer) so long as the impurities do not interfere in the analysis. Walls can be purified if necessary (e.g. to see if a particular polymer is firmly wall-bound, or to determine the absolute polymer composition of walls) but purification is often difficult.

Exponentially-growing cell suspension cultures are an excellent source of primary cell walls uncontaminated with secondary walls. Other rapidly growing tissues can also be used, e.g. young stems, roots, leaves and coleoptiles. Relatively little work has been done on the walls of specific cell types, but would be extremely useful, perhaps especially work on the epidermis in view of the likelihood that this tissue limits the growth of young stems and petioles (Section 10.1.5).

To minimise degradation of polymers, wall isolation should not involve heating, pH values outside the range 4-7, or prolonged incubation in aqueous solutions (in which wall-bound enzymes may cause autolysis).

Fig. 1.1.4: Paradermal sections through palisade layer of young Phaseolus leaf (from [265]). **a:** Cell plates (P) forming at 90° to mother cell walls (MCW) [x 27 800]. **b:** Cell plates have become daughter cell walls (DCW) and fused with mother walls; a 'central zone' (C) becomes electron-dense (C) [x 83 800]. **c:** Central zone becomes larger and adjacent mother wall (*) shows signs of degradation [x 58 200]. **d:** Air space formation [x 37 600].

1.2.1: Isolation of walls by washing in organic solvents

A simple cell wall preparation, which meets the above criteria and is adequate for many purposes, consists of washing the tissue in 70 % ethanol (Panel 1.2.1 a). This extracts low-M_r sugars, amino acids, organic acids and many inorganic salts, leaving behind an **alcohol-insoluble residue** (AIR) containing polymers. The polymers will include intracellular proteins, RNA and starch as well as wall polymers, but the contaminants can often be ignored. Simple stirring in 70 % ethanol is adequate for finely dispersed cell-suspension cultures; prior homogenisation in 70 % ethanol is preferable for tissues from intact plants.

Starch is present in very large amounts in some plant tissues, and can be a problem even in simple operations such as total acid hydrolysis since it results in chromatograms being overloaded with glucose. The most satisfactory method (Panel 1.2.1 b) for removal of starch is stirring overnight at 25°C in 90 % dimethylsulphoxide (DMSO) [444]. [DMSO may extract a small proportion of certain hemicelluloses, and if this is critical it should be investigated — see Panel 1.2.1 b, footnote [1]].

PANEL 1.2.1 a: Preparation of alcohol-insoluble residue (AIR)

Sample: 20 g fresh spinach leaves

Step 1:   To 20 g fresh weight, add 140 ml ice-cold absolute ethanol + 40 ml H_2O to give a final solvent composition of 200 ml 70 % ethanol (the tissue being assumed to contain about 20 ml H_2O).

Step 2:   Liquidise the tissue by 5 1-min bursts at full power in a liquidiser e.g. Polytron, Ultra-Turrax or Sorvall 'OmniMixer', with cooling to 0°C between each burst.

Step 3:   For complete extraction of low-M_r cell components, stir the suspension at 0°C with a magnetic stirrer bar for several hours.

Step 4:   Collect the AIR on a filter (e.g. a sintered glass funnel or a fine nylon gauze). Wash with several 100-ml aliquots of 70 % ethanol. To monitor progress of removal of low-M_r compounds, measure the $\underline{A}_{280}$ of a sample of each filtrate.

Step 5a: (for samples from which wall polysaccharides will be isolated: see Section 3.4.1) — Resuspend the final residue in more 70 % ethanol, and store in a tightly-stoppered bottle preferably at about 0°C.

Step 5b: (for other samples, e.g. to be subjected to acid- or base-catalysed hydrolysis: see Sections 4.2, 4.3 and 4.4) — Wash the final residue several times with 100 % acetone, and air-dry. Store in a tightly-stoppered bottle to prevent water-regain.

The routine use of amylases to de-starch cell walls is not favoured because **(a)** many commercial amylase preparations are contaminated with enzymes that cleave wall polysaccharides (especially bacterial, fungal or plant amylases) or with proteases (especially pancreatic amylases) — human saliva is one the safest sources [388];   **(b)** during the treatment with amylase, which may require days for completion, endogenous wall enzymes (see Ch 6) may partially hydrolyse wall polysaccharide and glycoproteins.

Removal of extraneous **protein** from cell walls is more difficult, but may be useful as it appears to render wall polysaccharides more accessible to exogenous enzymes used analytically (Section 4.2.5). During wall isolation walls may pick up intracellular protein both by ionic binding and by co-precipitation with vacuolar tannins;   and in AIR, essentially all the cellular protein will be present with the walls. The walls of living cells

---

PANEL 1.2.1 b: De-starching cell walls with DMSO

Sample:	Alcohol-insoluble residue (Panel 1.2.1 a) from <u>Lemna</u> fronds.
Step 1:	Stir AIR (10 g) in 100 ml of 90 % DMSO at 25°C overnight.
Step 2:	Filter on sintered glass. Rinse with 50 ml 90 % DMSO.[1]

Step 3: (Tests for unextracted starch) <u>Either</u>: to a small portion of DMSO-insoluble residue (washed in H_2O) add a drop of 0.33 % iodine/0.67 % KI: starch goes blue-black (use microscope if necessary). <u>Or</u>: remove two 50-mg (wet weight) portions of DMSO-insoluble residue, resuspend each in 5-10 ml H_2O, and centrifuge (5 min, 2,500 **g**). Reject supernatants; add 0.2 ml 0.1 M NaOH to each pellet. Heat at 100°C for 10 min to swell starch grains. Cool. Adjust pH to 7 (add 40 µl 1 M NaH_2PO_4). To one sample, add 0.1 ml human saliva[2]; to the other add 0.1 ml H_2O. Incubate at 25°C for 2 h. Centrifuge (5 min, 2,500 **g**). Assay supernatants for maltose (reducing sugar) by the PAHBAH test (Panel 3.7 a). An increase in maltose content above the amylase-free control indicates that starch remained.

Step 4: Repeat steps 1-3 until the sample is judged starch-free.

Step 5: Wash walls 6 times with 70 % ethanol to remove DMSO.[3] Store as in Panel 1.2.1 a.

[1]Filtrate contains starch and any DMSO-soluble wall-polysaccharide: to recover these, add 5 volumes acetone, stand at 0°C for 1 h, collect the precipitate, wash in pure acetone, and dry.

[2]Readily available, highly effective, and uncontaminated by enzymes that degrade wall-polysaccharides. An alternative is pig pancreas amylase.

[3]For more complete removal of DMSO, dialyse against 70 % ethanol.

contain a complement of specific glycoproteins, which ideally would be
retained during wall purification. However, most of them (other than the
covalently wall-bound kinds e.g. extensin) are removed during rigorous de-
proteinisation since, post-mortem, wall proteins do not behave differently
from extraneous proteins. Probably the most generally applicable solution
is to remove all non-covalently bound protein by treatment with **phenol/HOAc
/H$_2$O (2:1:1, w/v/v)** (PAW) [444], and to accept that this will remove many
genuine wall glycoproteins too. Wall glycoproteins can be studied on a
separate sample by alternative techniques (Sections 1.2.3, 3.4.2 and 6.4).

PAW is an excellent extractant for proteins and glycoproteins (except
extensin), but is a very poor solvent for polysaccharides. It can therefore yield very clean cell walls (Panel 1.2.1 c). The possibility should
be considered that PAW also removes some non-protein wall material
(tannins?). If PAW is to be used, there is little point in pre-treating
with 70 % ethanol, which merely coagulates the contaminating protein and
makes it less extractable. Instead, the tissue is initially homogenised in
PAW, which extracts low-M$_r$ sugars, amino acids etc. as well as proteins.
[PAW is a very powerful solvent and can cause poisoning by skin contact.]

PANEL 1.2.1 c: De-proteinisation by phenol/acetic acid/water (PAW)

Sample: 200 ml of rose cell suspension culture, PCV ca. 25 ml.

Step 1:   Collect cells on 4 layers of muslin in a filter funnel, and
wash in 100 ml of fresh culture medium.

Step 2:   Squeeze muslin gently to remove surplus medium.

Step 3:   Weigh the ball of cells (without muslin). Let weight = **W** g.

Step 4:   Plunge into 7 x **W** ml of solution **A** [100 ml glacial acetic
acid + 250 ml 80 % (w/w aqueous) phenol (the latter is a liquid, available
e.g. from BDH, and is easier to handle than solid phenol)].

Step 5:   Stir the sample at 25°C for at least 2 h; overnight
incubation is acceptable.

Step 6:   Filter on sintered glass [phenol dissolves nylon]. Mix 1
ml of filtrate with 50 µl 10 % ammonium formate + 5 ml acetone: a precipitate indicates protein.

Step 7:   Resuspend cell (wall) residue in 5 x **W** ml of solution **B**
[100 ml glacial acetic acid + 250 ml 80 % (w/w) phenol + 50 ml H$_2$O].

Step 8:   Repeat steps 5-7 until filtrate no longer contains protein.

Step 9:   Wash walls thoroughly with 70 % ethanol until washings no
longer smell of phenol. Store walls as in Step 5a or 5b of Panel 1.2.1 a.

1.2.2: Homogenisation of fresh tissue in aqueous buffers

An alternative approach involves homogenisation of the tissue in an aqueous buffer [444,229]. The cells are physically broken open and their cytoplasmic contents washed out to leave clean walls. Disadvantages of this approach include **(a)** the danger that wall autolysis will occur during homogenisation unless stringent precautions are taken to minimise the activity of wall enzymes, **(b)** loss of water-soluble components of the wall into the homogenisation medium, **(c)** polymerisation and wall-binding of phenolics by oxidases brought into contact with them upon cell disruption, and **(d)** the requirement for very rigorous physical homogenisation since the buffer has little disruptive effect on the tissue, unlike ethanol, DMSO or PAW. However, the approach does have advantages, e.g. cell-rupture allows PAW and DMSO to act more rapidly.

The aqueous homogenisation medium may contain some or all of:

(i) a buffer to stabilise wall polymers [20 mM HEPES, pH 7.5, is useful as most autolysing enzymes have pH optima of 4-5 and HEPES does not chelate Ca^{2+}; phosphate is not recommended as it may remove Ca^{2+} from pectin];

(ii) a reducing agent [e.g. 10 mM dithiothreitol, mercaptoethanol or sodium metabisulphite] to minimise the chance of native or extraneous phenols becoming oxidised and cross-linked ;

(iii) a detergent to disrupt membranes [e.g. 0.5-1 % sodium deoxycholate or 0.5-2.0 % sodium dodecyl (= lauryl) sulphate; however, detergents can extract some pectins and non-covalently bound wall proteins];

(iv) a Ca^{2+} salt [e.g. 3 mM $CaCl_2$] to ensure that endogenous Ca^{2+} is not removed from pectins by cytoplasmic chelating agents, e.g. citrate [Ca^{2+} cannot be used together with SDS because they precipitate on mixing], and

(v) 1-2 M sucrose or sorbitol so that the cytoplasm is withdrawn from the walls by plasmolysis before homogenisation begins [this only works with living tissue, not frozen or detergent-treated].

Plant material should be fresh, or, if this is impossible, stored frozen. The temperature of the homogenisation medium should be kept as low as possible, preferably just above freezing point (a slightly higher temperature is needed to keep SDS in solution).

Various methods of homogenisation are in use, including:

(a) Grinding with a pestle in a mortar in the presence of a little acid-washed sand. This is applicable to small amounts of any fresh plant tissue (previously chopped with a knife if necessary) and can be effective, but is laborious. The sand may be troublesome to remove from the walls.

   (b) Liquidising with a blender, e.g. the Sorvall 'OmniMixer' or the 'Polytron'. Blenders require a reasonable volume of liquid, which may be excessively large for small samples e.g. of highly radioactive walls (Ch 2). Liquidising often fails to break open <u>all</u> the cells in a sample, but the soup produced may be more amenable to methods (c), (d) or (e).

   (c) Wet ball-milling. A ball-mill is a pot (e.g. 1 l) partly-filled with porcelain, glass, agate or steel balls. It is either rotated steadily between a pair of rollers or physically shaken by an eccentric rotor (the latter is termed vibratory ball-milling). The tissue plus buffer are thus ground into a soup, typically overnight. Prolonged mechanical processing of any kind may cause some polysaccharide breakdown [254] but this is minimal in (non-vibratory) wet ball-milling for moderate lengths of time [444].

   (d) Sonication. To be susceptible to sonication, a sample has to be already fairly soup-like. Well disaggregated cell suspension cultures can be used directly, but tissues from intact plants require a pre-treatment such as (b) or (c). A metal probe is placed in the sample and made to vibrate at a high frequency, e.g. in an MSE 'Soniprep'. Slender probes are available such that very small samples (e.g. 1 ml in an Eppendorf tube) can be treated. The sample requires cooling on ice during treatment. Vigorous sonication will cause some polysaccharide breakdown [485].

   (e) French pressure cell. The sample is forced under high pressure, e.g. 3000 psi (200 Atm), through a narrow orifice [490]; as the cells emerge, they rupture. Again, pre-treatment by method (b) or (c) is necessary except for fine cell-suspension cultures.

   Cell rupture is checked by light microscopy. When rupture is complete the suspension is centrifuged at low speed; this pellets the pieces of wall, which are relatively large and dense, while the cytoplasmic fragments remain in suspension. Separation may be aided by centrifugation through a cushion of 50 % sucrose [205]. Filtration on fine gauze is an alternative. The wall pellet is resuspended in homogenisation buffer and the centrifugation repeated. This process is continued until the walls are free of cytoplasmic contamination. Cytoplasmic markers include green colour [in chlorophyllous tissue], microscopical evidence of cell contents [e.g. staining with Evans' Blue, acid fuchsin, Orange G or (under the UV-fluorescence microscope) ethidium bromide], and chemical evidence of phospholipids [either residual radioactivity in tissue previously grown for 1-2 h in the presence of a trace of [^{14}C]choline, or detection of phosphatidyl choline in a CHCl$_3$/MeOH extract (10 mg wall slurry + 35 μl MeOH followed by 35 μl CHCl$_3$) upon TLC on silica-gel in CHCl$_3$/MeOH/H$_2$O (65:25:4) and staining in iodine vapour — see Section 4.5.3].

Residual proteins can be removed by proteases or PAW (Section 1.2.1), and both proteins and RNA can be removed by washing with detergents or concentrated salts (e.g. 1 M NaCl). Residual starch can be removed by amylase

---

PANEL 1.2.2: Preparation of cell walls from potato tuber [444]

Sample: 50 g Fresh potatoes.

Step 1:   Peel potatoes, chop finely, plunge into liquid N$_2$. Grind to a powder in a pestle. Store frozen if step 2 is not performed immediately.

Step 2:   Liquidise in 150 ml of solution **A** [1.5 % SDS[1] containing 5 mM sodium metabisulphite (Na$_2$S$_2$O$_5$) and 20 mM HEPES, adjusted to pH 7.5 with 1 M NaOH] and a few drops of octan-1-ol (to suppress foaming) for 2 min in a Waring blender followed by 3 min in an Ultraturrax. Satisfactory results can also be obtained with one **or** the other rather than both.

Step 3:   Filter through fine nylon gauze, and wash with two 50-ml portions of solution **B** [0.5 % SDS in 3 mM sodium metabisulphite and 20 mM HEPES, pH adjusted to 7.5 with 1 M NaOH].

Step 4:   Resuspend the slurry in 200 ml solution **B**, and ball-mill with porcelain balls in a cold-room (2-4°C) overnight.

Step 5:   Separate walls from balls by filtration on a coarse sieve.

Step 6:   Centrifuge the filtrate (5 min, 2,500 **g**). Resuspend the pellet[2] and re-centrifuge as above — once more in 100 ml solution **B** and twice in 100 ml H$_2$O.

Step 7:   Treat the pellet in PAW (Panel 1.2.1 c). Resuspend the walls in water and centrifuge (twice).

Step 8:   Treat the pellet with 90 % DMSO (Panel 1.2.1 b). Resuspend the walls in water and centrifuge (6 times). [If it should be necessary to remove every last trace of DMSO, dialyse the wall slurry against 5 litres of water.] Store final pellet as a frozen slurry.

[1] Sodium deoxycholate (1 %) can be used instead of SDS, with the advantage that less foaming occurs and less pectic polysaccharide is extracted, but with the disadvantage that less intracellular protein is extracted.

[2] The supernatant contains some pectic polysaccharides, which can be precipitated by addition of 5 volumes of ethanol or acetone followed by storage overnight at 25°C.

or DMSO (Section 1.2.1), and residual lipids with MeOH followed by $CHCl_3$/ MeOH (1:1) and $CHCl_3$/MeOH/H_2O (10:10:3).

The final wall pellet is often washed with pure water to remove buffer salts, detergents, etc.; however, it is advisable to retain a trace of reducing agent (e.g. 2 mM mercaptoethanol) during the washing. The material can be freeze-dried and stored dry (mercaptoethanol is volatile), but it is preferable to store the walls as a frozen slurry since wall poly-saccharides are less readily extracted after drying.

1.2.3: Homogenisation in glycerol

A valuable collection of other methods for wall isolation is given by Harris [229]. One particular approach worth consideration is homogenisa-tion of tissue in glycerol [229,258]. Glycerol remains liquid at low temperatures (e.g. -20°C, although it is very viscous at that temperature and requires vigorous homogenisation). It does not seem to precipitate intra-cellular proteins on to the walls, nor does it appear to leach water-soluble glycoproteins out of walls. It thus has considerable potential as a tool for the investigation of endogenous wall glycoproteins.

1.2.4: Separation of different cell walls

Walls of a selected cell type can be obtained by preliminary dissection of the tissue of interest. Epidermal cells are often relatively easy to purify, by stripping, especially in certain species e.g. broad bean (Vicia faba) and lamb's lettuce (Valerianella), although the task becomes more difficult in young, rapidly growing specimens. Epidermal hair cells can be isolated if the plant part is rolled on 'Sellotape' adhesive tape; the hairs are subsequently released by treatment with methylcyclohexane, which dissolves the adhesive [399]. Other workers have achieved partial separation of the various cell walls present in a whole-organ homogenate by density gradient centrifugation in EtOH/CCl_4 mixtures [198].

2 Radioactive labelling of cell walls

2.1: THE PRINCIPLES OF RADIOACTIVE LABELLING

2.1.1: Analysis of picogramme quantities

Many of the analytical techniques described in this book are facilitated by radioactive labelling of the cell wall material. One reason for this is to aid in the tracing of metabolic pathways (Section 2.1.4 & Ch 5). However, a second reason is to aid the structural analysis of components present in very small amounts. This has 3-fold importance: **(i)** Some polymers are present in the wall in trace amounts. **(ii)** Even the more abundant polymers are usually composed of several different sugars and amino acids, some of which are present in small proportions. **(iii)** The plant material of interest may be available only in very small quantities, perhaps just a few cells.

The advantages of radioactive labelling in micro-analysis can be illustrated by comparison with the sensitivity (limit of detection) of various techniques commonly used in the analysis of non-radioactive sugars:

Separation technique (and detection method)	Limit
Paper chromatography ($AgNO_3$ detection)	0.1 µg
TLC (aniline hydrogen phthalate detection)	0.1 µg
GC (packed column; F.I.D. detection)	0.1 µg
GC (capillary column; F.I.D. detection)	1.0 ng
HPLC (refractive index detection)	0.1-10 µg
Colorimetric assay (anthrone reaction)	0.5 µg
Radioactive labelling (scintillation counting)	1-10 pg

Radioactive labelling of the molecules of interest therefore improves the limit of detection by 2-6 orders of magnitude.

A further advantage of radioactive labelling is that it makes substances detectable by non-destructive methods (e.g. scintillation-counting; see Section 2.8.2). Multiple analyses can therefore be performed on a single 1-10 pg sample (see Section 4.6).

Radioactive walls for micro-analysis are obtained by prolonged incubation of a radioactive precursor with the living cells, which incorporate it into newly-synthesised wall components. Useful radioactive precursors include the following [for a list of specific examples, see Panel 2.1.1]:

Precursor	Typical specific activity supplied	Qty detectable[*]
[U-^{14}C]Monosaccharides	300 mCi/mmol	6 pg
[1-^{3}H]Monosaccharides	10 Ci/mmol	½ pg
[U-^{14}C]Amino acids	300 mCi/mmol	6 pg
[s.c.-3-^{14}C]Cinnamic acid	40 mCi/mmol	40 pg

[*]The lower limit for routine, reliable measurement of ^{14}C by scintillation-counting is about 20 dpm, i.e. ca. 10 pCi, and that for ^{3}H is about 50 dpm. i.e. ca. 25 pCi (see Section 2.8). Background counts are ca. 10-20 cpm.

2.1.2: Specific versus non-specific labelling

A distinction is drawn between specific and non-specific labelling. **Non-specific** ^{14}C-labelling yields cell walls in which all the carbon atoms have an equal chance of being radioactive. It is achieved by growth of the tissue with its only carbon source in radioactive form. Typically [U-^{14}C]-glucose or [U-^{14}C]sucrose are used for tissue cultures, and ^{14}CO$_2$ for green plants. The main advantage of non-specific labelling is that it enables the detection of all organic components in a sample, regardless of chemical nature, whether or not their presence was expected. However, the inclusion of inositol in tissue culture media can prevent uniform labelling as it can be a significant carbon source for the synthesis of uronic acid, pentose and apiose groups [6,525]; if possible, inositol should be omitted. Similarly, many cells in green plants, especially seedlings, may obtain much of their carbon from endogenous carbohydrates or fats, rather than CO_2.

Specific labelling results in the radioactivity becoming incorporated preferentially into specific components of the cell wall. For instance, if [^{14}C]arabinose is fed to tissue cultures, radioactivity is readily incorporated into pentose residues (arabinose, xylose and 2-O-methyl-xylose), but hardly at all into hexose, uronic acid, amino acid or phenolic residues (Fig. 2.5). Specific labelling usually relies on uptake and incorporation via 'scavenger' pathways (= pathways that do not constitute the principal biosynthetic route to a particular product, but that come into play when an appropriate substrate is made available exogenously [173,155] — Section 5.1.2). These pathways may work at varying

16

efficiencies under varying growth conditions, or in diverse cell-types. Also, although a particular precursor may label specific wall components in one cell-type, it may label more indiscriminately in another: e.g., some tissues can metabolise arabinose, via glucose, to all wall components.

2.1.3: Maximising the specific activity of <u>in-vivo</u> labelled walls: Continuous labelling

The calculations of minimum detectable amounts (Section 2.1.1) are based on the specific radioactivity (**SA**) of labelled precursors available commercially. Brief feeding (i.e. **pulse**-labelling; see Section 2.1.4) of a radioactive monosaccharide will yield cell wall components with an SA much lower than that of the exogenous radioactive precursor (SA_{ex}). However, longer periods (i.e. **continuous** labelling), with a precursor giving <u>non-specific</u> labelling of walls, may yield walls approaching SA_{ex} (Fig. 2.1.3 a); for instance, if [^{14}C]glucose is fed to an exponentially-growing suspension culture which has glucose as its sole carbon source, for a period equal to 1 mean cell-cycle, cell walls will probably be produced in

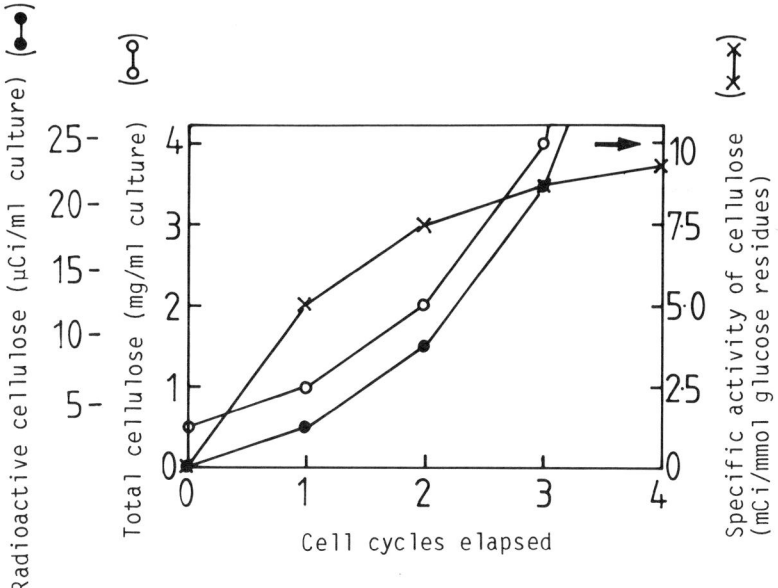

Fig. 2.1.3 a: Predicted result of growing a cell suspension culture in a medium containing [U-^{14}C]glucose as sole carbon source.
Arrow (→) indicates the specific activity of the supplied glucose.

17

Panel 2.1.1: To show which components of wall polymers are labelled by 17 precursors[1]

Radio-active prcrsr	GalA	GlcA	D-Gal	L-Gal	Glc	Man	Ara	Xyl	Fuc	Rha	MeXyl	MeFuc	Api	Cou	Fer	Hyp	Tyr	Idt	Lig	Cut
Glc(2,5)	+	+	+	+	+	+	+	+	+	+	+	+	+	+	+	+	+	+	+	+
Gal(2,5)	+	+	+	+	+	+	+	+	+	+	+	+	+	+	+	+	+	+	+	+
Fru(2,5)	+	+	+	+	+	+	+	+	+	+	+	+	+	+	+	+	+	+	+	+
Man(2,5)	+	+	+	+	+	+	+	+	+	+	+	+	+	+	+	+	+	+	+	+
Man(3)			+	+	+	+		+	+	+	+	+	+							
GlcA(5)	+	+																		
GlcA(4)	+	+																		
GalA(5)	+	+																		
Ino	+	+					+	+			+		+							
Ara							+	+			+									
Fuc									+		+									
Pro																+				
Tyr														(6)	(6)	+	+		(6)	(6)
Cinn														+	+				+	+
Cou														+	+				+	+
Fer															+				+	+
HDA(7)																				+

Abbreviations: **Api** = D-apiose, **Ara** = L-arabinose, **Cinn** = trans-cinnamic acid, **Cou** = p-trans-coumaric acid, **Cut** = cutin and suberin, **Fer** = trans-ferulic acid, **Fru** = D-fructose, **Fuc** = L-fucose, **Gal** = D-galactose, **L-Gal** = L-galactose, **GalA** = D-galacturonic acid, **Glc** = D-glucose, **GlcA** = D-glucuronic acid, **HDA** = hexadecanoic acid, **Hyp** = L-hydroxyproline, **Idt** = L-isodityrosine, **Ino** = myo-inositol, **Lig** = lignin, **Man** = D-mannose, **MeFuc** = 2-O-methyl-L-fucose, **MeXyl** = 2-O-methyl-D-xylose, **Pro** = L-proline, **Rha** = L-rhamnose, **Tyr** = L-tyrosine, **Xyl** = D-xylose.

[1] Potential precursors which have been tested but failed to give useful labelling of wall polymers include xylose, rhamnose, apiose and ribose.

[2] Glucose is readily taken up and converted to glucose 6-phosphate, a central metabolite; therefore radioactivity from it labels all the organic compounds of the cell. Similar results are obtained with any precursor that is readily metabolised to glucose 6-phosphate (e.g. sucrose, mannose, glycerol, etc.), or to UDP-glucose (e.g. D-galactose).

(3)The labelling pattern shown here is that obtained with [2-³H]mannose (available commercially), which loses its ³H atom as ³H₂O during conversion to glucose; cf. footnote (2).

(4)The labelling pattern shown here is that obtained with [6-¹⁴C]glucuronic acid, which loses its ¹⁴C as ¹⁴CO₂ upon conversion to a non-uronic acid sugar. Unfortunately [6-¹⁴C]glucuronic acid is no longer available commercially. However, [6-¹⁴C]glucose, which is available, could be oxidised to GlcA [346], and for an alternative route to [6-¹⁴C]GlcA, see [464].

(5)The labelling pattern shown is that obtained with readily available labelled versions of the compound, e.g. [U-¹⁴C]- or [1-³H]sugar; versions labelled elsewhere may show somewhat different labelling patterns (see footnotes 3 and 4).

(6)In Monocots only.

(7)HDA (hexadecanoic acid = palmitic acid) is water-insoluble, and must be converted into micelles by sonication in hot buffer prior to administration to plants [318].

which the mean SA is ca. ½ SA_{ex}, 2 cell-cycles ca. ³/₄ SA_{ex}, 3 cell-cycles ca. ⁷/₈ SA_{ex}, and so on. [The SA_{ex} is calculated taking into account any non-radioactive glucose in the medium.] Some wall components undergo turn-over (Section 8.1); their SA will reach ½ SA_{ex} in somewhat less than 1 cell-cycle. The conclusion is that, if the labelling period is long enough, cells can be produced with their walls labelled to nearly the same high SA as the commercially-available precursors, so that 1-10 pg of a wall component can be measured. Toxicity of high concentrations of high-SA precursors is not usually a problem with ¹⁴C or ³H, though problems may arise with higher energy β-emitters e.g. ³²P.

Cell walls with the highest SA will be produced by 'continuous labelling' of suspension-cultured cells for at least 3 mean cell-cycles (ca. 3-8 days for most cultures) in a medium in which, say, [U-¹⁴C]glucose is the sole carbon source. Added non-radioactive 'carrier' glucose should be kept to a minimum, and carry-over of non-radioactive sugars from the parent culture medium should be prevented by washing of the cells in sugar-free medium. The total glucose concentration should be kept high enough to sustain cell growth for the ca. 3 cell-cycles, but, in the interests of economy, should not be much higher or else some [¹⁴C]glucose will be left unused. Typically 100 µCi (cost ca. £15) of [U-¹⁴C]glucose at an SA of 330 Ci/mol is used in 0.5 ml of medium containing 15 mM non-radioactive glucose [thus final glucose concentration = 15.6 mM (0.28%) and final SA = 13 Ci/mol]. For even higher final SA, larger amounts of [^{14}C]-glucose are used, in smaller volumes of carrier-free medium.

Equivalent comments apply to the labelling of green plants with $^{14}CO_2$, although the maintenance of a constant supply of $^{14}CO_2$ to the atmosphere requires more sophisticated equipment.

The situation is different with specific labelling. Here the SA of the predominantly-labelled wall components may sometimes remain well below the SA_{ex} because the normal (non-scavenger) biosynthetic pathway, using non-radioactive glucose or CO_2, continues to operate, diluting 'hot' with 'cold'. Nevertheless, with optimisation of the following three factors, specific labelling can still result in highly radioactive cell wall components, often approaching SA_{ex} in ca. 3 cell-cycles:

(i) The precursor chosen should, if possible, be one that is taken up and incorporated by a pathway of sufficient capacity for the cells' total requirement of the wall component of interest. For instance, cultured cells are often able to take up both galacturonic acid and glucuronic acid, and to incorporate them into pectins; however, the cells' maximum capacity for uptake and incorporation of galacturonic acid is lower than that for glucuronic acid, and the latter would therefore be chosen for in vivo feeding experiments.

A simple test of sufficiency, which eliminates the need to buy radio-chemicals that are later found unsuitable, is to grow the tissue under continuous, non-specific labelling conditions (e.g. with [U-^{14}C]glucose) in the presence of various concentrations of the non-radioactive form of the potential precursor for ca. 3 cell-cycles: a good precursor, e.g. glucuronic acid, virtually stops the labelling of the wall component of interest under these conditions (Fig. 2.3.1 b, but not Fig. 2.3.1 c) because the scavenger pathway swamps and/or inhibits the endogenous biosynthetic pathway from [U-^{14}C]glucose. Clearly, it must also be shown that the precursor is non-toxic at the concentrations used; this can be done by checking that it does not block the incorporation of [U-^{14}C]glucose into an unrelated wall component — in this case galactose residues.

(ii) The precursor should be supplied at just above the ½-saturating concentration for the uptake + incorporation pathway. A ½-saturating concentration can be roughly estimated by quantitative use of the simple test already outlined (Figs. 2.3.1 b,c).

(iii) The specific radioactive precursor should be fed for at least 3 cell-cycles (see above).

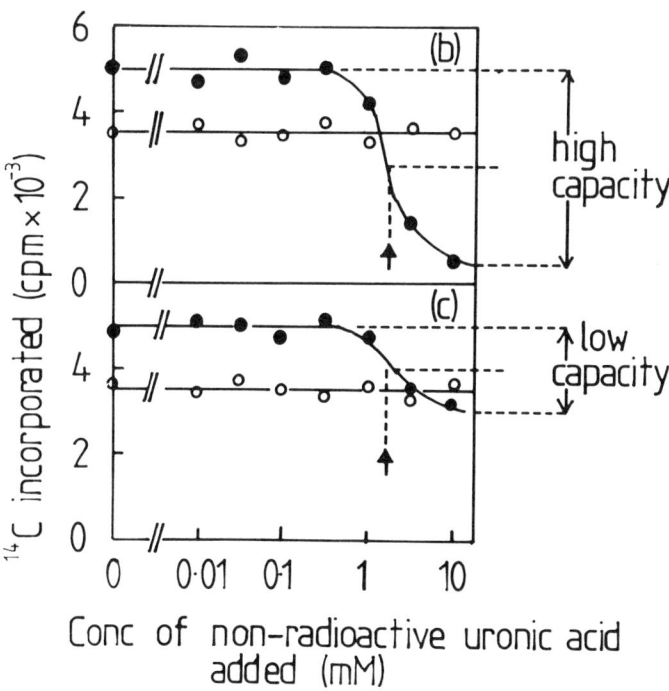

Figs. 2.1.3.b & c: Effect of added non-radioactive glucuronic acid (b) and galacturonic acid (c) on the labelling of pectin-bound GalA residues (filled circles) and Gal residues (open circles) in a suspension-culture incubated for several days in a medium containing 2 % [U-^{14}C]glucose (low SA) as sole carbon source.

The ½-saturating concentration (solid arrows) for both added uronic acids is ca. 1.7 mM.

The cells' capacity for utilisation of added glucuronic acid is essentially sufficient for synthesis of pectic GalA residues; that for utilisation of added galacturonic acid is not.

The added uronic acids are shown to be non-toxic at the concentrations tested since they do not affect the labelling of pectic Gal residues (for which they are not potential precursors; cf. Fig. 2.5).

2.1.4: Tracing of metabolic pathways: Pulse-labelling

Besides the preparation of high-SA cell walls to facilitate micro-analysis, the other major use of in vivo labelling is to answer questions about metabolic pathways, e.g: What is precursor of what? Which polymers are being synthesised at any given moment and in what amounts? Which polymers turn over and which are stable? Techniques to answer these questions are discussed in more detail elsewhere (Sections 5.2, 5.3, 8.1); they are introduced here to draw attention to **pulse-labelling**, where maximising the SA is not the main aim.

Again the details of labelling depend on whether the labelling is specific or non-specific. With specific labelling, it is undesirable to supply precursors at a concentration high enough to swamp the endogenous pathways as was recommended in Section 2.1.3. Instead, high-SA radioactive precursors are infiltrated into the metabolic pathways at 'tracer' concentrations so low that the experimental conditions themselves introduce no significant metabolic change. Normal synthetic pathways will continue to predominate over the scavenger pathways, diluting the radioactive material with a large excess of non-radioactive material. Wall components will therefore be labelled to an SA much lower than that of the precursor.

With non-specific labelling, the precursor itself is usually a major component of normal cells and radioactive precursor can thus be used at much higher concentrations (so long as they do not approach the concentrations present in vivo) without the risk of studying 'abnormal' pathways.

2.1.5: Simple pulse versus pulse-chase labelling

Simple **pulse-labelling** involves incubation of cells with a radioactive precursor for a time, after which the cells are killed and analysed. This method is used to determine which polymers are being synthesised over a given time interval. End-products, e.g. cellulose, become more radioactive the longer the cells are incubated [if the supply of precursor is not exhausted]. Intermediary metabolites (e.g. UDP-sugars) quickly reach a plateau of labelling where rate of synthesis of radioactive metabolite = rate of conversion to next substance in the pathway.

Pulse-chase labelling involves incubation of cells in the presence of radioactive precursor for a defined time, and then in its absence. This method is favoured in attempts to identify which compounds act as precursors for which others, and in studies of turn-over. It is a way of tagging a set of molecules, all synthesised during a particular time interval, and of then monitoring their fate, without the complication of continued formation of new tagged molecules. Stable end-products, e.g. cellulose, accumulate radioactivity steadily during the pulse period, and then enter a plateau during the chase; intermediary metabolites lose radioactivity during the chase (Fig. 2.1.5).

The 'chase' can be achieved by **(a)** washing of cells in medium lacking the radioactive precursor, and/or **(b)** addition of the non-radioactive form of the precursor to a concentration well above its ½-saturating concentration, thus swamping the pathway with non-radioactive material. Methods (a) and (b) are both suitable for specific labelling; method (a) should not be used alone in non-specific labelling experiments since it results in the cells starving during the chase period.

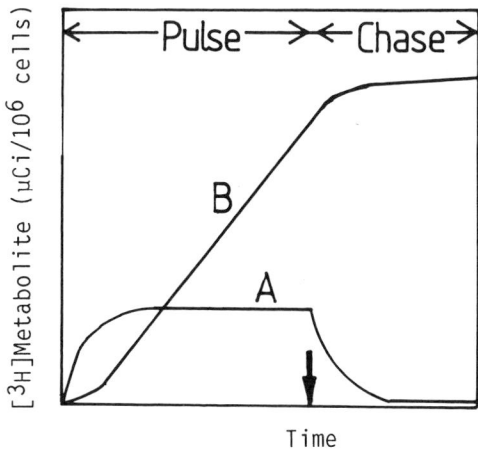

Fig. 2.1.5: Labelling of an intermediary metabolite (**A**, e.g. UDP-glucose) and an end-product (**B**, e.g. cellulose) during a pulse—chase experiment.
A ^{3}H-labelled precursor (e.g. glucose) was added at time 0 and later replaced with non-radioactive precursor (solid arrow).

23

2.2: METHODS OF FEEDING RADIOACTIVE PRECURSORS (TISSUE CULTURES)

Plant tissue cultures, and especially cell suspension cultures, are easier to label with radioactivity than are organs of whole plants. Cultured cells are often well disaggregated, so all the cells in the population are in contact with the medium; and there is no cuticle to act as a barrier to uptake. This means that all the cells in a culture start to take up an added precursor simultaneously, greatly simplifying kinetic studies. In addition, the precursors can be supplied without cutting or other wounding. Often all the cells in a culture appear similar, whereas stems, leaves and roots are composed of many different cell-types, each with its own characteristic metabolism. Tissue cultures can be kept aseptic, so prolonged feeding of precursors is possible without infection.

The details of the feeding method used will depend on the aims of the experiment: continuous, pulse or pulse-chase; specific or non-specific labelling; high or low SA. Pulse-labelling is described in more detail in Chs 5 and 8; this chapter focuses on continuous labelling as a background particularly for the micro-analytical techniques described in Chs 3 and 4. A sample protocol for continuous labelling of suspension-cultured cells with [14C]glucose is given in Panel 2.2 a. Methods for the necessary sterilisation of radiochemicals and other culture medium additives are given in Panel 2.2 b, and selected methods for killing the cells prior to analysis are given in Panel 2.2 c. A list of other precursors that can be fed to suspension cultures to label their walls is given in Panel 2.1.1 and Fig. 2.5.

When the aim is non-specific pulse- (Sections 2.1.4 & 2.1.5) rather than continuous-labelling, a hexose such as [14C]glucose or [14C]galactose is usually used as precursor. The [14C]hexose is supplied at high specific radioactivity, and it is helpful to have the cells growing in a hexose-free medium. This will make the cells 'hungry' for hexose, and they will there- fore take up a high proportion of the [14C]hexose very quickly. Non-hexose carbon sources on which some cell cultures will grow (though often rather slowly) include glycerol and occasionally D-xylose, L-malate or succinate [189,249,375].

Callus cultures, grown on agar-solidified medium, can be fed the same precursors as suspension cultures. The main difficulty with solid calli is the inhomogeneity of the tissue. For example, cells at the centre of a lump of callus are often different from the outer ones, and are screened to some extent from the added precursor. This means that the kinetics of wall-labelling are difficult to interpret and that the 'screened' cells may be receiving an unspecified metabolite of the added precursor, rather than the precursor itself. This may not be important in continuous labelling experiments, but it will be a serious problem in pulse-labelling work. An advantage of solid calli is that they more closely resemble the intact plant than do suspension cultures, for instance in synthesising more secondary products [4].

PANEL 2.2 a: Continuous labelling of cultured cells with [U-14C]glucose[1]

Sample: Suspension-cultured plant cells.

Step 1:   Wash the cells aseptically in sugar-free medium. This is to avoid carry-over of non-radioactive sugar, which would dilute the [U-14C]glucose. Allow the cells to sediment in sugar-free medium for 5-10 minutes, discard the medium, and resuspend the cells in fresh sugar-free medium; repeat this 2-3 times.

Step 2:   Autoclave 0.5 ml of radioactive medium in a 5-ml flat-bottomed glass vial. The medium sould be inositol-free, if the cells can possibly be grown without this additive. The sugar of the medium is partly replaced by 100 μCi of [U-14C]glucose, previously dried if necessary to remove any ethanol. The vial must be broad-based enough that the cells will stay in suspension in it with gentle shaking. It should be autoclaved with a tight-fitting lid, preferably a Teflon-lined screw cap, to avoid volume-changes due to evaporation or condensation. Alternatives to auto-claving are given in Panel 2.2 b.

Step 3:   Transfer <u>ca.</u> 25 mg fresh weight of washed cells into the autoclaved, cooled vial.   Well-disaggregated cells can be transferred as a suspension of known cell-density by pipette; alternatively, cells can be collected on a stainless steel mesh and scraped off with a spatula. Cap the vial with autoclaved cotton-wool.

Step 4:   Continue incubation under standard growth conditions for at least 3 cell-cycles; a cell-cycle is 20-60 h for typical suspension cultures. Note that the cells will evolve $^{14}CO_2$ by respiration, and precau-tions should be taken to avoid inhaling this. They will also lose water vapour, and the volume should be maintained if necessary by addition of sterile distilled water. The problem of drying can be reduced by keeping the culture vial in a 'humidity chamber' (beaker lined with wet paper).

Step 5:   Collect the cells. This can be done by filtration, e.g. on nylon or steel gauze, or by low-speed centrifugation. It is often worth collecting the filtrate or supernatant as these will contain some wall-related polymers that may be of interest. The method of killing the cells will depend on the purpose of the experiment: common methods are listed in Panel 2.2 c.

PANEL 2.2 b: Sterilisation of culture medium additives

Some radiochemicals are supplied as sterile solutions, and the required amount can then be dispensed with a disposable sterile hypodermic syringe. [It is important to ensure that any sterilant present (e.g. ethanol or HCl) will be diluted sufficiently not to interfere with the plant tissue: Final ethanol concentration should be kept below 0.05 %, and problems with HCl can be overcome with non-toxic buffers (Panel 2.7.1 b).] Compounds that are not supplied pre-sterilised can be made sterile as described here with special reference to additions of radiochemicals to cell suspension cultures; the methods are also suitable for many other applications requiring asepsis.

(1) Autoclaving

Application Heat-stable compounds e.g. amino acids, cinnamic acid, and many sugars.

Method 120°C for 15 min would be typical. The solution should be at a pH known not to affect the compound; if in doubt, autoclave a small amount and test quantitatively for loss, e.g. by paper chromatography or TLC (see Section 4.5).

(2) Use of self-sterilising solvents.

Application Low-M_r, heat-labile compounds e.g. certain sugars, phenols, gibberellins.

Method Dissolve the precursor as a concentrated stock solution in a self-sterilising, water-miscible liquid which when diluted into the cell suspension will be below the toxic level. Useful self-sterilising liquids are 90 % DMSO (which requires at least 50-fold dilution) and 80 % ethanol (requires at least 2,000-fold dilution).

(3) Use of ether or acetone.

Application   Heat-labile and/or insoluble materials, e.g. pectins, whole cell walls.

Method Oven-sterilise (e.g. 150°C for 4 h) a dry glass vessel fitted with a cotton wool bung and capped with foil; cool, and place in it the material, suspended in diethyl ether or acetone. Thoroughly dry off the solvent by leaving the vessel overnight, with the cotton wool still in place, on a sterile bench. (N.B. --- fire hazard!)

(4) Filter-sterilisation

Application Water-soluble labile polymers e.g. enzymes.

Method   A Millipore filter disk (type GS; nominal pore-size 0.22 µm) is fitted in a holder and autoclaved. The enzyme solution is taken up into a syringe barrel, the nozzle of which is then connected to the holder and the solution is forced slowly through the GS disk into a sterile vial.

PANEL 2.2 c: Methods for killing and extracting labelled tissue

(1) FREEZING ON SOLID CO_2 OR IN LIQUID N_2 followed by isolation of walls by homogenisation in aqueous buffer [see Panel 1.2.2]

Extracts: Unused precursor and most water-soluble cell components.
Recovery of Extracted Compounds: Freeze-drying (use a volatile buffer for wall-isolation). Or, polysaccharides and proteins can be precipitated (add 0.05 vols of 10 % ammonium formate followed by 5 vols EtOH or acetone; store overnight at 0°C — check this does not precipitate buffer salts).
Does not Extract: Water-insoluble components (walls, starch grains, chromatin, membranes); water-soluble components that bind to these structures.

(2) PLUNGING INTO 80 % ETHANOL[1] [see Panel 1.2.1 a]

Extracts: Unused precursor and most low-M_r metabolites.
Recovery of Extracted Compounds: Drying, either in bulk or by loading directly on to a paper chromatogram.
Does not Extract: Polysaccharides, proteins, lignin.

(3) PLUNGING INTO ICE-COLD 15 % TRICHLOROACETIC ACID[1] [see Section 3.5.6]

Extracts: Unused precursor, low-M_r metabolites, H_2O-soluble polysaccharides.
Recovery of Extracted Compounds: For polysaccharides, neutralise with NaOH [care] & de-salt (Sect 3.5.1). Low-M_r substances are difficult to recover.
Does not Extract: Proteins, lignin, insoluble (e.g. wall-bound) polysaccharides.

(4) PLUNGING INTO PHENOL/HOAc/H_2O (2:1:1, w/v/v)[1] [see Panel 3.4.2 a]

Extracts: Unused precursor, low-M_r metabolites, proteins, most glycoproteins.
Recovery of Extracted Compounds: Drying in vacuo. Or, proteins & glycoproteins precipitated with 0.05 vols 10 % ammonium formate + 5 vols acetone.
Does not Extract: Polysaccharides, lignin, covalently wall-bound glycoproteins.

(5) PLUNGING INTO 3 % SDS / 1 % MERCAPTOETHANOL / 1 % NH_4HCO_3 at 100°C[1]

Extracts: Unused precursor, low-M_r metabolites, proteins and most glycoproteins, and water-soluble & heat-sensitive polysaccharides.
Recovery of Extracted Compounds: Polymers pptd with EtOH or TCA. SDS is not removed by dialysis or GPC so recovery of low-M_r substances is difficult.
Does not Extract: Wall-bound polysaccharides (except some pectins), covalently wall-bound glycoproteins, and lignin.

(6) PLUNGING INTO CHLOROFORM/METHANOL/H_2O (10:10:3)[1] [see Section 5.3]

Extracts: Unused precursor, low-M_r metabolites including lipids.
Recovery of Extracted Compounds: Drying. Or, add 0.2 vols H_2O to give 2 phases; lipids partition into lower, most other cmpds into upper phase.
Does not Extract: Polymers.

----
[1]This is the **final** concentration or composition; the H_2O content of tissue + any culture medium should be taken into account.

2.3: METHODS OF FEEDING RADIOACTIVE PRECURSORS (INTACT PLANTS)

Under natural conditions, plants take in few exogenous precursors: CO_2 by photosynthesis, and H_2O and mineral salts through the roots. Thus, if the aim is for experimental conditions to mimic as closely as possible the natural situation and minimise wounding (which can trigger changes in wall metabolism) the three preferred radioactive precursors are inorganic.

2.3.1: Supply of inorganic precursors

(i) $^{14}CO_2$, supplied to photosynthesising leaves, is incorporated into all organic compounds synthesised by the leaf. This gives relatively non-specific labelling (see Section 2.1.2). If the aim is to label non-photosynthetic organs, the radioactivity has to be delivered via the phloem: this introduces a considerable lag. There is also more danger that stored carbon sources (e.g. starch) will be used in preference to the ^{14}C supplied from CO_2. A simple but effective method for feeding $^{14}CO_2$ to a potted plant is given in Panel 2.3.1.

(ii) Root-fed ^3H_2O becomes available to metabolism, as a result of which a small proportion of the ^3H atoms are incorporated into monosaccharides and hence into polysaccharides. One example of the use of ^3H_2O, in

---

PANEL 2.3.1: Feeding $^{14}CO_2$ to a potted plant (see also ref [284])

Step 1: Place the plant on a platform in a large glass vessel, e.g. a desiccator (without desiccant).

Step 2: Under the platform, place a stirrer bar with card 'wings' taped to it so that, when set over a magnetic stirrer, a draught will be created inside the desiccator. The draught helps to disperse the $^{14}CO_2$ to all leaves.

Step 3: Either generate $^{14}CO_2$ in situ by injection of $NaH^{14}CO_3$ solution (e.g. 10 µCi) into 1 ml of 1 M HCl in a test tube elsewhere in the desiccator, or bleed commercially available $^{14}CO_2$ into the desiccator from a gas cylinder. The former method, which has the advantage of simplicity, is adequate for short-term (pulse) labelling; the latter, supplemented with a means of precisely controlling the influx of $^{14}CO_2$ and of allowing the escape of O_2, ethylene and unused $^{14}CO_2$, is necessary for long-term (continuous) labelling. Both methods require the use of a fume cupboard to eliminate the escaping $^{14}CO_2$, which can also be quantified if the outlet gases are passed through a 1 M KOH 'trap'.

Step 4: Supply light from 2 floodlamps (e.g. 300 W); each light beam should pass through a trough of water to minimise its heating effect.

Step 5: Harvest the plant. Note that it will evolve $^{14}CO_2$ by leakage and respiration; continue to use a fume cupboard.

---

animals, is in the study of glycogen synthesis [405]. This emphasises the fact that incorporation of radioactivity from ^3H_2O into carbohydrates does not only occur by photosynthesis, although leaves do incorporate more ^3H atoms per glucose molecule than do non-photosynthetic cells, and this can make quantification difficult in tissues with some green and some non-green cells. However, there is considerable advantage to the fact that ^3H_2O is readily taken up by all cells without wounding and, once taken up, is not compartmentalised. Adequate precautions, e.g. involving the use of a condenser and containment within a fume cupboard, are necessary to avoid inhaling ^3H_2O vapour; the condenser could also serve the secondary function of recovering ^3H_2O lost by transpiration.

(iii) The third natural uptake system — of mineral salts — has limited use in cell wall studies because none of the radioactive minerals available is a major precursor of wall polymers. Possibilities exist with $^{35}SO_4^{2-}$, to label the (very few) methionine and cyst(e)ine residues in wall glycoproteins, and with $^{45}Ca^{2+}$, to label the pectin-bound calcium.

2.3.2: Infiltration of organic precursors

It is also possible to feed labelled organic precursors that plants do not naturally find in their environment (although individual cells in an intact plant may well receive some of them from neighbouring cells); see also Panel 2.1.1 and Section 2.5. One problem is that the precursor will often only reach the tissue of interest via other tissues; the latter could selectively withold certain precursors, or could transfer radioactive metabolites derived from the precursor originally fed. This screening effect complicates the kinetics of wall-labelling by introducing a long and unpredictable lag, it increases uncertainty as to the chemical nature of the labelled products, and it results in wastage of radioactivity to uninteresting tissues. A second problem is maintenance of asepsis: although it is unlikely that contaminating micro-organisms would seriously compete with a healthy plant for $^{14}CO_2$, ^3H_2O or $^{35}SO_4^{2-}$, plant surfaces usually bear micro-organisms that can rapidly metabolise organic precursors such as amino acids and sugars. Therefore, surface-sterilisation and use of aseptic technique is recommended for labelling experiments using organic precursors and lasting more than ca. 4 h.

Methods for supplying organic precursors to tissues of interest include:

(i) Whole, hydroponically-grown plants can, with minimal damage, be fed via their roots. Duckweeds (Lemnaceae) are particularly amenable and can be grown aseptically [454].

(ii) Leaves and fruits can be excised and incubated with their stalks in a solution of the precursor, which is taken up by the transpiration stream [399]; excision introduces some mechanical damage, but this may not be serious if the tissue to be studied is far enough away from the cut.

(iii) A step further is to excise a small portion of the tissue of interest and incubate it on wet filter paper or agar in the presence of the precursor; uptake will probably occur, but wound-responses become a problem. A refinement of this approach, which can keep the tissue alive for a considerable period, sometimes indefinitely, is to incubate the explanted piece of tissue as an organ culture (see Section 2.7).

A different type of approach is to work with whole, potted plants but to use a hypodermic syringe to inject the radioactive precursor close to the site of interest [163]; this cuts down loss of radioactivity to uninteresting tissues, but inevitably leads to wound responses. More sophisticated versions of this approach include the introduction of a wick into the wounded tissue so the precursor can be supplied slowly, avoiding to some extent any 'detoxification' reactions.

Two main methods are in use to surface-sterilise plants. [Surface sterilisation is adequate as the interior of a plant is usually free of microbes.] The first is to surface-sterilise the seeds, e.g. with 50 % ethanol for 30 min, wash them with sterile water, and then germinate them aseptically, e.g. in a glass pie-dish that has been lined with wet filter paper, covered with metal foil and autoclaved. The second method is to excise and surface-sterilise the aerial organs of a non-aseptically-grown plant. Stems and leaves can be sterilised as described in Panel 2.3.2.

PANEL 2.3.2: Surface sterilisation of a stem or leaf

Step 1: Swirl in 70 % ethanol for 10 seconds. This makes the cuticle wettable.

Step 2: Swirl in dilute sodium hypochlorite for 10-30 minutes, e.g. a 20-fold dilution of the household bleach 'Domestos'.

Step 3: Quickly transfer the tissue into a large volume of sterile (autoclaved and cooled) distilled water. Swirl for a few minutes. Repeat the water-washing several times to remove the last traces of hypochlorite.

2.4: UPTAKE OF PRECURSORS

A distinction is drawn between uptake and incorporation. **Uptake** is the passage of molecules from the culture medium, through the plasmalemma, into the cell. **Incorporation** is their conversion into defined metabolites, e.g. polysaccharides.

Plant cells will take up a wide variety of exogenous substances. Some pass freely through the plasma membrane because they are soluble in the hydrophobic phase of the membrane: such substances are low-M_r, non-glycosylated, non-polar compounds, e.g. acetate, ethanol, cinnamate, phenols, auxins, CO_2, etc. Uptake of weak acids, e.g. acetate, can be facilitated by use of a low-pH medium (e.g. pH 4.0-4.5). Other compounds, though not hydrophobic enough to dissolve in the lipid bilayer, can nevertheless penetrate the plasma membrane owing to the presence of specific proteins (permeases or translocators): examples are sugars, amino acids and certain inorganic salts. Other substances are unable to enter the living cell: examples are polysaccharides, proteins and phosphorylated compounds (sugar-phosphates, NDP-sugars, etc.).

It is valuable to measure uptake of a radioactive precursor because **(i)** it gives a quick indication of the progress of a labelling experiment, and **(ii)** it provides a way of correcting 'incorporation' data for the inevitable variation in the amount of radioactivity made available to the cell. Sources of variation in uptake rate are: pipetting errors, effects of the particular experimental conditions on the uptake system (e.g. the functioning of permeases, the provision of ATP for active transport, or fluctuation in pH), and variable accessibility of the cells to the medium (especially in studies with intact plant organs).

The two principal ways of measuring uptake are: **(i)** Subtraction of the radioactivity remaining in the culture medium from that initially supplied. As subtraction is involved, this method is most suitable for long-term work, where over 10 % of the precursor is being taken up. **(ii)** Assay of the radioactivity trapped within the tissue. This requires a washing step to remove radioactivity adhering to the cells but not yet taken up, and washing introduces the possibility that some intracellular radioactivity will be leached out; the washing should therefore be rapid. Also, some of the radioactivity taken up may have been incorporated into insoluble polymers and most techniques for determination of radioactivity do not detect insoluble radiochemicals as efficiently as soluble ones (Section 2.8). Method (ii) is thus best suited to short-term work, where incorporation is negligible.

Measurements of uptake often make two assumptions: **(a)** that the precursor does not bind directly to the cell walls (although it may do if, for example, it has a positive charge and can bind ionically to acidic polysaccharides) and **(b)** that the end-product metabolites, e.g. polysaccharides, are not released back into the medium (a proportion of them often are, e.g. pectins and hemicelluloses). Direct binding to cell walls can be tested with killed (e.g. frozen/thawed) tissue; the release of any end-product metabolites can be checked by paper chromatography of a sample of the spent medium. Another source of artefacts is the release of ^3H_2O by metabolism of ^3H-labelled compounds: any ^3H_2O soon becomes evenly distributed between the intra- and extracellular water. If this occurs rapidly, it may give the impression that the precursor is not taken up. An indication of the presence of ^3H_2O would be loss of radioactivity on drying.

An effective method for monitoring the uptake of $[^{14}C]$leucine over the short term is given in Panel 2.4.

PANEL 2.4:   Short-term assay of uptake of $[^{14}C]$leucine in cell cultures and its incorporation into protein [573]

Step 1:   Incubate a suspension culture (20 ml) in a 100-ml conical flask. It is helpful to leave the cells in the flask for about an hour to overcome 'transfer shock'. The flask should be shaken gently, under conditions as close as possible to the culture conditions normally used.

Step 2:   Add 100 µl of $[^{14}C]$leucine solution (0.1 µCi;   high specific activity) and start timing.

Step 3a: After timed intervals (0-60 min), remove and filter a 0.5-ml sample of cell suspension. For the filtration, quickly pass the sample through a 2.5-cm disk of Whatman GF/C glass fibre filter paper set up in a Hartley funnel or in a Millipore 12-port filtration assembly. Wash the cells with 2 x 10 ml of culture medium, using a moderate vacuum to expedite drainage. The entire filtration and washing should take less than 15 seconds.

Step 3b: After alternate timed intervals, transfer further 0.5-ml aliquots into tubes containing 5 ml of 15 % trichloroacetic acid (TCA) at 0°C. Incubate in an ice bucket for 1 h. Filter as in step 3a, washing the cells with 2 x 10 ml of 15 % TCA (0°C).

Step 4:   Place the cells, plus filter, in scintillation fluid and determine the ^{14}C content. Data from step 3a represent **uptake**; data from step 3b represent **incorporation** into protein.

2.5: ACTIVATION AND INTERCONVERSION OF PRECURSORS

Once taken up, the radioactive precursor must be activated before it can be incorporated into polymers. Sugars have to be phosphorylated and then attached to nucleotides, amino acids have to be esterified with tRNAs, and phenolic acids probably have to be thioesterified with CoA or converted to glucosyl-esters. It is sometimes useful to monitor these activation reactions during in vivo labelling experiments, in order to provide a more rigorous way of correcting incorporation data for fluctuations in the quantity of radioactivity actually available to the polymer-synthesising machinery (cf. Section 2.4). In addition to these activation reactions, some precursors can be interconverted — for instance [3H]arabinose acts as a precursor for both UDP-[3H]arabinose and UDP-[3H]xylose [179]. Thus the quantity of radioactivity available to the polymer-synthesising machinery (in this case xylan-synthesising) may depend not only on the capacity of the uptake and activation pathways, but also on that of the interconversion pathway.

Some activation and interconversion pathways relevant to cell wall synthesis are summarised in Fig. 2.5. Precursors that can be fed to cultured cells in order to label their walls are listed in Panel 2.1.1, which also shows the major interconversions undergone by each precursor.

Methods for analysis of [3H]sugar-phosphates, NDP-[3H]sugars, CoA-thioesters, glucosyl-esters and aminoacyl-tRNAs are given in Section 5.3.

2.6: DETERMINATION OF WALL POLYMER SYNTHESIS
2.6.1: Measurement of radioactive polymers

The most common way of determining radioactivity incorporated into polymers is based on differences in solubility between the precursor (plus its low-M_r metabolites) and the polymers. As a rule, precursors and low-M_r derivatives are soluble in 80 % ethanol, ice-cold 20 % trichloroacetic acid, and phenol/acetic acid/water (2:1:1 w/v/v) [PAW]. Most proteins and polysaccharides are insoluble in 80 % ethanol. Proteins are insoluble in ice-cold 20 % trichloroacetic acid (Panel 2.4), although water-soluble polysaccharides usually remain soluble in this. Polysaccharides are insoluble in PAW, whereas proteins (even membrane- and chromatin-bound) are soluble. Glycoproteins may exhibit intermediate behaviour. In addition, some polymers are water-insoluble, especially those that have become

33

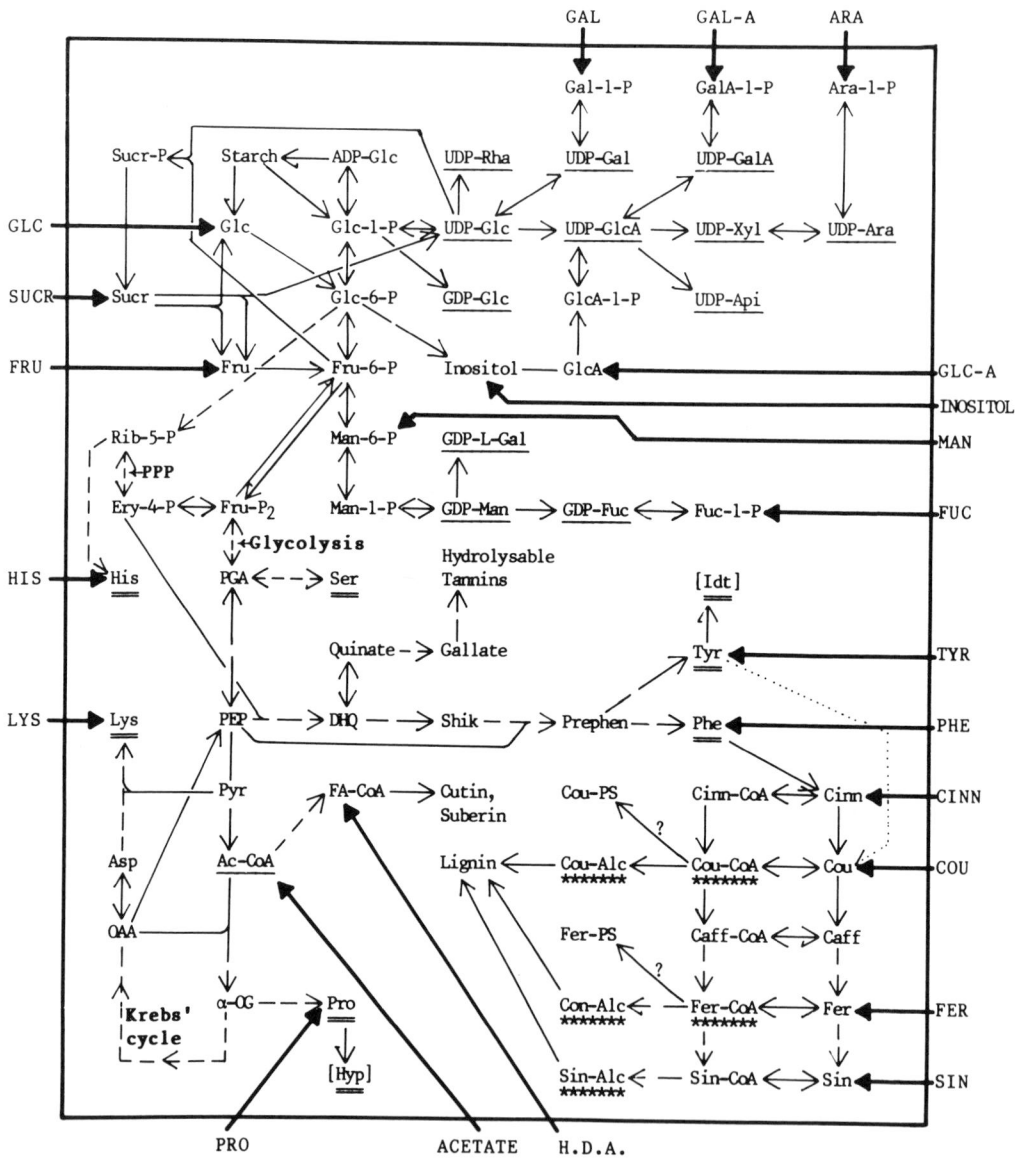

Fig. 2.5: Metabolic map. For legend see facing page.

34

integrated into the cell wall, whereas precursors and low-M_r metabolites are usually water-soluble.

A common technical problem is that, although a polymer may be insoluble in 80% ethanol (in the sense that dry polymer will not go into solution in it), an aqueous solution of the polymer may fail to form a collectable precipitate upon addition of ethanol. With dilute polymer solutions it is helpful to add a 'carrier' polymer (e.g. 0.5 % soluble dextran or bovine serum albumin) with which the radioactive polymer can co-precipitate. Flocculation can also be aided by addition of a salt [ammonium formate (0.1 M) is useful because it is highly ethanol-soluble], and/or by storage for several hours at 0°C.

In practice (Panel 2.4), radioactive cells are killed in a solvent (Panel 2.2 c) chosen **(i)** to extract the precursor and any low-M_r metabolites and **(ii)** to precipitate the polymers within the cells. Radioactive polymers present in the culture medium can also be precipitated by similar means. The precipitated polymer is collected either by filtration on Whatman GF/C glass fibre disks, or by low-speed centrifugation. It is washed with the same solvent to remove the last traces of low-M_r material; washing consists of either passage of more solvent through the filter paper until the filtrates are no longer radioactive, or thorough resuspension of the pellet in solvent and re-centrifugation (repeated until the supernatant

Fig. 2.5: [opposite] **Metabolic map showing the activation and inter-conversion of some precursors involved in wall synthesis.**
    The box represents the cell. Solid arrows show fate of selected useful exogenous precursors. Donors for wall polysaccharides are indicated by _____ ; precursors of the wall phenolics by************; and the main amino acids of extensin by ========.

Abbreviations used:
Caff = Caffeate, **Con-Alc** = Coniferyl alcohol, **Cinn** = Cinnamate, **Cou** = p-Coumarate, **Cou-Alc** = p-Coumaryl alcohol, **DHQ** = Dehydroquinate [13], **Ery** = Erythrose, **FA** = Fatty acids, **Fer** = Ferulate, **Fru-P₂** = Fructose 1,6-bisphosphate, **HDA** = Hexadecanoic acid (palmitic acid), **OAA** = Oxaloacetate, **PEP** = Phosphoenolpyruvate, **PGA** = 3-Phosphoglyceric acid, **PPP** = Pentose-phosphate pathway, **Prephen** = Prephenate, **PS** = Polysaccharide, **Pyr** = Pyruvate, **Shik** = Shikimate [see ref. 13], **Sin** = Sinapate, **Sin-Alc** = Sinapyl alcohol, **Sucr** = Sucrose, α**-OG** = α-Oxoglutarate.
[   ]   indicates components formed post-translationally.
............→   indicates a reaction probably restricted to Monocots.
----------→   indicates multi-step reactions (the number of dashes does not necessarily indicate the number of steps involved, and an irreversible arrow does not necessarily ·indicate that all the steps are irreversible).

contains negligible radioactivity). Some polymers, e.g. neutral arabinans, are somewhat soluble in 70-80 % ethanol. In such cases, dialysis or GPC (Section 3.5.1) give better recoveries than washing in solvents.

A convenient alternative method for measuring radioactive polymer in small volumes (less than 0.5 ml) of culture medium is to streak the sample on to Whatman 3MM chromatography paper, run it overnight in ethyl acetate/acetic acid/water (10:5:6) [which causes most precursors and low-M_r metabolites to migrate well away from the origin — Panel 4.5.2 d, page 164] and then to cut out and scintillation-count the origin, where the polymers will have remained. [For methods, see Section 4.5.2.]

2.6.2: Calculation of the rate of polymer synthesis: Pools

In prolonged non-specific labelling experiments (e.g. feeding of [^{14}C]glucose to glucose pre-grown cells for at least 2 h, so that the rate of incorporation becomes linear) it is reasonable to assume that the new polymer molecules synthesised have the same specific radioactivity (SA) as the radioactive glucose in the medium. For instance, if [^{14}C]glucose of SA 10 Ci/mol is fed for 5 h, after which time radioactivity is being incorporated into cellulose linearly at 4 µCi/h, then the rate of cellulose synthesis was approximately 0.4 µmol/h.

In contrast, in 'specific' labelling (Section 2.1.2), measurement of the incorporation rate does not readily give the rate of polymer synthesis. There is uncertainty in **(a)** the rate of uptake of the precursor (say, [^{3}H]-arabinose), **(b)** the rate of conversion of the supplied precursor into the immediate precursor of the polymer (e.g. UDP-[^{3}H]xylose), and **(c)** the size of the non-radioactive UDP-xylose pool already present in the cells, and continually being replenished, which will compete with the UDP-[^{3}H]xylose molecules for the attention of xylan synthase. Problems (a) and (b) have already been mentioned (Sections 2.4 and 2.5); pool problems (c) are discussed below.

One potential solution to problem (c) would be to measure the specific radioactivity (SA) of the UDP-[^{3}H]xylose, and to assume that the SA of the newly-synthesised polymer molecules was the same. However, measurement of SA is very difficult for most immediate precursors owing to the low concentrations at which they are present in vivo. In addition, the assumption mentioned may be invalid because the UDP-xylose may be present in two or more pools (e.g. cytosol and Golgi cisternae), some of which are more easily labelled than others but only one of which is utilised as precursor for the polymer under investigation. Measurement of SA of total extracted UDP-[^{3}H]xylose gives an average SA for the whole cell.

For this reason, it is best to restrict the use of specific labelling to the determination of <u>ratios</u> of the rates of synthesis of a group of related polymers. For instance, exogenous [^{3}H]arabinose will very efficiently label the arabinose and xylose residues of arabinans, arabino-galactans, xyloglucans, xylans, rhamnogalacturonans, feruloyl-arabinans, and extensin. Some of these are synthesised in extremely small amounts, and would be difficult to study by non-specific labelling (e.g. with [^{14}C]glucose) owing to the diversion of large amounts of radioactivity to other polymers. Specific labelling of the pentose residues allows the rates of synthesis of these particular polymers to be determined relative to one another. The absolute rate of synthesis of one major pentose-rich polymer (e.g. xyloglucan), measured separately by [^{14}C]glucose-feeding, would allow estimation of absolute rates for each of the minor pentosans.[*]

2.7: INTRODUCTION TO PLANT TISSUE CULTURE TECHNIQUES

Many parts of this book recommend the use of plant tissue culture. The present Section is a brief outline of the methods involved [see also H.E. Street (1977) 'Plant Tissue and Cell Culture', Blackwell, Oxford].

2.7.1: Initiation of a callus culture

The culture is initiated as wound callus at the cut surface of a plant organ. The organ, e.g. a stem, is surface-sterilised (Panel 2.3.2). Several 1-cm lengths are cut with a sterile scalpel and aseptically trans-ferred into a 9-cm Petri dish containing 10 ml of culture medium composed of the following (dissolved in double-distilled water):

a sugar (usually 2 % sucrose, but 2 % glucose is equally effective and preferable for many cell wall labelling purposes),
mineral salts (providing the elements N, P, S, B, I, Cl, K, Na, Mg, Mn, Fe, Zn, Cu, Mo and Co),

---

[*]This approach still makes an assumption — **either** that all pentose-containing polymers draw on a common pool of UDP-pentose precursor molecules **or** that they draw on different pools of equal specific radioactivity — but this assumption is clearly less risky than that mentioned in the previous paragraph.

growth regulators (an auxin e.g. 2,4-dichlorophenoxyacetic acid; and, if found necessary for growth, also a cytokinin e.g. kinetin), NaOH or HCl to give a suitable pH (usually 4-6), agar (1-2 %) to gel the mixture.

A useful general-purpose culture medium is given in Panel 2.7.1 a. Sometimes 'vitamins' (e.g. inositol, thiamine, pantothenic acid, nicotinic acid and pyridoxine), amino acids, yeast extract, or coconut milk are also supplied. However, they are often unnecessary and should be avoided if possible; this is especially true for inositol (see Section 2.1.2). It has been rare to include pH buffers, although they would be valuable in some biochemical studies: see Panel 2.7.1 b.

PANEL 2.7.1.a: Composition of recommended culture medium[1] (from [367])

Stock ref.	Compound[2]	Concn in in stock soln (g/l)	Amount of stock used (ml/l)	Final concentration of compound (mg/l)	Molar (approx.)
-	KNO_3	[added as solid]		1900	2×10^{-2}
-	NH_4NO_3	[added as solid]		1650	2×10^{-2}
A	$CaCl_2.2H_2O$	294	1.5	441	3×10^{-3}
B	$MgSO_4.7H_2O$	7.4	50.0	370	2×10^{-3}
	$MnSO_4.4H_2O$	0.45		22.5	1×10^{-4}
C	H_3BO_3	1.24		6.2	1×10^{-4}
	$ZnSO_4.7H_2O$	2.07		10.4	4×10^{-5}
	KI	0.17	5.0	0.83	5×10^{-6}
	$Na_2MoO_4.2H_2O$	0.05		0.25	1×10^{-6}
	$CoCl_2.6H_2O$	0.005		0.025	1×10^{-7}
	$CuSO_4.5H_2O$	0.005		0.025	1×10^{-7}
D	KH_2PO_4	181	0.94	170	1×10^{-3}
E[3]	$Na_2EDTA.2H_2O$	7.4	1.0	7.4	2×10^{-5}
	$FeCl_3.6H_2O$	5.4		5.4	2×10^{-5}
F[4]	2,4-D	0.2	5.0	1.0	5×10^{-6}
G[5]	Kinetin	0.2	2.5	0.5	2×10^{-6}
-	Glucose	[added as solid to 1.8 %]			1×10^{-1}
-	NaOH or HCl	[added to bring pH to 4.5][6]			

[for footnotes see opposite]

38

PANEL 2.7.1 b: Buffers suitable for use in culture media

[When additions are made to suspension cultures it is important to maintain
the pH. Many common buffers are toxic to plant cells. Little systematic
attempt has been made to define which are non-toxic; some that have proved
successful in our laboratory are listed here. However, it would be wise to
check toxicity against the particular culture under investigation; a convenient short-term test is to measure incorporation of [14C]leucine into
protein over a period of 1 h (Panel 2.4), and compare it with a buffer-free
control at about the same pH. Buffers are generally useful at pHs within
0.5 pH units of their pK_a(s). The buffer acid is used at a final concentration of 10-25 mM, and brought to the correct pH by addition of NaOH.]

BUFFER	pK_a at 25°C
Found sutiable	
MES	6.1
PIPES	6.8
Maleic acid	2.0, 6.2
Phthalic acid	3.0, 5.4
Tartaric acid	3.0, 4.4
Found unsuitable[1]	
Acetic acid	4.8
Formic acid	3.8
Citric acid	3.1, 4.8, 6.4
Malic acid	3.4, 5.1
Fumaric acid	3.0, 4.4
Succinic acid	4.2, 5.6

[1]These compounds caused loss of chlorophyll and a decrease in the rate of
protein synthesis in cultured spinach cells.

[footnotes to Panel 2.7.1 a]

[1]The ingredients are prepared as 7 concentrated stock solutions (**A-G**)
which are stored at 4°C. To prepare 1 litre of medium, the stock solutions
plus the solid ingredients are added with vigorous stirring to 0.9 litre of
water, the pH is adjusted, and the volume made up to 1 litre with water.

[2]All ingredients except 2,4-D and kinetin are AnalaR grade. Water should
preferably be double-distilled.

[3]Dissolve the $Na_2EDTA.2H_2O$ and $FeCl_3.6H_2O$ separately. When completely
dissolved, mix the two solutions with stirring.

[4]2,4-D (20mg) is dissolved in 1 ml DMSO. This solution is then added,
slowly and with very vigorous stirring, to 100 ml water.

[5]Kinetin (20 mg) is dissolved in 1 ml 1 M NaOH. This solution is then
added, with stirring, to 100 ml water.

[6]pH 4.5 is lower than traditionally used, but often optimal for growth.

Typical incubation conditions are 25°C under continuous dim light; the Petri dish is sealed with 'Parafilm'. Callus is sub-cultured whenever necessary by aseptic transfer of a small portion on to a new Petri dish. This is preferably done at regular intervals (usually every 2-4 weeks or after a 5- to 10-fold increase in size).

2.7.2: Characteristics of callus cultures

Callus cells show limited similarity to any cells in the intact plant. They divide rapidly, but usually differ from meristematic cells in having large vacuoles. Typical callus cells are white or pale brown (rarely green), rich in amyloplasts, and lack secondary cell walls. In some calli the cells are firmly held together, forming hard lumps; others are friable, a property often associated with rapid growth and very large vacuoles. Calli from stems, leaves and roots usually show no consistent differences from each other. However, replicate calli derived from a single source often differ, both structurally and metabolically; the cause of this 'somaclonal variation' is unclear, but it makes it desirable to use one particular line of callus throughout any set of experiments.

2.7.3: Establishment and maintenance of a suspension culture

Callus can be converted to a cell-suspension culture by aseptic transfer of a 2-g portion into a 250-ml conical flask containing 50 ml of autoclaved liquid medium (usually the same composition as for callus but without agar). The flask is sealed with sterile aluminium foil and incubated with constant shaking (90-120 rpm) at 25°C in dim light. Friable calli are more likely than lumpy ones to form fine suspensions. Suspensions are sub-cultured at 1-3-week intervals by transfer of 5 ml of old suspension, e.g. via an autoclaved ARH pipetting unit [A.R. Horwell, London], into 50 ml of fresh medium. [It is useful to have cannulae of internal diameter 4 mm custom-made for the ARH units, as the cannulae routinely supplied are too narrow for some lumpy suspension cultures.] Unfortunately, suspension cultures never consist of single cells, but, for the sake of easy pipetting, the aggregates should be of less than 100 cells. Disaggregation can, if necessary, be encouraged by sieving through sterile gauzes (typically a sieve with ca. 1-mm holes would be suitable) at each sub-culture. Sieving can also be used immediately before a particular experiment, so that only the smaller aggregates are used.

The growth of suspension cultures is conveniently monitored by measurement of packed cell volume (PCV): 10 ml of the suspension is centri fuged at low speed in a graduated tube. The volume of packed cells is ex pressed as a % of the 10 ml. PCV is often used to measure cell expansion. Dry weight is measured after washing in water and heating the cells at 95°C for 24 h. Cell number is measured in a haemocytometer after maceration in CrO_3 to disaggregate the clumps [484]. Viability of the cells can be quickly assessed prior to an experiment by staining with Evans' Blue or fluorescein diacetate.

2.7.4: Differentiation

Both callus and suspension cultures can sometimes be induced to undergo cell differentiation, organogenesis or embryogenesis. This is achieved by empirical manipulation of the composition of the medium, and can provide a useful system for study of cell wall changes associated with these developmental processes. Cultures that have been recently initiated from the intact plant are more likely to respond than are old cultures.

2.7.5: Synchronisation of cell division

Suspension cultures can be induced to undergo synchronous cell division, and this is useful for study of changes occurring in the cell wall during the cell cycle. The culture is grown to stationary phase in a medium modified such that the supply of an essential component (e.g. nitrate, phosphate or cytokinin) will run out prematurely. This may arrest most of the cells in one particular phase (e.g. G1) of the cell cycle. The cells are then replaced in standard medium, whereupon cell division resumes in the whole population simultaneously. Division is partially synchronous for 1-2 cell cycles thereafter. Occasionally the synchrony is perpetuated, or even reinforced, for 4-5 cell cycles.

2.7.6: Isolation of protoplasts [153]

Isolated protoplasts (cells with the walls removed) are useful for demonstration that an enzyme was located in the wall (Section 6.1). Also, isolated protoplasts may regenerate a new cell wall in vitro; they can thus be useful for study of the synthesis, secretion and cross-linking of wall polymers, uncomplicated by a large excess of existing wall material.

The cell wall is removed by enzymic digestion, which can be achieved in callus and suspension cultures, as well as in leaves and other organs. The tissue (stripped of any epidermis or finely chopped) is incubated at 25-30°C for 1-6 h, with very gentle shaking (20-30 rpm), in a wide-bottomed vessel containing a shallow layer of enzyme solution made up in solution **A** [1 mM $CaCl_2$, 5 mM KH_2PO_4 and 20 mM MES, adjusted to pH 5 with 1 M NaOH] supplemented with an osmoticum (typically 0.4-0.6 M mannitol) so that the protoplasts do not burst. The enzymes are commercial preparations from fungal cultures [161], and contain a wide variety of hydrolases. Cellulase activity, assayed viscometrically with carboxymethylcellulose (Panel 6.5 part 2a), gives little indication of the effectiveness of an enzyme mixture because cellulase alone will not readily breach cell walls; cellobiohydro-lase activity is also required (Section 11.2). One very effective preparation, which can be used alone, is 'Driselase' [sold by Sigma Chemical Co.]; others are 'Onozuka R-10' plus 'Macerozyme', or 'Micellase' plus 'Pectolyase'. These enzyme mixtures are impure and often incompletely soluble. It is advisable to eliminate insoluble matter by centrifugation before use of the enzyme, and it is a simple matter to remove low-M_r solutes [added to the enzymes by the manufacturers to enhance stability] by GPC on Bio-Gel P-2 or Sephadex G-25 (Section 3.5.1) followed by freeze-drying; the enzymes can then be used at much higher concentrations, which may be beneficial in tissues with resistant walls. Tissues vary widely in the time taken for protoplast isolation, and in yield.

To free the isolated protoplasts from remaining whole cells and from enzyme, the suspension is layered on to an equal volume of solution **A** in a sucrose solution equimolar to the mannitol (sucrose is denser), and then centrifuged at low speed (about 50 **g** for 2-5 min). Living, isolated proto-plasts collect at the interface whereas walls, walled cells and dead proto-plasts pellet. Last traces of enzyme solution are removed by repeated resuspension and pelleting of the protoplasts in 0.5 M mannitol in soln **A**.

2.8: INTRODUCTION TO THE USE AND DETECTION OF RADIOACTIVITY [199]

N.B. The use of radioactivity is controlled by law in most countries. New users should obtain instruction from their Radiation Supervisor.

2.8.1: Basic concepts

A radioactive atom is one whose nucleus is liable to decay, emitting radiation. Such atoms have the same chemical properties as the non-radioactive form (isotope) of the same element. They can therefore be used as 'tracers', the emission of radioactivity providing a way of distinguishing them from other atoms of the same element, and also making them detectable in very small quantities. The radioisotopes discussed here are β-emitters, i.e. their decaying nuclei emit β-particles. The rate at which a sample emits β-particles is a measure of the number of radioactive atoms it contains. Quantity of a radioisotope is usually measured in dpm (disintegrations per minute), mCi (millicuries) or MBq (megabecquerels), 1 Bq being the amount that emits 1 β-particle per second. Conversions are:

1 Bq $= 10^{-6}$ MBq $= 10^{-9}$ GBq $= 60$ dpm.

1 mCi $= 10^{-3}$ Ci $= 10^3$ µCi $=$ ca. 37 MBq $=$ ca. 2.2×10^9 dpm.

The number of radioactive atoms that make up 1 Bq depends on the **half-life** ($t_{\frac{1}{2}}$, a physical constant equal to the time taken for 50 % of the radioactive atoms in any sample to dacay) — see Panel 2.8.1.

Another constant is the maximum energy of the β-particles (E_{max}: Panel 2.8.1). Isotopes of high E_{max}, e.g. ^{32}P, are easier to detect but more hazardous than those with low E_{max}, e.g. ^3H.

Radioisotopes, in the form of radioactive compounds, are available e.g. from Amersham International (U.K.), New England Nuclear (USA), and Sigma Chemical Co (USA). The main specifications when ordering are:

(a) Chemical position of radioactive atom(s). Compounds can either be

---

PANEL 2.8.1: Properties of commonly used radioisotopes

Isotope	Half-life	$E_{max.}$ (MeV)	Product of decay	Max. theoret. SA[1] of element (Ci/mmol)
^3H	12.4 y	0.0186	^3He	28.8
^{14}C	5730 y	0.156	^{14}N	0.0624
^{35}S	87.4 d	0.167	^{35}Cl	1494
^{32}P	14.3 d	1.709	^{32}S	9131

[1]The maximum theoretical specific activity (SA) of a compound depends on how many atoms of the element in question are present per molecule.

labelled uniformly, meaning that all the atoms of the element in question are equally likely to be radioactive (designated, for instance, [U-^{14}C]glucose); or the radioactivity can be restricted to one or a few specific positions (say, carbon atom nos. 1 and 6, designated [1,6-^{14}C]glucose). Uniform labelling is usually chosen when it is important to have a high specific activity. Restricted labelling may be preferred for economy or if it results in the radioactivity being lost when a particular metabolic pathway is followed: e.g. the ^{3}H from [2-^{3}H]mannose is 'lost' as ^{3}H$_2$O when the molecule is isomerised to fructose, but retained during incorporation into mannan.

(b) Specific activity (SA). This is a measure of the average number of radioactive atoms per molecule of the compound. It is expressed in Ci/mmol or GBq/mmol. It depends not only on the ratio of radioactive : non-radioactive atoms, but also on the number of atoms of the element present in the molecule. Thus, the maximum possible SA of [^{14}C]glucose (a 6-carbon compound) is 374 Ci/mol and that of [^{14}C]glycerol (a 3-carbon compound) is 187 Ci/mol. The maximum theoretical SA for common radioisotopes is shown in Panel 2.8.1.

(c) Quantity of radioactivity. This is specified in Ci or Bq. The amount required can be deduced from consideration of the sensitivity of the detection method to be used (see Section 2.8.2).

Radiochemicals should be stored under the conditions recommended by the suppliers to minimise microbial and radiation-induced decomposition.

2.8.2: Methods for the detection of radioactivity

Several methods are in use (Panel 2.8.2 a): the choice is determined by the importance attached to cost, spatial resolution (the ability to say where the radioactivity is within the specimen), sensitivity (the ability to detect low levels), and ability to recover of the sample after measurement. Sensitivity can be assessed as counting efficiency, [β-particles detected (measured in cpm)] / [β-particles emitted (measured in dpm)]. The principal methods available are:

(a) Autoradiography. The sample, which must be dry, is pressed flat against X-ray film (e.g. Cronex 4) in complete darkness for several days or even months. The film is then developed and fixed according to the manufacturer's recommendations, and blackening caused by the β-particles [cf. (b) below] shows where the radioactive material had been. Autoradio-graphy is often used for paper and thin-layer chromatograms, whole plants (after pressing), and microscopic specimens, for which special techniques are needed [574]. Autoradiography is at best only semi-quantitative, but gives excellent spatial resolution with ^{14}C. Resolution is even better with ^3H, but ^3H has such a low E_{max} as to be virtually undetectable on a TLC spot at normal working levels.

  (b) <u>Fluorography</u> is similar to autoradiography, but gives quicker results. The specimen, impregnated with fluor (e.g. PPO — see (d) below),

---

PANEL 2.8.2 a: Comparison of methods for detection of radioisotopes

Method	Cost	Spatial resolution	Counting efficiency		Quantification	Recovery of sample	Approx. time required to detect reliably 1 nCi[1]	
			^3H	^{14}C			^3H	^{14}C
Autoradiography	£100	+++++	−	++	±	easy	−	1 dy
Fluorography	£150	++++	+	++	±	often easy[3]	2 wk	10 h
GM-cntr	£350	+	−	2%	+	easy	−	1 mn
RITA	£15000	++[2]	1%	10%	+++	easy	10 mn	1 mn
Radio-isotope camera	£9000	++	+	++	±	easy	30 mn	10 mn
LSC (aqueous)	£20000	−	35%	75%	++++	hard	1 mn*	1 mn*
LSC (non-aq)	£20000	−	6%	55%	+++	easy	1 mn*	1 mn*

[1]I.e. approximately 2200 dpm, dried as a 1 cm^2 spot on Whatman no. 1 paper (except LSC aqueous method).

[2]Resolution is essentially one-dimensional only.

[3]The PPO could be hard to separate from some phenolic wall components.

*Or less.

is exposed to a film to detect light that is secondarily emitted, rather than the β-particles themselves. This method makes it feasible to see ^3H on TLC plates and paper chromatograms. If the radioactive compounds of interest are insoluble in ether (sugars and amino acids, but not phenols), the whole chromatogram is dipped in a 7 % PPO solution in diethyl ether and dried [409]; if not, the PPO solution is sprayed on. For the most rapid results, the film is 'pre-flashed' for less than 1 ms [313] before exposure to the chromatogram at about -80°C (e.g. on a bed of solid CO_2).

(c) Instruments based on the ionisation of a gas. A basic Geiger-Müller (GM) counter is cheap, and can give quantitative data. The sample, preferably dry, is placed close to the detector. β-Particles pass through a thin 'window', into the gas, which is thereby ionised. Standard GM counters detect ^{14}C with ca. 1 % efficiency, but will hardly detect ^3H at all because its low-E_{max} β-particles cannot penetrate the window. More sensitive versions, which can detect ^3H, use an 'open' window, in which case the gas is constantly replenished from a cylinder.

Specialised radiochromatogram scanners incorporate a moving detector or a long detector [e.g. the 'RITA' — Raytest Instruments, Sheffield], permitting continuous measurement of radioactivity along a 1-dimensional TLC or paper chromatogram.

A different refinement uses a broad array of individual detectors to map qualitatively the distribution of radioactivity on a 2-dimensional TLC plate. This gives pictures similar to, but even quicker than, fluorography (typically ½ h rather than 1-2 weeks for a 0.001 μCi spot of ^3H) though with considerable loss in spatial resolution. Such an instrument is the 'Radioisotope Camera' [Birchover Instruments, Letchworth].

(d) Liquid scintillation-counting (LSC). LSC is the method of choice for quantitative work. The sample is mixed with a scintillant, i.e. a fluid that absorbs β-particles and re-emits their energy as light. The scintillation-counter registers the number of pulses of light produced per minute (cpm). This technique has no spatial resolution, but provides efficient detection of total radioactivity within any one sample.

A recommended scintillant is **0.5% PPO + 0.05% POPOP in toluene.** The sample + scintillant are mixed in 5- or 20-ml scintillation vials made of glass or translucent plastic. Certain widely used reagents, especially acetone and any coloured substances, reduce the counting efficiency of LSC, and should be avoided (Panel 2.8.2 b). TLCs and paper chromatograms should whenever possible be stained after LSC. Four types of sample are frequently met:

(i) Specimen dissolved in MeOH (or other toluene-miscible solvent). Sample is added directly to scintillant in the ratio sample/scintillant = 1:10.

(ii) Specimen dissolved in H_2O (or other toluene-immiscible solvent). Triton X-100 is added to increase miscibility. The sample is mixed with a modified scintillant [**0.33 % PPO + 0.033 % POPOP in toluene/Triton X-100 (2:1)**] in the ratio 1:10 (sample:scintillant), and shaken until a single, clear phase forms. Deviation from the 1:10 ratio in either direction results in immiscibility and alters the counting efficiency.

(iii) Insoluble specimen e.g. cell walls. These are either mixed directly with non-Triton scintillant (in which case counting efficiency is low and rather variable) or else hydrolysed by the 72 % H_2SO_4 method (Panel 4.2.1 b), neutralised, and mixed with Triton scintillant (in which case the counting efficiency is high, but the specimen cannot be analysed further).

(iv) Chromatogram strips. [Cf. Section 4.5.2.] Dried strips from TLCs or paper chromatograms can be soaked in non-Triton scintillant (1 ml is enough for a 2 x 4 cm strip of Whatman no. 1 paper) and assayed with low but reproducible efficiency. Most cell wall substances (sugars and amino acids) are insoluble in toluene and remain on the paper; some phenols tend to leach into the toluene and are thus counted at higher efficiency. [To test for leaching, remove the strip and re-count the scintillant.] The ability of sugars to remain on the strip during LSC is useful, as it means that they can subsequently be stained on the strip (Section 4.5.2), or else re-covered by elution with water, after washing in toluene and drying.

A higher and more consistent counting efficiency is obtained if the sugar is eluted from the strip in water (Panel 4.5.2 h) and then assayed in Triton scintillant. If enough sample is available, a known proportion of the eluate can be assayed in Triton scintillant while the rest is used for further analysis. Alternatively, if only small amounts of radioactive material are present, the whole operation can be performed in one scintillation vial: H_2O (0.4 ml) is added to the dry paper, then Triton scintillant (4 ml) is added, shaken, and left to stand overnight in a capped vial. The sample, including paper, is then shaken again and read in the scintillation counter. Despite its reproducibility and simplicity, this approach has the disadvantage that recovery of the sugar, after assay, is difficult.

PANEL 2.8.2 b: Effect of some common laboratory reagents on the efficiency of liquid scintillation-counting
Identical samples of [^{14}C]lactose or [^{3}H]arabinose were prepared as indicated, and assayed by liquid scintillation-counting.

Method of preparation of sample	Approx. counting efficiency as % of that obtained[1] with sample in distilled water	
	^{3}H	^{14}C
4 ml of TS[2] added to sample dissolved in 0.4 ml of:		
Water	(100)	(100)
2 M TFA	60	80
2 M HOAc	100	100
2 M NH$_3$	100	100
1 M NaOH	70	90
10% Pyridine	70	90
5% Pyridine + 5% HOAc	70	90
2 M CaCl$_2$	100	100
80 % Ethanol	100	100
80 % Acetone	15	80
Dimethylsulphoxide	100	100
Dimethylformamide	100	100
Ethyl acetate	110	100
Sample dried on to 1 x 4 cm paper and added to 4 ml of NTS[3]		
Whatman no. 1	15	80
Whatman no. 3	15	80
Sample dried on to 1 x 4 cm paper and then treated with 0.4 ml H$_2$O followed by 4 ml of TS[2]		
Whatman no. 1	90	95
Whatman no. 3	75	95
	[→85[4]]	

[1]Actual counting efficiencies: ^{3}H = 35 %; ^{14}C = 75 %.

[2]TS = Triton scintillant [0.33 % PPO + 0.033 % POPOP in toluene / Triton X-100 (2:1, v/v)].

[3]NTS = Non-Triton scintillant [0.5 % PPO + 0.05 % POPOP in toluene].

[4]The relative efficiency increased from 75 % to 85 % on standing overnight, presumably owing to gradual leaching from the paper.

3 Wall polymers: extraction and fractionation

Growing plant cell walls contain polysaccharides, glycoproteins and phenolics. Since information on protein chemistry and methodology is widely accessible, the emphasis here is on carbohydrates and phenolics.

3.1: Polysaccharides

Plant cell wall polysaccharides are built up of about a dozen main sugars (Fig. 3.1.1), which can be linked in an enormous variety of ways. The types of linkage, as well as the sugars themselves, dictate the properties of polysaccharides, and thus the uses to which the plant can put them.

3.1.1: Chemistry of sugars

A sugar is a polyhydroxy alcohol with a C$=$O group. Sugars in which the C$=$O group is an aldehyde are **aldoses**; those in which it is a ketone are **ketoses**. Almost all cell wall sugars are aldoses, and have 5 or 6 carbon atoms (i.e. are **pentoses** or **hexoses**). The single exception is KDO, an 8-carbon ketose present in very small amounts (Fig. 3.1.1 s).

Straight-Chain and Ring Forms. Most free monosaccharides can exist in 1 straight-chain and 4 different ring forms (Section 3.1.2). The 5 forms rapidly interconvert in aqueous solution, setting up an equilibrium mixture. The straight-chain form has a $=$O group whereas the ring forms do not. The ring forms are the building blocks of wall polysaccharides.

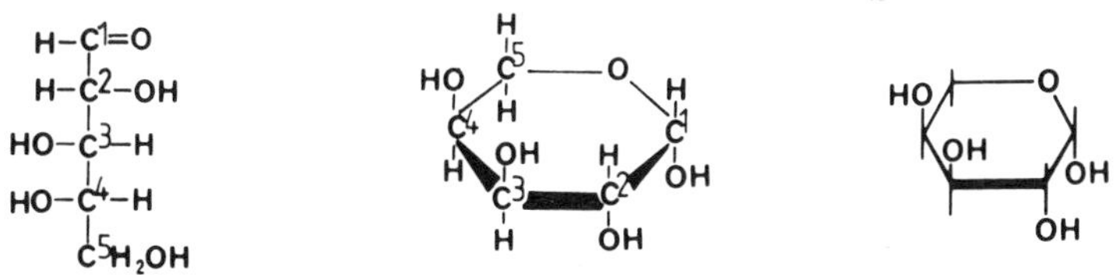

Arabinose: straight-chain form, one of the four ring forms,
and the same ring-form in abbreviated representation.

Numbering of carbon atoms. The end C atom nearest the potential $=$O group is no. 1. Others are numbered in sequence from that. In ring-forms, the potential C$=$O is shown as the right-hand extremity.

<u>Naming the sugar</u>.   Straight-chain pentoses and hexoses have 3 and 4 asymmetric centres respectively.   There are thus 2^3 pentoses (D-ribose, D-arabinose, D-xylose and D-lyxose, and the four corresponding L-sugars) and 2^4 hexoses (D-allose, D-altrose, D-glucose, D-mannose, D-gulose, D-idose, D-galactose and D-talose, and the eight corresponding L-sugars).   Sugars are named by reference to the configuration of the —H and —OH groups at these centres.   Changing the configuration at an asymmetric centre is **epimerisation**, which can in specific cases be enzyme-catalysed.

Arabinose and its 4-epimer, xylose.

<u>Optical isomers</u>   are named from the configuration at the <u>penultimate</u> C atom (= no. 4 in a pentose).   If this configuration resembles D-glyceraldehyde it is a D-sugar; if not it is an L-sugar.   A D-sugar (e.g. D-arabinose) is the mirror image of the corresponding L-sugar (L-arabinose) and thus differs in configuration at <u>all</u> 3 of its asymmetric centres.   The fact that D- and L- forms are mirror images can be seen in the ring form if it is remembered that the ring is supposed to be viewed edge-on, with the —H, —OH and other substituents pointing up and down.   Although D- and L-arabinose are identical in chemical reactions, any given enzyme will only act on one or the other, so optical isomerism is biologically important. The two forms can be distinguished by the direction in which they rotate plane-polarised light, and more sensitively by enzyme-based assays, but cannot normally be separated by chromatography.

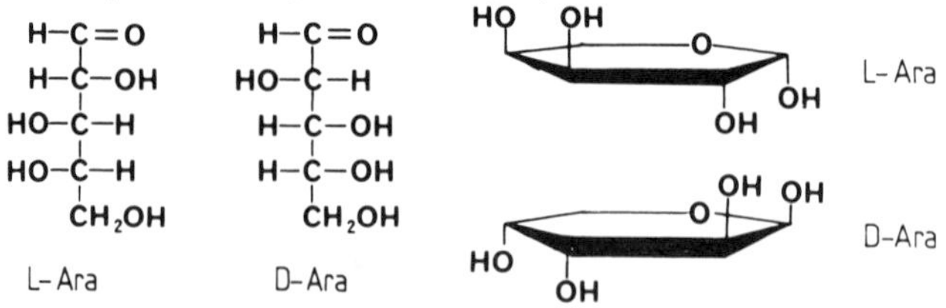

D- and L-arabinose are mirror images.

<u>Modified sugars</u> (Fig. 3.1.1).   Wall polymers contain several modified sugars which can alter the properties of the polysaccharides in which they occur;   they require identification in a complete description of a polymer.

50

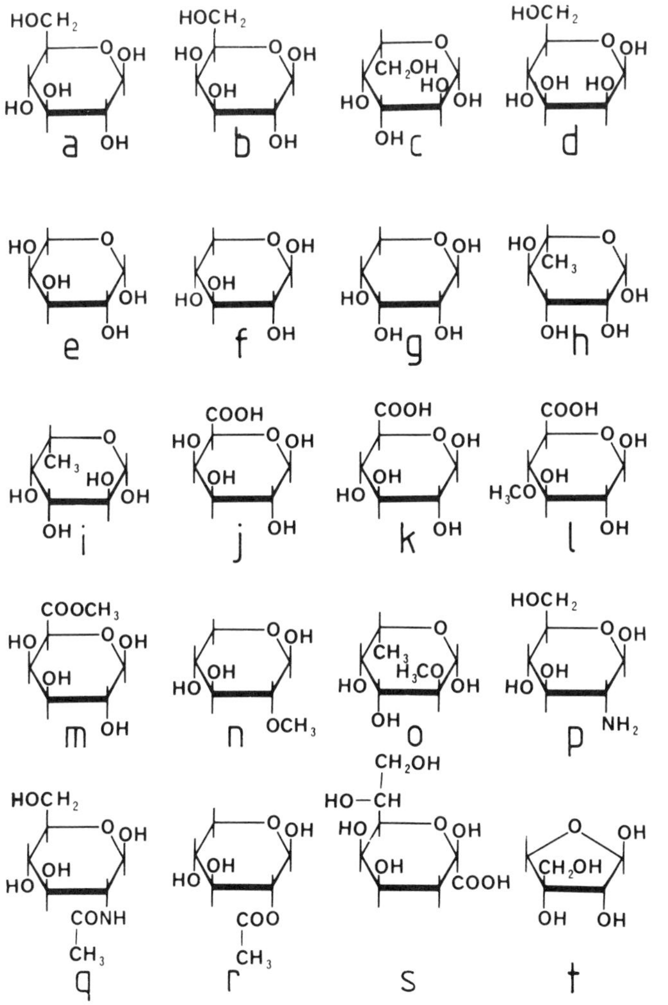

Fig. 3.1.1 a-t: Structures of the major sugars of the plant cell wall, and some of their naturally-occurring derivatives. All the sugars are shown in their β-pyranose form, except apiose, which can only adopt the furanose form. (a) D-Glucose, (b) D-Galactose, (c) L-Galactose, (d) D-Mannose, (e) L-Arabinose, (f) D-Xylose, (g) D-Ribose, (h) L-Rhamnose, (i) L-Fucose, (j) D-Galacturonic acid, (k) D-Glucuronic acid, (l) 4-O-Methyl-D-glucuronic acid, (m) D-Galacturonic acid methyl ester, (n) 2-Ō-Methyl-D-xylose, (o) 2-O-Methyl-L-fucose, (p) D-Glucosamine, (q) N-Acētyl-D-glucosamine, (r) 2-O-Acetyl-D-xylose, (s) Ketodeoxyoctulosōnic acid (KDO), (t) D-Apiose. [N̄.B. — D-ribose is probably not a wall component, but is often encountered from contaminating RNA during analysis.]

51

Examples are **uronic acids**, in which the —CH$_2$OH group is replaced by —COOH; **6-deoxy-hexoses** e.g. fucose and rhamnose, in which the —CH$_2$OH group is replaced by —CH$_3$; **2-amino-sugars** (= 2-amino-2-deoxy-sugars) e.g. glucosamine, in which the —OH group on carbon atom no. 2 is replaced by —NH$_2$; O-**methyl-sugars**, in which an —OH group is replaced by an —O—CH$_3$, and a few others (Fig. 3.1.1). In addition to these modified sugars, which are recovered intact after hydrolysis (Ch 4), there are others (e.g. methyl esters of uronic acids, acetyl and feruloyl esters of various sugars, and acetamides of 2-amino-sugars) which may not be.

3.1.2: Glycosidic Linkages

'-ose' Units and '-yl' Units. Monosaccharides ('glyc**ose**s') spontaneously equilibrate between the straight-chain form and 4 different ring forms. In a poly- or oligosaccharide, the single sugar unit (usually shown at the right-hand end of the structure) with no other sugar attached to its potential C=O group can undergo a similar equilibration. Such sugar units, since they can form a C=O group, will reduce Fehling's solution and are thus known as **reducing termini.** A polysaccharide cannot have more than one reducing terminus. When a monosaccharide (**X**) is linked via its potential C=O group to another molecule (**Y**), **X** is a **glyc<u>osy</u>l ('non-reducing')** **residue,** and is fixed in one of the 4 ring forms. **Y** is the **aglycone of X.**

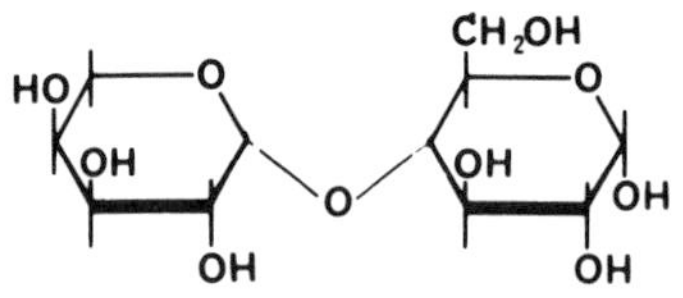

Arabinos**yl**-gluc**ose**

α- and β-Linkages (Anomerism). In a glycosyl residue, the carbon bearing the potential =O group is fixed in one particular configuration. Two different glycosyl linkages are therefore possible, called **α** (where the configuration is the same as at the penultimate carbon atom) and **β** (where it is opposite). The α- and β-forms (**anomers**) of otherwise identical molecules differ greatly in physical and biochemical properties (compare starch and cellulose) and are not interconvertible.

α-L-Arabinosyl-**Y** and $\underline{\beta}$-L-arabinosyl-**Y**.

Ring-size   A glycosyl residue is also fixed in either the 5-membered or the 6-membered ring-form (**furanose** or **pyranose** respectively).   These forms can be distinguished by mild acid hydrolysis, since furanose forms are much less stable.

α-L-Arabinopyranosyl-**Y** and α-L-arabinofuranosyl-**Y**.
[The ring oxygen atom 'belongs' to the carbon atom to its left. For the purposes of defining α-/β- and D-/L-, the bond between these two atoms is regarded as pointing 'down' (from C to O) if the —CH$_2$OH group is pointing up;   and vice versa.]

Carbon Atoms Involved in the Linkage   When two sugars are linked as part of a polysaccharide, the linkage is between the carbon bearing the potential $=$O group of one sugar and an —OH group on a different carbon atom of the other.   These linkages, in the case of aldoses, are thus described as $(1\rightarrow2)$, $(1\rightarrow3)$, $(1\rightarrow4)$ etc.   The arrow points from the glycosyl residue towards the reducing terminus.

α-L-Arabinofuranosyl-   and α-L-Arabinofuranosyl-
$(1\rightarrow4)$-D-galactose $(1\rightarrow6)$-D-galactose

3.1.3: Properties of polysaccharides

Name.   A polysaccharide composed of only one type of monosaccharide may be systematically named by replacing the '-ose' of the monosaccharide by '-an'. For example, cellulose is a glucan, or more specifically a β-$(1\rightarrow4)$-D-glucopyranan. When the polysaccharide possesses two or more different types of monosaccharide, the name is taken from the major sugar of the polysaccharide backbone, e.g. 'xylan' can be used as a general term for any polysaccharide with a backbone rich in xylose. The name can be elaborated to draw attention to other monosaccharide residues present, either as side-chains (e.g. arabinoxylan) or as minor backbone components (e.g. rhamnogalacturonan).

Size. Whereas a protein can be ascribed an exact molecular weight (M_r), this is not the case for a polysaccharide, because **(a)** M_r can only be measured after the polysaccharide has been solubilised, a process which may depend on some depolymerisation (see Section 3.4.1), and **(b)** there is no unique M_r value since polysaccharides are **polydisperse** (i.e. the individual molecules in a sample of 'a' polysaccharide are not all the same length). Estimates of M_r should therefore state the method of solubilisation, and the range of M_rs as well as the mean. The size of a polysaccharide can be quoted as M_r or as **degree of polymerisation** (DP = the number of monosaccharide units per polysaccharide molecule).

Shape [73,415]. **Linear** polysaccharides consist of a chain of monosaccharide residues. The chain can be rigid and rod-like (e.g. cellulose) or flexible and more globular (e.g. β-$(1\rightarrow3),(1\rightarrow4)$-glucan). **Branched** polysaccharides generally also have a clearly defined main chain ('backbone'), some of whose sugar residues have extra sugar residues attached to them, forming side-chains. Depending on the size of the side chains and on the flexibility of the backbone, branched polysaccharides can also be either rod-like or globular.

Charge. Cellulose, β-$(1\rightarrow3),(1\rightarrow4)$-glucan, callose, most xyloglucans, and a few arabinoxylans and arabinogalactans are **neutral**, consisting largely of uncharged sugar residues. Other wall polysaccharides, especially pectins but also many xylans and arabinogalactans, contain uronic acid residues, making them **acidic** (negatively charged). No plant cell wall polysaccharides have a net positive charge: most fungal walls contain chitin, a polymer of 2-amino-glucose, in which however the amino groups are largely neutralised by acetylation.

Solubility. A distinction should be drawn between **solubility** and

54

extractability. The polysaccharides in an intact wall are extracted into cold water to a very limited extent. However, all polysaccharides except cellulose can be extracted from the wall by other aqueous extractants (Section 3.4.1), and most of them thereafter are soluble in cold water. This may be because the extractant has caused some degradation.

Extracted polysaccharides that are difficult to re-dissolve in cold water can often be brought into solution by addition of 0.2 M EDTA, pH 6.5, and/or 0.2 M imidazole-HCl, pH 7.0 (for pectins) [385] or 0.1 M NaOH containing 0.1 % $NaBH_4$ (for hemicelluloses). Dried pectin, e.g. commercial citrus pectin, can be slow to dissolve because it forms gummy lumps; this can be prevented by making the pectin into a slurry with a little ethanol or DMSO and then pouring the slurry slowly into a large volume of rapidly stirred water. Polysaccharides are insoluble in most organic solvents, but a few (e.g. starch) will dissolve in 90 % DMSO, and heavily acetylated xylans are solubilised by 100 % DMSO; most neutral polysaccharides (even cellulose) will dissolve in hot MMNO (Section 3.4.1).

Binding to proteins and antigenicity. A useful property of polysaccharides is their affinity for certain proteins (lectins). Lectins are often specific for minor sugars of the plant cell wall that would be hard to study by other methods. Commercially available lectins with potential in cell wall work include:

<u>Ptilota plumosa</u> lectin	binds terminal α-D-Gal;
<u>Viscum album</u> lectin	binds terminal β-D-Gal;
<u>Canavalia ensiformis</u> lectin[*]	binds terminal α-D-Glc & α-D-Man;
<u>Dolichos biflorus</u> lectin	binds terminal α-D-GalNAc;
<u>Ulex europaeus</u> lectin I	binds D-GlcNAc-β-(1→4)-D-GlcNAc;
<u>Ulex europaeus</u> lectin II	binds α-L-Fuc.

[*]known as 'Concanavalin A'.

Carbohydrate-specific antibodies can also be produced. The polysaccharide is usually coupled to a protein, e.g. ovalbumin, to increase its antigenicity [276,360]. Lectins and antibodies can be labelled (e.g. fluorescently) and used histochemically to locate particular sugars within the wall [240].

Binding to other polysaccharides. Certain polysaccharides can interact with each other as shown, for instance, by the adsorption of hemicelluloses to cellulose [241] and the precipitation of β-(1→3),(1→4)-glucan on mixing with arabinoxylan [532]. In addition, pure xylans will self-aggregate whereas the side-chains present in arabinoxylans prevent this [16].

Binding to inorganics. Borate binds reversibly to certain sugar
residues via readily dissociable esters [542]. As a result of this
binding, a neutral polysaccharide will acquire a negative charge (Sections
3.5.5 & 4.5.4). Binding depends on the sugar's ability to present 2 —OH
groups at suitable positions for interaction with a borate ion:

Reversible interaction of borate with L-rhamnose

This ability may be abolished if one of the two —OH groups is substituted
by another sugar residue, either as part of the polysaccharide backbone or
as a side-chain. Binding is generally favoured at pH values above 8.5.
Several other substances also have the useful property of binding to
specific polysaccharides (Section 3.5.6).

Stability. Acids degrade polysaccharides by hydrolysis, especially
of furanosyl bonds. Apiose residues are exceptionally labile to warm mild
acid. Alkalies degrade monosaccharides and polysaccharides, especially at
the reducing terminus, and can cause step-wise 'peeling' of sugar units
from the reducing terminus. Alkalies can also cause internal cleavage of
acidic polysaccharides by an elimination reaction (Section 4.2.4), and will
very rapidly remove any ester-substituents present on the polysaccharide.
Both acid- and alkali-catalysed degradative reactions become more severe
at higher temperatures; heat also widens the pH range over which hydrolysis and (especially) elimination occur; thus the heating of polysaccharides at any pH value is never completely safe: pH 3-4 is usually the
best compromise if heating is essential. Polysaccharide chains are also
broken by physical treatments e.g. ultra-sonication [485] and by vigorous
ball-milling or other mechanical processing [254]; these treatments should
be avoided if M_r is subsequently to be measured.

Polysaccharides are particularly susceptible to microbial attack,
which can occur very quickly and without visible microbial growth, e.g.
during a prolonged column chromatography run. Aqueous solutions of polysaccharides should thus be stored cold and sterile, at pH 4-7. As a short-
term alternative, they can be stored in cold solution at pH 4-7 with an

56

adequate antimicrobial agent e.g. 0.05 % NaN_3, 0.05 % chlorbutol or 20% ethanol. The best long-term way of storing polysaccharides is in dry form: freeze-drying often gives a fluffy product that readily goes back into solution but an alternative drying method is precipitation with ethanol, washing with acetone, and drying in a desiccator.

3.2: CHEMISTRY OF PROTEINS AND GLYCOPROTEINS

Cell walls contain insoluble structural proteins, soluble mucilages, numerous enzymes, and sometimes one or more lectins. All these are glyco-proteins, i.e. polypeptides with carbohydrate side-chains which can be a mono-, oligo- or polysaccharide. In mucilages, the side-chains can be much bulkier than the polypeptide backbone to which they are attached. Wall polypeptides are built up of the 20 L-amino acids specified by the genetic code, plus at least two others (hydroxyproline and isodityrosine) produced by post-translational modification. The amino acids are joined by secondary amide (peptide) linkages between the —COOH group of one amino acid and the —NH_2 group of another. There is no equivalent of α-/β-, or pyranose-/furanose-, or $(1\to2)/(1\to3)/(1\to4)/(1\to5)/(1\to6)$-linkages.

The linkage between the carbohydrate side-chain and the polypeptide backbone is often a normal (O-) glycosidic bond, in which what would have been the reducing terminus of the carbohydrate is attached to an —OH group on the polypeptide. Specific examples of linkage points are:

L-Arabinofuranosyl-β-L-hydroxyproline   [20]
D-Galactopyranosyl-α-L-serine   [20]
D-Xylosyl-L-threonine   [201]

In addition, there are bonds between N-acetylglucosamine and the side-chain of L-asparagine; the linking atom here is N rather than the usual O, and the linkage is thus an N-glycosidic bond [308].

The polypeptide backbone, being genetically-encoded, presumably has a precisely defined initial size. However, post-translational proteolysis could introduce variation; the extent to which this occurs in the wall is not known. The number and size of the carbohydrate side-chains is also variable. The shapes of wall glycoproteins span the extremes: extensins are rod-like structures [469], whereas the arabinogalactan-proteins (AGPs) adopt a much more globular 'wattle blossom' shape [157]. Wall enzymes are likely to be globular, in common with other enzymes. Charge also spans the extremes. Extensins are highly basic (pI ca. 11; rich in lysine and histidine) and AGPs are acidic (pI <u>ca</u>. 1.5; rich in uronic acids). Wall enzymes occupy a wide range of pI values.

As with polysaccharides, <u>solubility</u> should be distinguished from <u>extractability</u>. Extractability is discussed in Section 3.4.2. Once extracted from the wall, most glycoproteins are soluble in cold water; where this is not the case, addition of salt (e.g. NaCl to 0.4 M) may help. More powerful solvents, useful for membrane proteins, include detergents (e.g. Triton X-100, SDS or CHAPS) and phenol/acetic acid/H_2O (2:1:1, w/v/v) [444]; SDS and phenol have the disadvantage of inactivating enzymes. Enzyme glycoproteins can usually be precipitated from solution by addition of $(NH_4)_2SO_4$ or trichloroacetic acid (TCA), but the concentration required varies from one glycoprotein to another. AGPs and extensin require relatively high TCA concentrations for precipitation: this can be exploited in their purification. Most proteins and glycoproteins are insoluble in EtOH.

<u>Stability</u> The carbohydrate moieties of glycoproteins are as acid- and microbe-labile as analogous polysaccharides (Section 3.1.3). Alkali-stability depends on the sugar—amino acid linkage present (Section 4.3.1). Peptide bonds are more acid-stable than glycosidic bonds, but less alkali-stable. They are also susceptible to microbial attack. Enzymic glycoproteins depend on a defined secondary and tertiary structure for activity, and can be 'lost' by denaturation upon warming and at extremes of pH; however, wall enzymes are generally among the most stable of enzymes because they have evolved to work in the harsh extracellular environment.

3.3: WALL PHENOLICS

3.3.1: Chemistry of phenolics (see Fig. 3.3.1)

A phenolic is a compound with an —OH group attached directly to a benzene ring; the simplest is phenol itself, C_6H_5OH. Walls of some <u>non-growing</u> cells, e.g. in wood, contain up to 30 % lignin [381], a phenolic polymer formed from p-coumaryl-, coniferyl- and sinapyl-alcohols (Sections 4.1.8, 5.1.8). Growing walls contain little or no lignin, but the traces of phenolics that are present may nevertheless act as important cross-linking sites between polysaccharides (Ch 7). Examples are ferulic acid, p-coumaric acid and p-hydroxybenzoic acid, which are linked to wall polysaccharides [174]. Tyrosine is a phenolic component of wall glycoproteins and is especially abundant in extensins [97]. Walls may contain condensed tannins (proanthocyanidins), which are oligomers of flavan-3-ols [234], and possibly also hydrolysable tannins, which are composed of several residues of gallic acid [and sometimes also its dimer, hexahydroxybiphenic (ellagic) acid] esterified to glucose. It has yet to be shown that tannins are wall components in living cells; it is possible that they bind there post-mortem. Other phenolics are present in cutin (in the cuticle of epidermal walls) and suberin (in cork and endodermal walls), both of which are aliphatic polyesters mixed, meshed or bonded with lignin-like material [253, 296]. Cutin, at least, is present in growing epidermal cell walls.

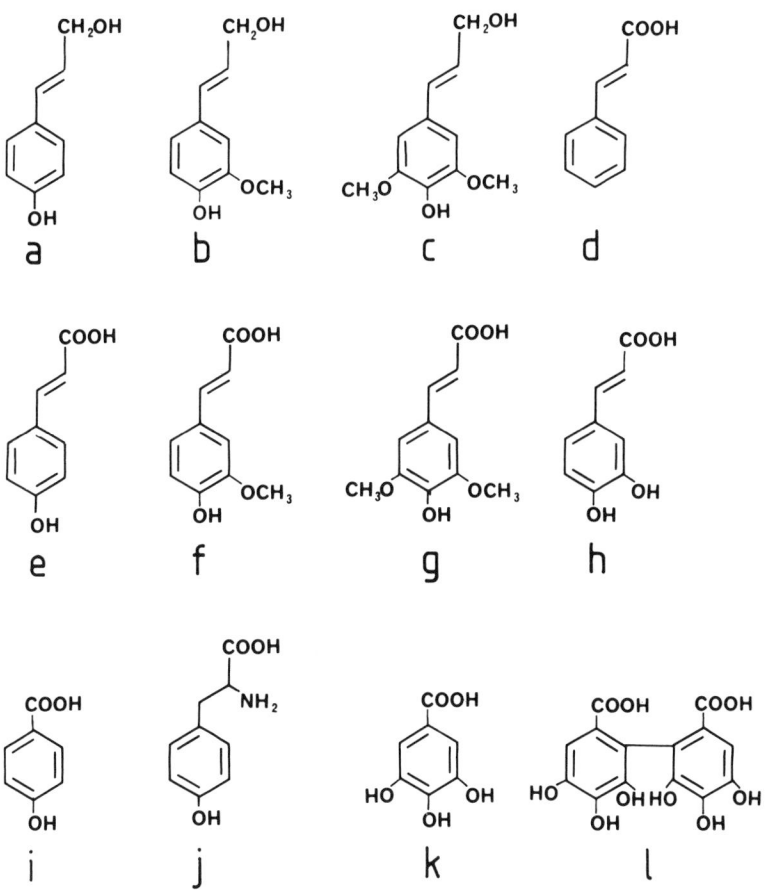

Fig. 3.3.1 a-l: Structures of some major wall phenolics. (a) p-Coumaryl alcohol, (b) Coniferyl alcohol, (c) Sinapyl alcohol, (d) Cinnamic acid, (e) p-Coumaric acid, (f) Ferulic acid, (g) Sinapic acid, (h) Caffeic acid, (i) p-Hydroxybenzoic acid, (j) L-Tyrosine, (k) Gallic acid, (l) Hexahydroxybiphenic acid [which spontaneously lactonises to ellagic acid].

3.3.2: Linkages involving phenolics

Phenol—polymer bonds.   The best-characterised bonds are ester
linkages between ferulic acid and wall polysaccharides [168,170,282].
These link the —COOH group of ferulic acid to specific —OH groups on
particular sugars of certain polysaccharides (Fig. 3.3.2 a).   The older
view that ferulic acid was rather randomly attached to polysaccharides is
not valid.

As well as ester bonds, there may also be some ether bonds between
ferulic or p-coumaric acid and wall polymers [50,92,171], especially lignin
[437] (Fig. 3.3.2 b).   There are also reports of ferulic acid being amide-
linked via its —COOH group to the NH_2-terminus of a protein.   Proanthocyanidins may possibly be linked to wall polysaccharides via ether bonds
[234].   Lignin itself may also be covalently linked to wall polysaccharides, possibly via ether bonds, and possibly via phenol—uronic acid
ester bonds [164,277,316].

Phenol—phenol bonds.   The hydroxycinnamyl alcohol building blocks
of lignin are linked by both C—C and C—O—C bonds in a variety of
positions (Fig. 3.3.2 c; [164]), reflecting the fact that the last steps in
lignin synthesis are non-enzymic (Section 5.1.8).   The bonds that hold
lignin together suggest some possible linkages between the phenols present
in growing walls.   Dimerised phenolics have been found in hydrolysates of
growing walls, but few of these have been characterised, and many others
probably await discovery.   Two which have been purified illustrate possible
linkages: the C—C (biphenyl) bond of diferulic acid (Fig. 3.3.2 d;
[335]), and the C—O—C (diphenyl ether) bond of isodityrosine (Fig. 3.3.2
e; [169]).

3.3.3: Properties of pure phenolics

Charge.   Phenolic —OH groups acquire a negative charge at high pH
(pK_a = 8-10).   The pK_a is lower in feruloyl esters (ca. 8.6) than in
ferulic acid (ca. 9.2).   The pK_a of dityrosine is very low (ca. 6.7) for
the 'first' —OH group, but much higher for the second.   Many naturally-
occurring phenolics also possess other ionisable groups e.g. —COOH.

Solubility.   Hydroxycinnamic acid derivatives (ferulic, diferulic,
p-coumaric), p-hydroxybenzoic acid, tyrosine and isodityrosine, have low
but significant solubility in water at neutral or acid pH; they become much

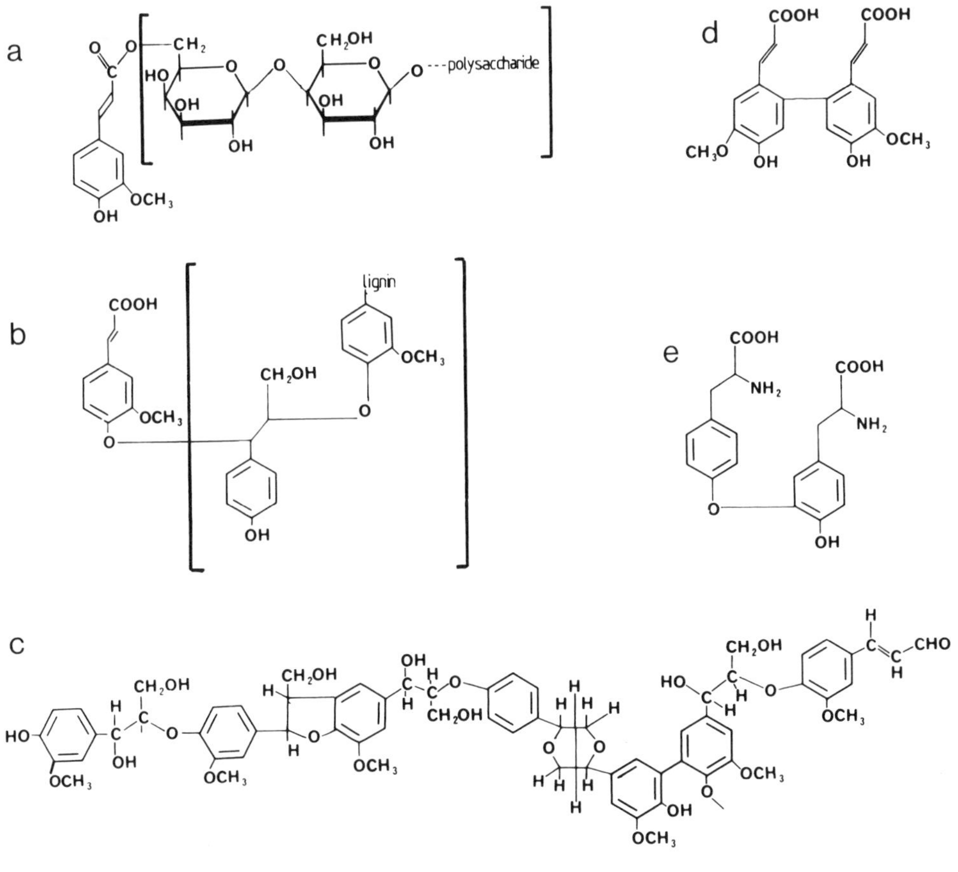

Fig. 3.3.2:
(a) Ester linkage of ferulic acid (at left) to a polysaccharide (shown in brackets).
(b) Ether linkage of ferulic acid (left) to lignin (in brackets).
(c) A hypothetical portion of a lignin molecule (from [2]), showing the main types of linkage.
(d) Diferulic acid.
(e) Isodityrosine.

more soluble as the pH rises and the phenolic —OH group ionises. Tyrosine and isodityrosine are not readily soluble in organic solvents; 1 M NH$_3$ is a suitable volatile solvent for them. Hydroxycinnamic and benzoic acid derivatives are soluble in many organic solvents e.g. DMSO, methanol, ethanol, acetone, and butan-1-ol. Such phenolics can be partitioned from aqueous solution (at pH 1-2, so that they carry no charge) into immiscible solvents e.g. butan-1-ol (Panel 4.4.1). Ferulic acid and related compounds also occur in vacuoles in the form of β-glucosyl esters (Section 5.1.4) which are much more water-soluble than the free acids. Polysaccharide-free lignin is soluble in 80-90 % dioxane, dimethylformamide, DMSO, glacial acetic acid, chloroform and pyridine/acetic acid/H$_2$O (9:1:4), but not in water, aqueous salt solutions or diethyl ether.

Stability. Phenolics are less stable in hot acid than are sugars; they can be partially recovered from acid hydrolysates, but losses should be taken into account. Alkali, used for liberation of esterified phenolics from wall polymers (section 4.4.1), renders phenolics highly O$_2$-labile, especially in the light. Phenolics with two or more —OH groups (arranged ortho- or para- to each other) on the benzene ring, e.g. caffeic and gallic acid, are particularly susceptible to alkali-induced oxidation, rapidly turning yellow or brown. Phenolics are also oxidised by O$_2$ on dried silica-gel TLC plates, where the colourless spots of compounds like caffeic acid and to a lesser extent ferulic acid gradually turn yellow or brown. Besides promoting oxidation, light can interconvert the geometrical isomers of phenolic acids. Hydroxycinnamic acid derivatives e.g. ferulic acid are biosynthesised as the trans-isomers; these are converted to cis/trans mixtures upon exposure to light, especially in UV light [568]. Phenolics can be stored for future analysis in solution in acetone (in a tightly capped tube, preferably filled with N$_2$) in the dark at 4°C.

UV absorbance. Phenolics absorb UV light at wavelengths below 240 nm but also (more diagnostically) in the 250-380 nm range. The absorption maximum and shape of the absorption spectrum depends on the chemistry of the phenol´ and the pH of the solvent. Simple phenolics, e.g. tyrosine or p-hydroxybenzoic acid, absorb maximally at about 250-280 nm at pH 7 or below, and at 275-300 nm in alkali owing to ionisation of the phenolic —OH group (Fig. 3.3.3). Phenolics with a conjugated side-chain, e.g. the cinnamic acid derivatives, absorb at longer wavelengths (typically 308-340 nm), shifting to 330-380 nm on addition of alkali (Fig. 3.3.3). Lignin absorbs strongly at about 280 nm, although some lignins exhibit a shoulder rather than a peak at 280 nm.

UV fluorescence. Many phenolics, when excited by absorption of UV, re-emit the energy as visible light. This fluorescence can be helpful in locating and identifying phenolics in chromatography and also within the cell by UV microscopy. Further information can be obtained by addition of a trace of NH_3 (solution or vapour), which often increases the intensity and alters the colour of the fluorescence [226].

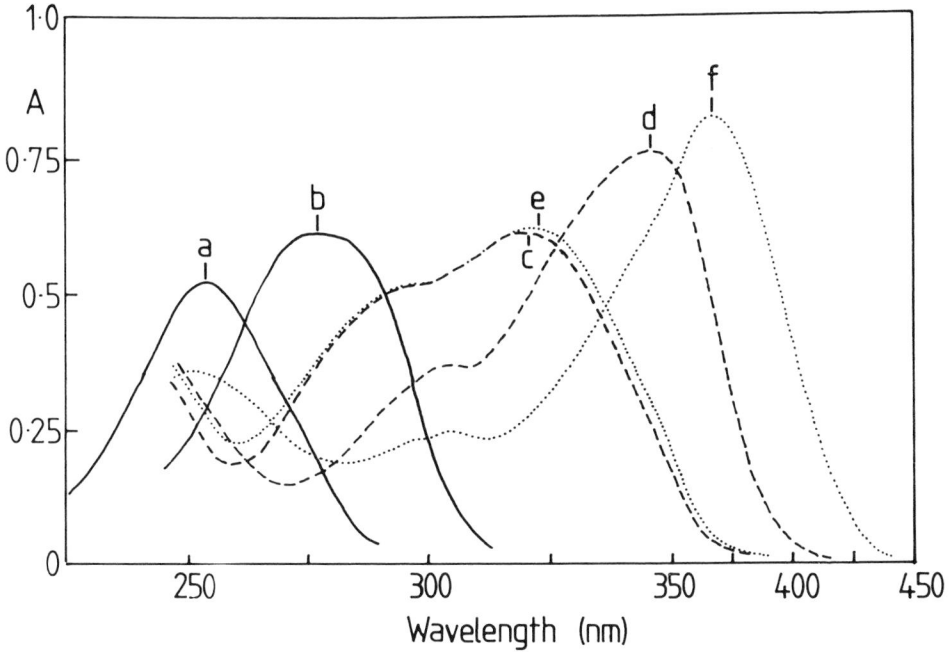

Fig. 3.3.3: Absorption spectra.
   (a) p-Hydroxybenzoic acid at pH 3,
   (b) p-Hydroxybenzoic acid at pH 11,
   (c) Ferulic acid at pH 3,
   (d) Ferulic acid at pH 11,
   (e) A feruloyl ester at pH 3 [the ester is pure synthetic 1-feruloyl-ethane-1,2-diol, but the spectrum is very similar to that of a feruloyl-polysaccharide],
   (f) A feruloyl ester at pH 11 [see (e)].

63

3.3.4: Properties of phenol—polymer conjugates

Many of the generalisations about the properties of polysaccharides (Section 3.2.3) apply equally to phenol—polysaccharide conjugates. Although phenolics are very different from sugars (Section 3.3.3), the few feruloyl, p-coumaroyl or p-hydroxybenzoyl groups present on polysaccharides of primary walls may have little net effect on the M_r, shape, charge (at pH below 7.5) or stability of the polymer. However, ester-linked phenolics are rapidly removed by cold alkali; feruloyl-ester bonds have an acid-lability intermediate between pyranose and furanose glycosidic bonds. The water-solubility of a polysaccharide may be reduced by attached phenolic groups. If the phenolics undergo oxidative coupling, the polysaccharides may become cross-linked, increasing their M_r, and perhaps decreasing their extractability [174]. Coupling of tyrosine residues in extensin similarly reduces the extractability of that glycoprotein [169].

A further property, unique to phenol-bearing polymers, is their intense UV absorbance (absorption maximum ca. 250 to 380 nm, depending on structure and pH: Section 3.3.3), and, in some cases, UV-fluorescence. The absorption spectrum of ferulic acid is considerably altered by esteri-fication, e.g. with a polysaccharide (Fig. 3.3.3) [168].

Some of the wall's lignin occurs bound to polysaccharides and can be extracted in the form of lignin—carbohydrate complexes. These molecules vary between very wide extremes in their carbohydrate:lignin ratio, and their physical and chemical properties vary likewise [31].

3.4: EXTRACTION OF WALL POLYMERS
3.4.1: Extraction of polysaccharides

Molecules can most satisfactorily be purified and investigated if they are in solution.

Growing walls of intact plants contain water-extractable polysaccha-rides that are lost by homogenisation in aqueous media (Ch 1). Only the 'alcohol-insoluble residue' (AIR) method recovers water-extractable wall polymers, which are precipitated by ethanol. Some of these polysaccharides can then be extracted from the AIR with aqueous buffer, e.g. 100 mM HEPES (pH 7), but extracts prepared in this way will contain, besides wall poly-saccharides, some intracellular polymers e.g. proteins and RNA.

In suspension cultures, water-extractable wall polymers accumulate in the culture medium and can be recovered from culture filtrates by GPC or dialysis (Section 3.5.1). The equivalent of the culture medium in an intact plant is the system of intercellular spaces and cell surface films (apoplast), from which soluble molecules can be obtained by vacuum-infiltration/centrifugation (Panel 6.3.1).

Most wall polysaccharides cannot be extracted with cold water. Two alternative approaches have therefore been used to extract these: **(a)** more powerful extractants, and **(b)** deliberate partial degradation.

Polysaccharide extractants

No extractant is perfect. The following account describes for each extractant: its presumed mode of action, the types of polysaccharide solubilised, and any side-effects (e.g. bonds broken). A group of extractants, often used sequentially, is described in Panel 3.4.1 a.

Dimethyl sulphoxide (DMSO), if not already used for de-starching, will extract a small proportion of the water-inextractable wall polymers, mainly hemicelluloses, and especially those that are heavily acetylated [69]. 100 % DMSO may be a more effective solvent for hemicelluloses than the 90 % DMSO used for starch. Treatment with DMSO is very mild, and unlikely to cause polymer degradation. Its chief disadvantage is that it only extracts a small proportion of the total wall polysaccharides.

Aqueous chelating agents completely extract Ca^{2+} from the wall at room temperature and moderate pH (e.g. 50 mM CDTA, pH 6.5 [262]). During and after this treatment, some pectic polysaccharides are gradually solubilised. The proportion varies greatly between cell types: typically it is over 50 % in fruit tissue, 20 to 50 % in the vegetative tissues of intact plants, but less than 10 % in tissue cultures. Treatment of walls with chelating agents at 25°C is unlikely to have any side effects, except that the prolonged treatments required may allow endogenous wall enzymes to act on wall polysaccharides and cause partial autolysis; this can be prevented by phenol-extraction during the isolation of walls (Ch 1).

Since chelating agents at 25°C extract only a proportion of the pectins, many attempts have been made to increase effectiveness by heating. Heating solubilises more polysaccharide, but with degradation, even at the least destructive pH values (pH 3-5) [455]. For instance, when cultured sycamore cell walls were heated at 90°C with oxalate (a chelating agent) at pH 3-6, there was a gradual solubilisation of pectins which continued for at least 24 h.   However, treatment under the same conditions with formate, which is not a chelating agent instead of oxalate gave approximately equal solubilisation. This suggests a process unrelated to the removal of Ca^{2+}.

PANEL 3.4.1 a: Sequential extraction of polysaccharides from primary walls — with minimal degradation until step (5)

Sample: Cell walls of fast-growing suspension-cultured spinach cells, prepared as in Panel 1.2.1 c; not dried.

Step 1:   **DMSO.** (Extracts starch + traces of native hemicellulose). Suspend cell walls (equivalent to 1 g dry wt, but still moist with 70 % EtOH) in 90 % DMSO (100 ml).   Stir 16 h at 25°C.   Centrifuge (5 min at 2,500 **g**).   Carefully pour supernatant from pellet. Assay supernatant for polymer by anthrone[2] or EtOH tests (Panel 3.7 a). Resuspend wall pellet in 50 ml 90 % DMSO, centrifuge, and pool the supernatants.   Repeat DMSO extraction until supernatants contain negligible polymer.   Remove residual DMSO by resuspending final pellet in water and centrifuging.

Step 2:   **CDTA.** (Extracts some pectins).   Re-suspend pellet from step 1 while still wet in 100 ml 50 mM CDTA, pH adjusted to 7.5 with 1 M NaOH, and chlorbutol added to 0.05 %. Shake or stir 16 h at 25°C. Centrifuge.   Pour supernatant from pellet.   Assay polymers by m-hydroxybiphenyl test[2] (Panel 3.7 a).   Repeat CDTA extraction until the supernatants no longer contain polymer. Resuspend final pellet in water and centrifuge.

Step 3[1]:   **Urea.** (Extracts small proportion of native hemicellulose). Resuspend pellet from step 2, while still wet, in 100 ml of 8 M urea (buffered at pH 7.5 with 50 mM HEPES/NaOH). Shake/stir 16 h at 25°C. Centrifuge.   Pour supernatant from pellet.   Assay polymers by EtOH or anthrone test[2] (Panel 3.7 a).   Repeat urea extraction until the supernatants no longer contain polymer. Resuspend final pellet in water and centrifuge.

Step 4[1]:   **Guanidinium thiocyanate (Gdm).** (Extracts some mannose-rich hemicellulose). Repeat step 3 with 4.5 M Gdm in place of 8 M urea.

Step 5:   **NaOH.** (Extracts high proportion of total hemicellulose, but with de-acylation; some pectins, with de-esterification). Repeat step 3 with 6 M NaOH containing 1 % $NaBH_4$ in place of buffered urea.   Neutralise the extract(s) with HOAc and assay by anthrone[2] or EtOH test.

Step 6:   **$NaOH/H_3BO_3$.** (Extracts small proportion of de-acylated hemicellulose; some de-esterified pectin). Repeat step 5, supplementing the extractant with 4 % H_3BO_3.

Step 7:   **Residue.** (Contains cellulose; some resitant pectins and hemicelluloses; and extensin).   Wash final pellet extensively with H_2O; freeze-dry; weigh.

Next steps: Removal of extractant; fractionation & characterisation of polymers (Sections 3.5; 3.6 and Ch 4).

[1]Steps 3 & 4 may be omitted; the polymers which they extract will then be recovered in steps 5 & 6, although with loss of acyl groups.

[2]Dilute the extracts with H_2O before applying these tests.

Degradation was occurring, as shown by GPC of the extract: fragments as small as monosaccharides (galacturonic acid) were detected.

Aqueous chaotropic agents extract hemicelluloses, but only a small proportion of the total. Urea (8 M), an excellent solvent for proteins, extracts very little wall polysaccharide [39]. Guanidinium thiocyanate (5.5 M) is slightly more effective, and may show the useful property of selectively solubilising mannose-containing polysaccharides [444]. As with DMSO, treatment with chaotropic agents at room temperature and neutral pH is unlikely to damage the covalent structure of the polysaccharides. However, pH requires careful buffering, especially with urea, which turns alkaline on warming. Again, the chief disadvantage is poor effectiveness.

Alkalies are traditionally used on the insoluble residue obtained after treatment with a chelating agent [501]. They solubilise hemicelluloses very effectively, and may also solubilise some of the pectic polysaccharides not extracted by the chelating agent. Their mode of action is unclear. They may act partly as chaotropes; however, alkalies act so much better than other chaotropes that the situation may be more complicated. They certainly hydrolyse ester-linked substituents, and if these act as cross-links (ch 7) then this process may help to solubilise the polysaccharide. They can also cause elimination degradation of certain glycopeptide bonds e.g. sugar—serine or sugar—threonine (Section 4.3.1).

Different hemicelluloses are extracted by different concentrations of alkali. For example in grass cell walls, three classes of hemicellulose are sequentially extracted by progressively increasing KOH concentrations (0.01-4.0 M) [91]: two classes of arabinoxylan and a β-(1→3),(1→4)-glucan (Panel 3.4.1 b). Further hemicellulose (especially mannans) can often be extracted if the 4 M KOH is supplemented with 4 % H_3BO_3 [501].

Alkalies cause 'peeling' of many polysaccharides, especially in the presence of O_2 (Section 4.2.4). This type of degradation can be partly suppressed by addition of the strong reducing agent $NaBH_4$ (0.1 %). Once the hemicellulose has been extracted into alkali + $NaBH_4$, it does not undergo significant decerease in M_r on storage. This argues against extensive mid-chain cleavage of the backbone.

Cadoxen is a solvent that complexes with, and solubilises, cellulose. It is a highly basic reagent, and may thus cause peeling (Section 4.2.4), but is a useful solvent in which to estimate the M_r of cellulose [76].

Cadoxen is prepared by stirring together 1,2-diaminoethane (310 ml), H_2O (720 ml) and cadmium oxide (100 g) at 20°C for 3 h followed by 4°C for 18 h. After bench centrifugation, the clear supernatant is used; it can be stored at 4°C [560]. Cellulose is also reversibly converted into a soluble derivative upon heating in dry DMSO + paraformaldehyde; free cellulose is regenerated as a precipitate on addition of H_2O or methanol [268].

Methylmorpholine-N-oxide (MMNO), used as a 1:1 (mol/mol) mixture with H_2O, is an extremely powerful chaotropic agent. A solid at room temperature, it is routinely used at 85°C to 120°C. It can solubilise virtually all the neutral polysaccharides, including cellulose, from the plant cell wall; acidic pectic polysaccharides are less effectively solubilised [96]. Any treatment at 85-120°C must be suspected of degrading polysaccharides, and indeed it has been found that MMNO treatment does cause some mid-chain cleavage, e.g. of cellulose [272]. However, once the polysaccharides have been dissolved in hot MMNO, the solution can be diluted into 5 volumes of DMSO at 25°C and the material will remain in solution for a few hours. Unfortunately, DMSO/MMNO mixtures at 25°C do not readily extract wall polysaccharides. In addition to slight backbone cleavage, hot MMNO seems to cause extensive removal of ester-linked substituents. MMNO is thus a promising new extractant, but it requires careful study.

PANEL 3.4.1 b: Sequential extraction of maize hemicelluloses ([91])

Sample: Maize coleoptiles, ground in liquid N_2, extracted with PAW and 90 % DMSO, washed with water and freeze-dried (Panels 1.2.1 b & c).

Step 1: Suspend the cell walls (1 g dry wt) in 50 ml 0.01 M KOH,[1] stir for 1 h, centrifuge (5 min at 2,500 **g**). Carefully retrieve the supernatant and bring it to pH 4.7 with acetic acid [care — H_2 evolved].

Steps 2 - 16: Resuspend the pellet from the previous step in 50 ml of a higher concentration of KOH[1] and repeat step 1. The progression of concentrations used by Carpita [91] is: 0.02, 0.03, 0.045, 0.06, 0.08, 0.1, 0.2, 0.3, 0.45, 0.6, 0.8, 1.0, 2.0, 3.0, 4.0 M KOH.

Step 17: Repeat step 16 with 4 M KOH[1] supplemented with 4 % H_3BO_3.

Next steps: The major polymers extracted are arabinoxylans and β-$(1\rightarrow3),(1\rightarrow4)$-glucans. These can be assayed by the orcinol and anthrone tests respectively (Panel 3.7 a). The polymers are recovered, fractionated and characterised as in Sections 3.5, 3.6 and Ch 4.

[1]Note: all the KOH solutions are supplemented with 0.3 % $NaBH_4$. The lowest concentrations of KOH do not significantly remove feruloyl ester groups; higher concentrations do.

Extraction by partial degradation

The second approach to the extraction of wall polysaccharides is deliberate partial cleavage of their backbones. This has mainly been applied to the pectic polysaccharides, which, as already mentioned, are often difficult to extract with chelating agents. Treatment of walls with purified pectinase will cleave pectin wherever there are 3-4 contiguous unsubstituted galacturonic acid residues, at wall sites to which the enzyme can physically penetrate [it may be excluded by tight cross-linking or by the presence of hydrophobic fillers such as lignin, cutin or suberin]. Limited enzymic cleavage with pectinase renders water-soluble much of the wall's pectic polysaccharides (Panel 3.4.1 c and Fig. 3.5.2 b) [118,343]. This approach clearly yields degraded polysaccharides, but so long as this is borne in mind and the products (rhamnogalacturonans) are not regarded as individual polysaccharides in their own right, the method is very useful, and has the great advantage that labile substituents (e.g. ester-linked groups) are not removed from the polysaccharide chain. Other enzymes that have been used include pectate lyase [261], β-$(1\to3)$,$(1\to4)$-endoglucanase [280], endoxylanase [281], and a novel endoglucanase that appears to be useful as it attacks β-$(1\to3)$,$(1\to4)$-glucan at a few highly specific loci [235].

PANEL 3.4.1 c: Use of pectinase to extract rhamnogalacturonans I and II [118]

Sample: **Walls** of suspension-cultured sycamore cells prepared by PAW extraction (Panel 1.2.1 c), H_2O-washed, and freeze-dried. **Pectinase** activity, purified either from culture filtrates of <u>Colletotrichum lindemuthianum</u> [144] or purified from a commercial source of pectinase (e.g. 'Driselase') by 'Chromatofocusing' (Pharmacia).

Step 1: Suspend cell walls (0.1 g dry wt) in 10 ml of a solution of the enzyme (40 Units/ml) made up in buffer **A** (25 mM pyridine, pH adjusted to 5.2 with acetic acid), containing 0.05 % chlorbutol. Incubate at 30°C with gentle shaking.

Step 2: At 1-2 h intervals, remove 50 μl of the suspension, add 450 μl of H_2O, centrifuge in an Eppendorf tube, and assay 200 μl of the particle-free supernatant for uronic acids by the m-hydroxybiphenyl test (Panel 3.7 a). Also assay 20 μl of the whole suspension (uncentrifuged; added to 180 μl H_2O) to indicate the total uronic acid content of the walls. Hence calculate % of polymer-bound uronic acid extracted. If necessary, add more enzyme during the reaction.

Step 3: When more than 50 % of the uronic acid is in solution, pellet the remaining walls (5 min at 2,500 **g**). Freeze-dry the supernatant.

Next steps: see Panel 3.5.2 a.

A simpler version of this approach is the use of hot aqueous buffers to effect the partial degradation of the pectic polysaccharides (Panel 3.4.1 d) [41]. The widespread practice of using hot solutions of chelating agents to extract pectins is an example; as already mentioned, the chelating agent is not required. Heating at moderately acidic pHs (e.g. 3-5) will cause partial hydrolysis of the pectic backbone since galacturonosyl bonds are more labile to <u>mild</u> acid than are neutral sugar glycosidic bonds, in contrast to the situation at pH less than 1 [455]. At pHs in the order of 6-8, heating causes elimination degradation (rather than hydrolysis) of the galacturonic acid-rich pectic backbone, especially if it is highly methyl-esterified, but has the same overall effect — backbone cleavage [36]. Thus, heating at any pH will solubilise much of the wall's pectic polysaccharides. Lower pHs (e.g. pH 3.5, which can be used with autoclaving to speed the effect) maximises the chance of retaining any ester-substituents, although apiosyl and a small proportion of arabinosyl linkages are likely to be lost; higher pHs (e.g. pH 7) retain sugar side-chains at the expense of some loss of ester-linked substituents.

PANEL 3.4.1 d: Rapid extraction of polysaccharides by autoclaving — with partial degradation

Sample: Alcohol-insoluble residue (AIR) of young tomato stems (see Panel 1.2.1 a)

Step 1:   Suspend AIR, ex 50 g fresh weight of stems, in 250 ml of pyridine/HOAc/H_2O (1:1:23), pH <u>ca.</u> 4.5. Autoclave for 30 min at 120°C in a tightly screw-capped bottle.

Step 2:   Cool. Filter the suspension through a sintered glass funnel, and rinse the wall residue with 2 x 50 ml H_2O; pool the filtrate + washings.

Step 3:   Freeze-dry the filtrate in a tared vessel to yield the **pectin.** For removal of the last traces of pyridinium acetate, re-dissolve the pectin in water and freeze-dry again. Weigh the total pectin extracted; assay uronic acid residues by <u>m</u>-hydroxybiphenyl test.

Step 4:   Recover the wall residue from the sintered glass and treat with NaOH (Panel 3.4.1 a, step 5) to extract **hemicellulose.**

Next steps: Ion-exchange chromatography (Panel 3.5.3 a), gel-permeation chromatography (N.B. the pectic material will contain some very low-M_r pectic fragments; these are recovered owing to use of freeze-drying for de-salting) (see Panel 3.5.2 a), fractionation and characterisation of polymers (Sections 3.5.2-3.5.7, 3.6 and Ch 4).

3.4.2: Extraction of glycoproteins

The diverse glycoproteins of the plant cell wall require diverse extractants. Arabinogalactan-proteins (AGPs) and some enzymes are water-extractable, and many of the comments about water-extractable wall polysaccharides (Section 3.4.1) apply to these glycoproteins, except that enzymes may be <u>irreversibly</u> insolubilised by alcohol. Other enzymes, lectins and newly-deposited extensin are ionically-bound to the acidic polysaccharides; these glycoproteins can be extracted with salt (Section 6.4). Further wall enzymes and mature extensin cannot be extracted from the wall except by degradation, and are said to be covalently bound. Some covalently-bound glycoproteins (not extensin) can be extracted by enzymic digestion of the wall polysaccharides (Section 6.4); mature extensin is extracted by degradation of isodityrosine with $NaClO_2$ (Panel 7.4 a). The extractant chosen will depend on the class of glycoprotein under investigation, on the methods used to isolate the walls, and on the method to be used for detection of the glycoprotein. Four recommended methods are:

(a) <u>Extraction of ionically-bound wall proteins with salt.</u> Living cells are treated gently with a non-toxic salt (see Section 6.4.2). The use of living cells is helpful because the salt leaches glycoproteins from the cell surface without significant contamination by intracellular proteins. The method is especially suitable for cell suspension cultures [458], where the cells are in direct contact with the medium, but could also be applied to intact tissues by vacuum infiltration [133]. The extraction is mild, and any enzyme activity is usually retained.

(b) <u>Extraction of total non-covalently-bound protein with phenol</u> (Panel 3.4.2). Phenol/acetic acid/water (2:1:1, w/v/v) (**PAW**) is a potent solvent for proteins, but not polysaccharide. As plants often have a high polysaccharide:protein ratio, PAW is an excellent choice for protein extraction (whether or not the interest is in cell walls) because a poly-sac-charide-free product is obtained [33,444]. Polysaccharides interfere during electrophoresis of plant proteins, but PAW-extracts give excellent electrophoretograms. Another extractant, hot SDS + mercaptoethanol, is less good because co-extracted pectins may gel in SDS extracts on cooling.

PAW-extraction can be performed on intact fresh (or frozen or freeze-dried) tissue; physical homogenisation may not be required in the case of finely dispersed cells such as suspension cultures. The treatment can be at

```
┌─────────────────────────────────────────────────────────────────────────────┐
│ PANEL 3.4.2:  Extraction of total non-covalently bound cell protein           │
│                                                                                │
│ Sample:          5 ml of exponentially-growing suspension-cultured spinach    │
│                  cells, previously grown aseptically for 24 h in the           │
│                  presence of 10 µCi of DMSO-sterilised (see Panel 2.2 b) L-    │
│                  [35S]methionine.                                              │
│                                                                                │
│ Step 1:          Add the whole suspension (5 ml) to 35 ml of freshly          │
│ prepared solution A [= 100 ml HOAc + 250 ml 80 % (w/w) phenol — care:         │
│ risk of poisoning by skin contact], and mix thoroughly.  Stir rapidly, and    │
│ incubate in a fume cupboard on a magnetic stirrer/heater at 70°C for ½ h.      │
│                                                                                │
│ Step 2:          Cool.  Filter on a pad of GF/C glass fibre paper in a         │
│ Hartley funnel or Millipore 12-port filtration unit, collecting the           │
│ filtrate.  Rinse the cell residue with a further 2 x 5 ml of solution B       │
│ [= 35 ml sol'n A + 5 ml H2O].  Pool and retain the filtrates.                  │
│                                                                                │
│ Step 3:          Wash the cell residue with 2 x 5 ml H2O, 2 x 5 ml 80 %        │
│ acetone, and 2 x 5 ml ether.  Dry;  assay 35S-labelled material (presumably   │
│ protein) not extracted.                                                        │
│                                                                                │
│ Step 4:          To a 1-ml portion of the PAW-filtrate, add 50 µl of 10 %      │
│ ammonium formate followed by 5 ml acetone to precipitate the protein.         │
│ Stand at 0°C for 1 h.  Swirl gently, pass the suspension through a new pad     │
│ of GF/C, and rinse with 2 x 5 ml 80 % acetone and 2 x 5 ml ether, and dry.    │
│ Assay 35S by LSC in non-Triton scintillant (Section 2.8.2).     Hence          │
│ calculate extracted [35S]protein in whole sample.                              │
│                                                                                │
│ Step 5:          To the remaining ca. 49 ml of PAW-filtrate, add 2.5 ml 10 %   │
│ ammonium formate followed by 250 ml acetone.  Stand at 0°C for at least        │
│ 1 h.  Centrifuge down the protein precipitate (5 min at 2,500 g).              │
│                                                                                │
│ Step 6:          Resuspend the precipitate thoroughly in 10 ml 80 % acetone,   │
│ and centrifuge again.  Repeat once more.                                       │
│                                                                                │
│ Step 7:          Resuspend the final pellet in 10 ml water, and freeze-dry.    │
│                                                                                │
│ Next steps:  SDS [305] or acid-urea [468] gel electrophoresis.                 │
└─────────────────────────────────────────────────────────────────────────────┘
```

25°C, when extraction of total [methionyl-35S]protein from cultured spinach cells was ca. 75 % after 16 h, or at 70°C, which extracted ca. 90 % after ½ h without degradation of proteins as judged by SDS gel electrophoresis. Slightly higher extraction was achieved at 100°C, but with some breakdown of the proteins. PAW extraction of whole tissue requires a method of distinguishing wall glycoproteins from others. A disadvantage of PAW extraction is the complete loss of any enzymic activity that the glycoproteins may have possessed.

(c) Extraction of extensin by degradation of isodityrosine. After exhaustive extraction of glycoproteins by salt, PAW and SDS, much of the extensin remains in the cell wall. Some of this can be extracted [389], with little if any cleavage of the polypeptide backbone [51], by treatment with warm acidified $NaClO_2$, which acts primarily by breaking isodityrosine cross-links between extensin molecules (Section 7.3.2).

(d) Solubilisation of covalently-bound glycoproteins by enzymic digestion of wall polysaccharides. Some otherwise inextractable wall enzymes (e.g. certain isozymes of peroxidase) can be brought into solution by treatment of the salt-washed wall with polysaccharide-degrading enzyme preparations, e.g. Driselase (see Sections 4.2.5 and 6.4.3). Note that Driselase, although containing little protease activity, will probably hydrolyse off the carbohydrate side-chains of a glycoprotein.

3.4.3: Extraction of phenol—polymer conjugates and lignin

Many of the most effective polysaccharide extractants described in Section 3.4.1 are not applicable to phenol—polysaccharide conjugates because the ester-linked phenolic group would be lost during treatment. Thus NaOH, MMNO and hot water are unsuitable. Extractants that can be used safely include DMSO, chaotropic agents and chelating agents (all at 25°C and buffered at pH 3.5 to 7.5). These extractants, which are poorly effective for simple wall polysaccharides, may be even less effective for phenol—polysaccharide conjugates, which may be cross-linked in the wall. Perhaps the most promising approach is deliberate partial degradation of the polysaccharide backbone by treatment with enzymes (e.g. Panel 3.4.1 c).

For polysaccharides that may bear phenolics linked via ether bonds, more vigorous extractants can be considered, e.g. cold aqueous NaOH [171]. However, some sugar—phenol ether bonds are relatively labile to acid and/or alkali [146], unlike most other ether bonds. Maintenance of samples under N_2 and in the dark during alkali-treatment is recommended to retard oxidation of the phenolic moieties.

Lignin can be obtained in a number of ways [436]. Most of the work has been done on wood or straw lignin and application of the methods to the small quantities of lignin that may occur in certain growing cells will require adaptation of methods. **Klason** lignin is the insoluble residue after dissolution of total wall polysaccharide in 70-72 % (w/w) H_2SO_4 at 25°C for 2-4 h; it is highly modified from the natural structure, and liable to be heavily contaminated with protein unless this is first removed from the walls by a pre-hydrolysis in hot dilute acid.

Björkman lignin (milled wood lignin) is extracted into dioxane after mechanical disruption of the microfibrils: the specimen (e.g. saw-dust) is suspended in toluene and subjected to 2 or more days' vigorous milling in a vibratory ball mill (Section 1.2); subsequent stirring at 25°C with 80-90 % dioxane extracts a proportion (30-50 %) of the lignin [436]. Some of the material solubilised remains in solution when transferred into water — this material is rich in lignin—carbohydrate complexes (LCCs) [31]; some of the dioxane-insoluble lignin can subsequently be extracted from the residue as LCCs in DMSO or 50 % aqueous acetic acid.

'Enzyme lignin' is obtained by treatment of ball-milled cell walls with commercial cellulase (e.g. 1 % 'Onozuka' cellulase or Driselase, pH 4.5, with 0.05 % chlorbutol, at 37°C for 3 days); this removes most of the carbohydrate and breaks down LCCs, and the residual lignin is extracted with 80-90 % dioxan [437].

'Alkali lignin' can extracted from grass cell walls with 2 M NaOH under N_2 at 37°C for up to 24 h. It is separated from hemicelluloses by its

---

PANEL 3.4.3: Extraction of lignin and its micro-scale assay by the acetyl bromide/acetic acid method [267]

Sample: 0.5 mg of finely ground Douglas fir wood, thoroughly extracted with ethanol/toluene (1:1) until the extracts no longer absorb UV light at 280 nm.

(care — hazardous reagents!):

Step 1: Mix 0.5 mg wood with 1 ml acetyl bromide/acetic acid (1:3) in a loosely capped glass tube.

Step 2: Incubate in a water bath at exactly 70°C with occasional shaking for 30 min.

Step 3: Cool to about 15°C. Transfer the solubilised sample into a mixture of 0.9 ml of 2 M NaOH + 5 ml glacial acetic acid (to hydrolyse excess acetyl bromide). Rinse out the tube with a little more acetic acid and pool with the bulk solution.

Step 4: Add 0.1 ml 7.5 M hydroxylamine-HCl [to destroy bromine and polybromide].

Step 5: Dilute the solution to exactly 10 ml with more acetic acid.

Step 6: Within 5 min of starting step 3, read the A_{280}. Lignin at 10 µg/ml gives an absorbance of about 0.24. An observed absorbance of 0.343 would indicate that the diluted solution contained lignin at 14 µg/ml, and therefore the original wood sample was 28 % (w/w) lignin.

solubility properties: e.g. the lignin is precipitated upon addition of HCl to pH 1 at 0°C, re-dissolved in dioxane, re-precipitated with 1 M acetic acid, re-dissolved in glacial acetic acid, and re-precipitated in a large volume of diethyl ether.   Although alkali is an efficient extractant for grass lignins (ca. 50 % solubilised [437]) it is little use for Dicot or Gymnosperm wood lignins (less than 10 % solubilised).

Perhaps the most useful lignin 'extractant' for small samples is acetyl bromide/acetic acid (1:3), which almost completely dissolves wood (and other cell wall) samples by acetylating the free —OH groups of poly- saccharide and lignin to form derivatives that are soluble in organic solvents.   [The reagent may also cause some degradation of polysaccharide backbones by acetolysis (Section 4.2.3).]   If the wall sample had been pre- extracted so that lignin was the only aromatic material left in the wall, the A_{280} of the AcBr/HOAc extract can be taken as a quantitative measure of lignin content [267].   The technique is outlined in Panel 3.4.3.

3.5:   SEPARATION OF WALL POLYMERS

Extracted polymers are separated, either **analytically** to characterise the polymers present, or **preparatively** to purify a chosen polymer in quantities adequate for further study.   Many but not all separation techniques can be used both preparatively and, by scaling down, analytically.   Section 3.5 emphasises preparative applications, and Section 3.6 analytical, but the distinction is rather arbitrary.   One important 'separation' method, not further discussed, is the differential extraction of polymers achieved by specific extractants (Section 3.4).

3.5.1:   Removal of extractant: de-salting

It is often necessary to separate polymer from extractant.   Since most extractants are low-M_r compounds, this is usually easy.   The four main methods are described below.   [Note: 'de-salting' is taken to mean removal of any low-M_r contaminant of a polymer sample.]

De-salting by dialysis (Panel 3.5.1 a) is slow but cheap, simple and effective.   Dialysis is excellent for removal of DMSO, urea, guanidinium thiocyanate, alkalies (after neutralisation with acetic acid and allowing time for H_2 bubbles to cease evolving from any $NaBH_4$), MMNO, chlorite (after neutralisation with NH_3 to prevent further evolution of ClO_2),

CaCl$_2$, LaCl$_3$, and other inorganic salts. Dialysis does not remove detergents, and removal of sodium hexametaphosphate (phosphate glass) and some other chelating agents may be slow.

Sample [useful range 1-500 ml] is enclosed in a tightly knotted sac of water-washed Visking tubing [range ca. 1 to 6 cm diameter], which is incubated in 10-50 volumes of stirred water (or dilute aqueous buffer, preferably volatile — see Section 4.5.4) at 0°C. Low-M$_r$ solutes slowly diffuse from the sac into the water; polymer molecules are trapped inside the sac. Several changes of the water are necessary. Since the process may take many hours it is useful to add an anti-microbial agent e.g. 0.05 % chlorbutol (chosen because it is volatile). It is also important to ensure

---

PANEL 3.5.1 a: De-salting hemicellulose by dialysis

Sample: 200 ml of total hemicellulose extracted by 6 M NaOH / 1 % NaBH$_4$ (see Panel 3.4.1 a — step 5).

Step 1: Prepare the dialysis sac by soaking a 65-cm length of 5 cm wide Visking (dialysis) tubing in water. A 5-min treatment is adequate; but soaking also gives an opportunity to wash the tubing (e.g. in 1 M EDTA, pH 6.5, followed by 1 M NaOH and finally copious water), for work requiring high purity. Tie two knots in one end of the tubing.

Step 2: Neutralise the sample (200 ml), and adjust the pH to 5 by addition of acetic acid (requires ca. 120 ml). Add acid gradually to avoid heating and so that bubbles of H$_2$ given off by NaBH$_4$ do not cause excessive frothing. [It is possible at this stage to centrifuge down and collect the precipitate of 'hemicellulose A' (flocculation of which is favoured by incubation at 37°C); 'hemicellulose B' is the polysaccharide that remains in solution on neutralisation. Alternatively dialyse the whole suspension.]

Step 3: Using a funnel, pour the sample into the tubing. Expel air, and tie two more knots at the other end of the sac. [It is unnecessary to leave room for osmotic expansion of the sample as Visking tubing is very resistant to bursting.]

Step 4: Place sac in a 2-litre measuring cylinder almost filled with 0.05 % chlorbutol. Stir gently for 1 h.

Step 5[1]: Change chlorbutol solution; stir for a further 1 h. Repeat hourly for 6 h. Finally stir overnight in fresh chlorbutol.

Step 6: Open sac with scissors (cautiously, in a large beaker — the sac will be under pressure). Estimate sodium acetate concentration of sac contents by use of a conductivity meter. If low enough, freeze-dry.

Next steps: See sections 3.5.2-3.5.7, 3.6 and Ch 4.

[1]As an alternative to frequent changing, running tap water can be used, followed by dialysis against distilled water to remove tap water solutes.

that unwanted enzymic reactions will not occur during dialysis, e.g. cata-lysed by enzymes co-extracted with the polymers of interest. Commercial Visking tubing contains small amounts of impurities: in work requiring very high purity, an extractant-only control is dialysed in parallel with the sample so that any significant contaminants can be detected.

De-salting by gel-permeation chromatography on Sephadex G-25 (Panel 3.5.1 b). GPC separates on the basis of M_r (see Section 3.5.2) and is often used preparatively to remove low-M_r extractant from a polymer. It is quicker than dialysis, but has a lower capacity and has the disadvantage that if de-salting causes precipitation, the polymer of interest may be lost in the gel bed. The gel (e.g. Sephadex G-25) is suspended in water or a buffer (preferably volatile) and poured into a column to form a bed about 3 times longer than wide, and total volume at least 5 times that of the sample to be de-salted. The sample [workable range ca. 0.1-200 ml] is applied evenly to the gel bed [1-1000 ml], and allowed to soak in. The column is then eluted with water or buffer. No pump is needed, as adequate flow occurs under gravity. Polymers (M_r greater than 5000) emerge when ca. 0.25 to 0.5 bed volumes have eluted (including the volume eluted during sample application), followed by the low-M_r solutes at ca. 0.6-1.0 bed volumes. The process usually takes ca. 10-60 min. After each run, the column is washed with at least one column-volume of pure eluent prior to re-use. A mixture of 0.2 % Blue Dextran and 0.5 % $CoCl_2$ provides coloured high- and low-M_r markers for a preliminary run to calibrate the column using the same volumes as will later be used for the sample of interest. GPC is suitable for all the low-M_r extractants used.

Precipitation of the polymer. (See also Section 2.6.1). Polysaccha-rides and glycoproteins can usually be precipitated from aqueous solution or from PAW by addition of 0.05 volumes of 10 % ammonium formate followed by 5 volumes of ice-cold ethanol. Glycoproteins (except those with a very high carbohydrate content), but not polysaccharides, can be precipitated from aqueous solution at 0°C by addition of trichloroacetic acid (TCA) to a final concentration of 20 %. Ethanol and TCA do not precipitate most low-M_r compounds (this should be checked on a sample of the pure extractant), and can therefore be used to separate polymer from extractant. Once the polymer has formed a flocculent precipitate, either in ammonium formate + ethanol or in TCA, the precipitate is washed several times (e.g. by the

PANEL 3.5.1 b: 10-min de-salting of glycoprotein or polysaccharide by GPC

Sample: 1 ml of $LaCl_3$-leachate from living, cultured spinach cells (Panel 6.3.2).

Step 1: Soak 4 g dry Sephadex G-25 (medium grade) in 100 ml H_2O for 2 h at 100°C or 100 ml 0.05 % chlorbutol for 2 days at 25°C. Do not stir.

Step 2: Plug the tip of a dry 10-ml pipette with a small wisp of glass wool, and fit a 3-cm length of narrow-bore flexible tubing to the outlet. Close this off with a clip.

Step 3: Swirl Sephadex, let the gel settle out, and pour off most of the free liquid, leaving a 5-ml head. Swirl gently to form a thick slurry, taking care not to trap bubbles. By use of a long-tipped Pasteur (or ARH) pipette, transfer the slurry into the 10-ml pipette, and clamp this vertically. Open clip and let liquid drain through. Do not let the gel bed run quite dry. Add/remove gel until the bed is at the 10-ml mark.

Step 4: Pass <u>ca.</u> 20 ml H_2O[1] through the column to wash the gel.

Step 5: Open the clip and let the meniscus almost run down to the gel bed. Close the clip, and position a 10-ml measuring cylinder under the outlet. Carefully layer 1 ml of 0.2 % Blue Dextran / 0.5 % $CoCl_2$ on to the gel, avoiding disturbance of the bed surface. Open the clip.

Step 6: When the coloured solution has just run into the gel, gently apply 0.5 ml water. (This can be made to rinse down any of the solution that may be adhering to the walls of the pipette.) Repeat step 6.

Step 7: Fill up the head of the column with water, and let flow continue. Blue Dextran & $CoCl_2$ will separate, forming within a few minutes a fast-running blue and a slow-running pink zone. Record volume in measuring cylinder when **(a)** blue solution starts, and **(b)** stops emerging; **(c)** pink solution starts, and **(d)** stops emerging. [ca. 2.5, 4.5, 6.5 and 10 ml respectively.] This calibrates the column, indicating the 'windows' in which high- and low-M_r compounds will be found.

Step 8: Repeat steps 4-5 using 1 ml $LaCl_3$-leachate in place of the coloured solution.

Step 9: Apply [**(a)** minus 1] ml of water, and let the bed just run dry, rejecting the eluate. [If flow is very slow, material has probably precipitated upon de-salting: speed the elution by very gentle pressure from a Pasteur pipette teat; in future use a higher pyridine/HOAc concentration as eluant (see[1]).] Apply [**(b)** minus **(a)**] ml of water and collect the eluate (high-M_r fraction, containing glycoprotein). Freeze-dry.

Step 10: Apply at least a further [**(d)** minus **(b)**] ml of water before re-use of the column. [Steps 8-10 can be performed on many columns at once, with the advantage that many samples can be de-salted in parallel.]

Step 11: If column is to be left overnight, pass 15 ml 0.05 % chlorbutol through it. Clip outlet tightly & seal top of pipette with Parafilm.

Next steps: see Sections 3.5.2-3.5.7, 3.6, 6.5 and Ch 4.

[1]In these and subsequent steps, pyridine/HOAc/H_2O (1:1:23; pH <u>ca.</u> 4.5) can be used instead of water to minimise ionic binding to the gel.

resuspension—centrifugation method) with pure 80 % ethanol to remove the ammonium formate or TCA, re-dissolved or re-suspended in water, and freeze-dried. The disadvantage of precipitation is that it is sometimes hard to get proteins back into aqueous solution after they have been precipitated. A few polysaccharides, especially neutral arabinans, are not precipitated by 80 % ethanol; higher concentrations, or acetone, would be needed, which would however be more likely to co-precipitate the low-M_r solute.

Drying. Volatile extractants, e.g. pyridinium acetate, can readily be removed by freeze-drying, or in a 'Speed-Vac' [Savant Instruments, Farmingdale, New York], or under a jet of air or N_2. Drying has the great advantage over dialysis, GPC and precipitation methods that any (non-volatile) low-M_r degradation products of the cell wall, generated during extraction, are retained. A special case of de-salting by drying is the removal of borate, which, although not itself volatile, can be removed as its volatile methyl ester (Section 4.5.1).

Where the 'extractant' was an enzyme e.g. pectinase, different techniques are needed for its removal. Possibilities are: precipitation of the enzyme by TCA [see above] or 80 % saturated $(NH_4)_2SO_4$, or treatment at 100°C to coagulate the enzyme (not recommended owing to the heat-lability of polysaccharides). Often, however, the extracted polymer is so different from the enzyme that any separation technique used for other purposes will have the side effect of removing the enzyme.

3.5.2: Separation of polymers on the basis of size

Separation on the basis of M_r is excellent for proteins but less useful for polysaccharides because they tend to be highly heterogeneous in size. Thus, in a crude hemicellulosic extract of a Dicot, the two main polysaccharides (xyloglucan and arabinoxylan) may considerably overlap in M_r, and separation on the basis of size would not be a useful preparative step. It would, however, be needed analytically when M_r was one of the molecular parameters being studied (see Ch 8).

Gel-permeation chromatography. The most widely used method of separation based on M_r is **gel-permeation chromatography** (GPC) [294,572]. The sample is passed through a gel bed, usually Sephadex, Sepharose or Bio-Gel P or A, packed in a glass column. Molecules above a certain size limit elute rapidly, as a group, at the void volume (V_0, usually ca. 0.25 to 0.3

bed volumes). Molecules smaller than a certain limit are retarded, eluting together, as a second group, in the included volume (V_i, usually <u>ca.</u> 0.7 to 0.8 bed volumes). Molecules of intermediate M_r are said to lie within the fractionation range of the gel; they elute between V_0 and V_i (say, at volume V_e). Different gels have different M_r fractionation ranges (Panel 3.5.2 b). For partially excluded molecules, there is an inverse relationship between V_e and log M_r (shown by the dashed line in Fig. 3.5.2). However, globular molecules elute later than rod-shaped molecules of the same M_r, and certain compounds, especially phenolics, adsorb to the gel, sometimes so strongly that they do not elute at all. Also the gels may have a small number of fixed negative charges and therefore behave to a slight degree as ion-exchange resins; the effect of this can be minimised by inclusion of a salt in the eluent [pyridine/acetic acid/H_2O (1:1:23), pH <u>ca.</u> 4.5, is useful because it is volatile].

Sephadex and Bio-Gel P are most suitable for relatively small polymers (M_r <u>ca.</u> 10^3 to 10^5). These gels can also separate higher-M_r polymers, but the grades of gel required (e.g. Sephadex G-200 or Bio-Gel P-300) are very soft: they tend to clog, and elution is then very slow unless great care is exercised. For polymers of M_r 10^5-10^7, Sepharose-CL and Bio-Gel A gels are recommended.

The gels are swollen in water or buffer according to the manufacturers' instructions, and poured into glass columns (pre-washed internally with 'Repelcote' or 2 % dichlorodimethylsilane in CCl_4, and dried). For high-resolution work the columns should be long and narrow (typically <u>ca.</u> 50 x 2 cm; sometimes as long as 150 cm). A booklet [400], produced by Pharmacia Fine Chemicals, is an excellent source of practical hints on Sephadex and Sepharose, most of which are also applicable to Bio-Gel.

The <u>volume</u> of sample applied to the column should not be greater than 4 % of the gel bed volume, but use of much less than 4 % will not significantly improve resolution. The <u>concentration</u> of the sample to be loaded should not be excessive. As a guide, under 1 % (w/v) total solutes (including any sugars and salts) should be aimed for; less for higher-M_r polymers. It is helpful to adjust the pH of the sample as close as possible to that of the elution buffer, to avoid precipitation during chromatography. It is useful, especially during analysis of radioactive material, to add <u>internal markers</u> to the sample e.g. 0.5 % Blue Dextran (M_r <u>ca.</u> 2,000,000;

detected by A_{620}) and 1 % glucose [M_r = 180;   detection by any convenient assay —— see Panel 3.7 a for quantitative assays;   alternatively, dry 20 μl of each column fraction on to filter paper and stain this with aniline hydrogen phthalate (Panel 4.5.2 f)].   Internal markers provide exact information about the V_0 and V_i in each individual analysis.   For columns with fractionation ranges extending above M_r 2,000,000, viruses or autoclaved cells of E. coli have sometimes been used as high-M_r marker.

The use of pectinase to digest native (wall-bound) pectin, and the analysis of the digestion-products by GPC is shown schematically in Fig. 3.5.2;   experimental details are given in Panel 3.5.2 a.

PANEL 3.5.2 a: **Separation of rhamnogalacturonan-I and -II, and oligo-galacturonides, by GPC**[1]

Sample: 50 mg of pectinase-digestion products (Panel 3.4.1 c).

Step 1:   Soak 20 g of dry Sephadex G-75 in 400 ml water for 2 h at 100°C or 400 ml 0.05 % chlorbutol for 2 days at 25°C.   Do not stir.

    Step 2:   Rinse a dry 100 x 1.5 cm 'Econocolumn' [Bio-Rad Labs Inc.] with Repelcote (or 2 % dichlorodimethylsilane in CCl_4), and re-dry.   Attach 40 cm of very fine-bore, flexible tubing to the outlet.   Close this off with a clip.

Step 3:   Pour the slurry into the column, following the precautions recommended [400].   Keep topping up the gel until the bed level is about 6-10 cm from the top of the column.

Step 4:   Wash the gel by passing 200 ml of buffer **A** [pyridine/acetic acid/water (1:1:18)] through the bed at ca. 30 ml/h.   Flow rate is adjusted by raising or lowering the eluent reservoir.

Step 5:   To calibrate the column, apply 5 ml of 0.5 % Blue Dextran / 0.5 % galacturonic acid (GalA) in buffer **A**.   Elute with buffer **A** at 30 ml/h, collecting 75 2.5-ml fractions with a fraction collector.   Assay Blue Dextran by A_{620}, and identify the fractions containing GalA by the m-hydroxybiphenyl assay (Panel 3.7 a) [or by drying 25 μl of each fraction on to filter paper and staining it with aniline H-phthalate (Panel 4.5.4 f)].

Step 6:   Repeat step 5 with sample (50 mg digestion-products in 5 ml buffer **A**) in place of Blue Dextran / GalA.   Assay the 75 fractions with m-hydroxybiphenyl.   Three peaks should be observed: RG-I ($K_{av.}$ = 0), RG-II ($K_{av.}$ = 0.6) and GalA-rich oligogalacturonides ($K_{av.}$ = 1.0).

Step 7:   Pool the fractions corresponding to these peaks.   Freeze-dry to remove pyridinium acetate.

Next steps: See Sections 3.5.3 to 3.5.7, 3.6 and Ch 4 for analysis of pectic polysaccharides.   Use paper chromatography in EtOAc/HOAc/H_2O (10:5:6) to resolve oligouronides (Panel 4.5.2 d).

[1] For further relevant precautions, see Panel 3.5.1 b.

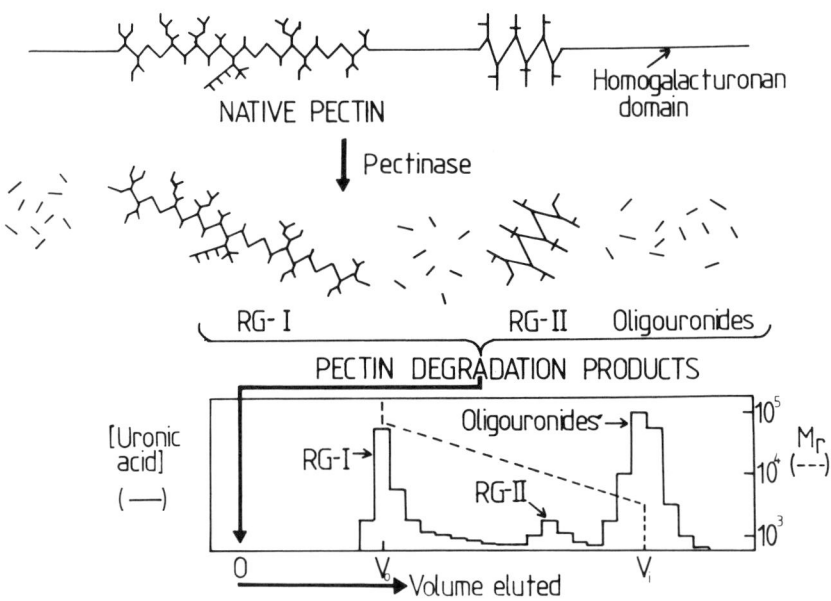

Fig. 3.5.2: **One interpretation of the action of pectinase on wall-bound pectins.** The enzyme splits the chain at sites of contiguous α-(1→4)-GalA residues, solubilising rhamnogalacturonans (RG-I and RG-II) and generating GalA-rich oligosaccharides from connecting homogalacturonan domains. The digestion-products are resolved according to M_r by GPC on Sephadex G-75. The caricatures of RG-I and RG-II are not to scale.

PANEL 3.5.2 b: **Fractionation ranges[1] of chromatography gels**

Name of gel	M_r Fractionation range reported for		Cost for 100-ml bed[2]	pH stability at 25°C
	linear dextrans	globular proteins		
Sephadex G-25	2×10^2–5×10^3	1×10^3–5×10^3	£ 8	2-14
Sephadex G-50	5×10^2–1×10^4	2×10^3–3×10^4	£ 5	2-14
Sepahdex G-75	1×10^3–5×10^4	3×10^3–8×10^4	£ 4	2-14
Bio-Gel P-2	not given	1×10^2–2×10^3	£14	2-10
Bio-Gel P-4	not given	8×10^2–4×10^3	£10	2-10
Bio-Gel P-6	not given	1×10^3–6×10^3	£ 9	2-10
Bio-Gel P-10	not given	2×10^3–2×10^4	£ 7	2-10
Bio-Gel P-30	not given	3×10^3–4×10^4	£ 6	2-10
Bio-Gel P-60	not given	3×10^3–6×10^4	£ 5	2-10
Sepharose CL6B	1×10^4–1×10^6	1×10^4–4×10^6	£11	3-14
Sepharose CL4B	3×10^4–5×10^6	6×10^4–2×10^7	£11	3-14
Sepharose CL2B	1×10^5–2×10^7	7×10^4–4×10^7	£10	3-14
Bio-Gel A-0.5m	not given	ca. 10^4–5×10^5	£22	4-13
Bio-Gel A-1.5m	not given	ca. 10^4–1.5×10^6	£22	4-13
Bio-Gel A-5m	not given	1×10^4–5×10^6	£17	4-13

[1]Manufacturers' data. [2]Estimated from 1987 prices.

Dialysis (Cf. Section 3.5.1). Sophisticated dialysis sac materials are available which can discriminate between polymers of different sizes. For instance, M_r–20,000 cut-off dialysis tubing can be used so that only polymers above this size are retained in the sac. The cut-off values quoted by manufacturers are only rough guides and depend on the chemistry of the compounds being analysed. An advantage of dialysis over GPC is the large sample-volume and -concentration that can be handled.

3.5.3: Separation of polymers on the basis of charge

The charge on a wall polymer is one of its most characteristic features, and is therefore the basis of some useful separation methods. The main **negatively** charged groups are the ―COOH groups (charged at pHs above 3-5) of galacturonate, glucuronate, aspartate and glutamate residues. Phenolic ―OH groups, e.g. of ferulate and tyrosine residues, also acquire a negative charge at pHs above 8-10. There are no reports of phosphate groups in plant cell wall polymers, and sulphated polysaccharides are restricted to algae. The main **positively** charged groups are the side-chains of lysine (charged below pH 11), histidine (charged below pH 6) and arginine (charged below pH 12.5) residues. A polymer's charge:mass ratio at a given pH is usually reproducible, but the ratio of neutral methyl-esterified galacturonate residues to acidic galacturonate residues in a pectin can be decreased by hydrolysis with either alkali or enzyme (pectin methylesterase); pectins should therefore be protected from these agents.

Ion-exchange chromatography involves the reversible binding of the polymer to a gel which bears a fixed charge of opposite sign. As an example of **anion-exchange chromatography**, negatively charged pectins in a solu-tion of pH ca. 7 are passed through a column of gel which bears fixed posi-tive charges. Suitable gels include Q Sepharose Fast Flow, QAE-Sephadex [41] and DEAE-Trisacryl [414]. Acidic pectins bind to the gel, whereas neutral and positively charged polymers pass through it unretarded. The bound pectins are then eluted from the gel by application of salt. The anion of the salt (e.g. HCO_3^-) competes with the pectin for positively charged sites on the gel, displacing the pectin. Different classes of pectin, differing in charge : mass ratio, are eluted by different concentrations of salt. For this reason, elution with a gradient of salt can give separation of related polymers (Panel 3.5.3 a; Fig. 3.5.3 a).

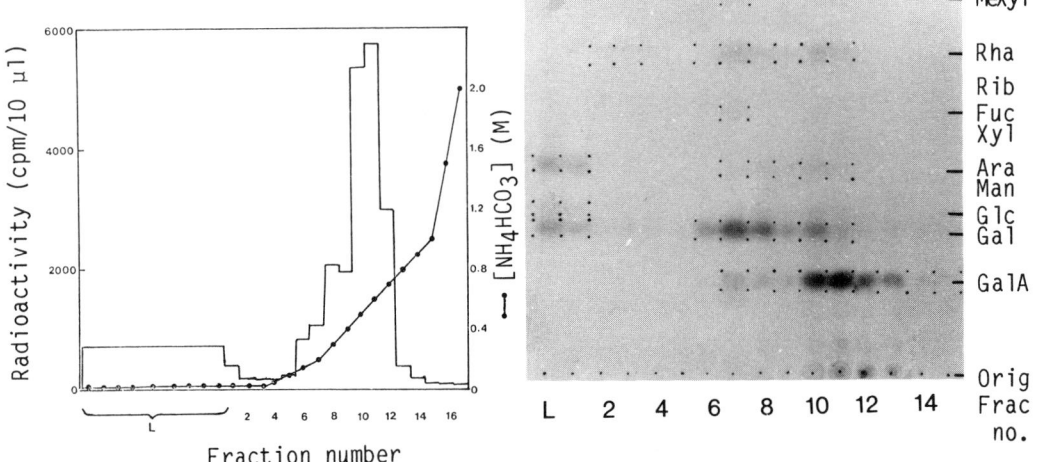

Fig. 3.5.3 a: Fractionation of [U-14C]pectins of rose by anion-exchange chromatography on QAE-Sephadex, by use of an ammonium bicarbonate concentration gradient. Left: elution of total polymer (cpm ^{14}C) [L = loading; **2-16** = fractions collected after completion of loading]. Right: Autoradiogram of acid-hydrolysis products separated by TLC [41] as in Fig. 4.5.3 (i). Areas in which blackening of the film was detectable are highlighted with dots.

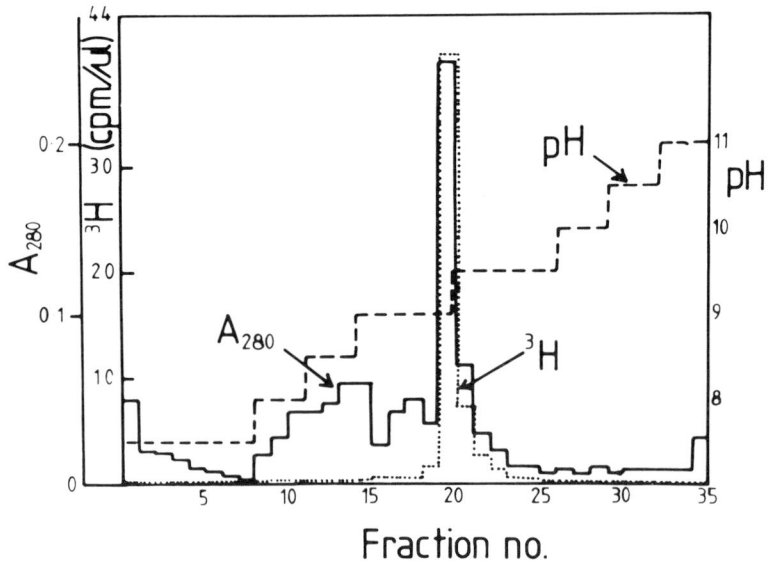

Fig. 3.5.3b: Cation-exchange chromatography of extensin (see Panel 3.5.3 b). Total protein and pentose (Ara) are monitored as $\underline{A}_{280}$ and ^3H respectively.

```
┌─────────────────────────────────────────────────────────────────────────┐
│ PANEL 3.5.3 a:  Anion-exchange chromatography of partially degraded pectins │
│                                                                           │
│ Sample:          0.2  g  freeze-dried  pectin  from  cultured  rose  cells │
│                  (extracted by the method of Panel 3.4.1 d).              │
│                                                                           │
│ Step 1:          Suspend 50 ml of 'Q Sepharose Fast Flow' or 10 g QAE-    │
│ Sephadex A-25 in 250 ml 2 M NH4HCO3 with occasional shaking for 1 h or 1  │
│ day respectively.  Let gel settle out, reject free liquid, & repeat step 1. │
│                                                                           │
│ Step 2:          Transfer gel into a sintered glass funnel, and wash with 2 │
│ 1 buffer A (10 mM NH4HCO3, pH adjusted to 7.5 with dilute NH3, and         │
│ containing 0.05 % chlorbutol).  Resuspend gel in buffer A, and pour into a │
│ 2.5 x 15 cm 'Econocolumn' [Bio-Rad Labs Inc].                             │
│                                                                           │
│ Step 3:          By passing buffer A through the bed, deduce height of     │
│ buffer-reservoir needed for a flow rate of 5 ml/min.                      │
│                                                                           │
│ Step 4:          Dissolve pectin (0.2 g) in 20 ml buffer A.  Re-adjust pH to │
│ 7.5 with dilute NH3.                                                      │
│                                                                           │
│ Step 5:          Connect column outlet to a fraction collector, and from now │
│ on collect in the form of 10-ml fractions all the material that eluates.  │
│ Pass sample(3) through gel.  Apply a further 50 ml of buffer A.  [Eluted:  │
│ neutral and very weakly acidic pectins, not ionically bound to gel.]      │
│                                                                           │
│ Step 6:          Apply a gradient of 10-500 mM NH4HCO3 (pH 7.5), made up in │
│ 0.05 % chlorbutol.  Gradients can be applied manually, in steps (e.g. 12  │
│ 50-ml portions of 20, 30, 40, 60, 90, 120, 160, 200, 250, 300, 400 and 500 │
│ mM).  Alternatively, they can be continuous, formed by a gradient maker   │
│ [439].  [Eluted:  progressively more acidic pectins.]                     │
│                                                                           │
│ Step 8:          Freeze-dry each fraction.  Redissolve residue in a minimum │
│ of water;  freeze-dry again.  Repeat until no NH4HCO3 remains.            │
│                                                                           │
│ Step 9:          Wash the gel bed extensively in 1 M NaOH before re-use.  │
│                                                                           │
│ Next steps:      See Sections 3.5.4-3.5.7, 3.6 and Ch 4.                  │
│                                                                           │
│ (1)The Mr of the polymer dictates the type of gel used:  a gel should be  │
│ chosen which the polymer can permeate, otherwise the binding capacity is  │
│ much reduced.  'Q Sepharose Fast Flow' and QAE-Sephadex A-25 have exclusion │
│ limits of Mr ca. 10^6 and 5x10^3 respectively for polysaccharides.        │
│                                                                           │
│ (2)The composition of the sample solution is carefully controlled with    │
│ respect to pH [chosen so that the gel and the polymer have opposite       │
│ charges] and to ionic strength (usually ca. 10 mM).                       │
│                                                                           │
│ (3)The volume of sample can be larger than the gel bed provided that the  │
│ binding capacity of the gel is not exceeded;  this is checked in a scaled- │
│ down preliminary run.                                                     │
└─────────────────────────────────────────────────────────────────────────┘
```

In **cation-exchange chromatography,** a positively charged polymer (e.g. soluble extensin) will bind to a negatively charged gel (e.g. S Sepharose Fast Flow or SP-Sephadex). Experimental details are given in Panels 3.5.3 b and results in Fig. 3.5.3 b (page 84). For further details, including the factors that determine choice of conditions, see [439].

PANEL 3.5.3 b: Cation-exchange chromatography to purify extensin

Sample: 1 mg de-salted (Panel 3.5.1 b), freeze-dried glycoprotein, LaCl$_3$-leached (Panel 6.3.2) from [^{3}H]arabinose-fed (Sections 2.1.3 & 2.2), cultured spinach cells.

Step 1: Soak 0.4 g of dry SP-Sephadex C-50 [Na$^+$ form, i.e. as supplied] in 100 ml buffer **A** [0.1 M NaOH, pH adjusted to 7.5 with solid HEPES] for 2 days at 25°C or 2 h at 100°C.

Step 2: Transfer the gel to a sintered glass funnel and wash with 2 x 100-ml portions of buffer **A**. Resuspend in a small volume of buffer **A** to form a slurry, and pack a 10-ml gel bed in a 10-ml pipette (Panel 3.5.1 b).

Step 3: Dissolve the glycoprotein in 10 ml buffer **A**. Remove exactly 0.1 ml for liquid scintillation-counting (LSC: see Section 2.8.2).

Step 4: Pass the remaining 9.9 ml through the column at 10 ml/h, collecting the eluate. Apply a further 9.9 ml of buffer **A**. Combine the eluates, and remove 0.2 ml for LSC. Hence demonstrate that the binding capacity of the gel has not been exceeded.

Step 5: Prepare buffer solutions of similar ionic strength[1] by titrating 0.1 M NaOH with solid HEPES to give pH 8.0, with solid TAPS to give pH 8.5 and 9.0, and with 0.1 M NaHCO$_3$ to give pH 9.5, 10.0, 10.5 and 11.0. Apply to the gel 7.5 ml at pH 8, 7.5 ml at pH 8.5, 15 ml at pH 9, 15 ml at pH 9.5, 7.5 ml at pH 10, 7.5 ml at pH 10.5, and 7.5 ml at pH 11 (flow rate ca. 10 ml/h). Collect 2.5-ml fractions throughout.

Step 6: Monitor each fraction for pH, ^{3}H, and protein (by simple UV absorbance — see Panel 3.7 a). [Extensin is eluted at pH ca. 9.0-9.5 — see Fig. 3.5.3 b.] Pool appropriate fractions, and de-salt by dialysis (Panel 3.5.1 a).

[1] Changes in ionic strength bring about large changes in bed volume with certain gels e.g. SP-Sephadex C-50; this can make the use of a pH gradient preferable over a salt gradient. The problem does not arise with S and Q Sepharose Fast Flow, which are in the form of rigid (cross-linked) beads.

Precipitation of acidic polymers by cationic detergents.   A second method of separating polymers on the basis of their charge involves the ability of acidic polymers to bind to, and form an insoluble complex with, cationic detergents e.g. cetyltrimethylammonium bromide (CTAB). Neutral and positively charged polymers do not bind and therefore remain in solution.

The thoroughly de-salted polymer solution (0.1-1.0 %) is mixed with an equal volume of 2 % CTAB containing 20 mM Na_2SO_4 and incubated at 37°C for about 1 h (or overnight for very weakly acidic or low-M_r polymers). Precipitated polysaccharides are collected by centrifugation (2,500 **g** for 5 min) or filtration. The precipitated acidic polymers can usually be re-dissolved by addition of more Na_2SO_4; the concentration of salt required depends on the properties of the polymer, and therefore gives a way of separating different acidic polymers. The polymers can subsequently be purified by precipitation with 80 % ethanol, in which CTAB is soluble, followed by de-salting (Section 3.5.1) [440,441].

3.5.4: Separation of polymers on the basis of density

Polysaccharides have a higher buoyant density than proteins, and this difference can be used to separate these polymers from each other and from glycoproteins, which have intermediate density. Separation is achieved by isopycnic centrifugation [52] in a caesium chloride (CsCl) density gradient, which is created in the ultra-centrifuge itself by prolonged centrifugation of the sample material in an initially uniform solution of CsCl (adjusted to a density of 1.48 g/cm^3). Recently, caesium trifluoro-acetate [Pharmacia] has been used for analysis of RNA, with the advantage over CsCl of being able to give a higher density (2.6 g/cm^3, compared with 1.9 g/cm^3 for CsCl).

Centrifugation is typically at 40,000 **g** [for an SW 50.1 rotor] and 25°C for 48 h, in a 15-ml polycarbonate tube [201]. After centrifugation, the solution is slowly withdrawn (by means of a peristaltic pump) via a hollow needle, from the bottom of the tube, into a fraction collector. Calibration of the density gradient is by measurement of the CsCl concentration in each fraction e.g. with a refractometer. Analysis of the separated polymers usually requires de-salting, but liquid scintillation counting is possible in Triton scintillant after 4-fold dilution of the CsCl solution.

3.5.5: Separation of polymers by their affinity properties

<u>Lectin-binding.</u> Polymers bearing certain sugar residues can bind to specific lectins (Section 3.1.3), and this property can be exploited in affinity chromatography. The lectin, immobilised on a column of agarose gel, will adsorb some polysaccharides and glycoproteins but not others. Many agarose-immobilised lectins are commercially available (e.g. Sigma). The polymer mixture is passed through the column in a buffer that favours sugar—lectin binding (typically 10 mM Tris-HCl, pH 8), and the column is washed with the same buffer to elute non-bound polymers; specifically bound polymers are subsequently eluted by application of a low-M_r sugar which competes for the binding sites. This sugar may be an oligosaccharide or (at a higher concentration) monosaccharide; the concentration required is determined empirically in a preliminary run by use of a concentration gradient: monosaccharides have been used in the range 10-500 mM. Sample volume can be larger than the gel bed volume, provided that the polymer of interest binds firmly. The binding capacity is low (e.g. 1 mg polysaccharide / ml gel bed; much less of an oligosaccharide). An example of the use of lectin affinity chromatography is the separation of a glycoprotein (peroxidase) from the corresponding unglycosylated protein [517].

<u>Cellulose-binding.</u> Hemicelluloses will bind to a column of powdered cellulose in neutral aqueous buffer, and can subsequently be eluted, e.g. by a gradient of urea (0-8 M) and/or NaOH (0-6 M) [29,241].

<u>Borate-binding.</u> The ability of some polysaccharides to bind borate (Section 3.1.3) can be exploited in 4 ways: (i) <u>Electrophoresis</u> — The polysaccharides are electrophoresed on GF/A paper (see Section 3.6.2) in 1.9 % borax ($Na_2B_4O_7.10H_2O$), pH 9.2. Neutral polysaccharides, which would be electrophoretically immobile at this pH in the absence of borate, move towards the anode if they can bind borate anions [278,542]. (ii) <u>Affinity chromatography</u> — an agarose gel carrying immobilised phenylboronate groups can be purchased (Sigma) and a column of this material will bind polymers that carry certain sugar groups. Bound polymers are later eluted with a solution of ethylene glycol or sorbitol, which bind borate strongly. (iii) <u>Ion-exchange chromatography</u> [279] — the polymers, after thorough de-salting, are dissolved in 5 mM borax, pH 9.4, and passed through a column of DEAE-Sephadex that had previously been equilibrated with a saturated solution of borax (pH 9.4) and washed with 5 mM borax. Those

polysaccharide that can bind borate will adsorb to the gel; unbound polymers are washed out with more 5 mM borax, and bound polymers are subsequently eluted by competition with 200 mM borax followed by 1 M NaOH. (iv) CTAB-precipitation — the (neutral) polymers, which would not otherwise be precipitated by CTAB (see Section 3.5.3), are supplemented with 0.33 % borax, pH adjusted to 8.5-10.0 [higher pHs being needed for polymers that bind borate weakly], and then treated with CTAB. Borate-binding polymers are precipitated [440,441].

3.5.6: Separation of polymers by differential precipitation and phase-partitioning

[See also use of CTAB — Sections 3.5.3 & 3.5.5.]

Eight unrelated techniques for the precipitation of polymers are summarised here. Such techniques do not separate on the basis of any single well-defined molecular property (e.g. charge, size or density), but they are often nevertheless very effective, giving qualitative separations of polysaccharides that would otherwise be difficult. They also have the advantage that they are easily scaled up for preparative use.

Ethanol. De-salted polymer is dissolved in aqueous buffer (e.g. 50 mM ammonium formate, adjusted to pH 4 with formic acid — for maximum reproducibility, the pH, ionic strength and polysaccharide concentration are kept constant between experiments). Pure ethanol is added dropwise to the rapidly stirred solution to give a final concentration of 20 %. After standing on ice for 10 min, the sample is centrifuged (3,000 **g** for 10 min), and the supernatant is carefully removed from any preciptate. More ethanol is added to the supernatant, with stirring, to give 30 %, and the next crop of precipitate is collected. The process is repeated, gradually increasing the ethanol concentration, until all the polymer has been precipitated. The precipitates containing the polymer of interest are pooled. If necessary, the process is fine-tuned by taking cuts at, say, 35, 40 and 45 % ethanol. The method works well for the polysaccharides liberated into the medium by cultured spinach cells: xylan, xyloglucan and arabinan precipitate at ca. 20, 50 and 90 % ethanol respectively (Panel 3.5.6). For polymers that do not precipitate until very high ethanol concentrations, e.g. arabinan, acetone may be preferred because lower concentrations suffice.

```
┌─────────────────────────────────────────────────────────────────────┐
│ PANEL  3.5.6:    Fractionation  of  secreted  polysaccharides  by  differential│
│ precipitation with ethanol                                            │
│                                                                       │
│ Sample:          Dialysed  spent  culture  medium  from  exponentially-growing,│
│                  [³H]arabinose-fed, suspension-cultured spinach cells.│
│                                                                       │
│ Step 1:          Dissolve  50  mg  ammonium  formate  into  10  ml  of  aqueous│
│ sample.  Add enough 100 % ethanol (1.765 ml) to give a final concentration│
│ of 15 % (v/v).                                                        │
│                                                                       │
│ Step 2:          Stand solution at 25°C for at least 15 min, and centrifuge│
│ (5 min at 2,500 g).  Carefully pour off (and keep) supernatant (S1).  Re-│
│ suspend  pellet  (if  any — may  be  very  difficult  to  see)  in  10 ml 15 %│
│ ethanol, and centrifuge again.  Reject supernatant (S2).  Redissolve pellet│
│ in 1 ml water, and freeze-dry.                                        │
│                                                                       │
│ Step 3:          Measure  exact  volume  of  S1  recovered  from  step 2.  Add more│
│ ethanol to give a final concentration of 25 %.  An equation for calculating│
│ the volume (V ml) of 100 % ethanol required to increase the concentration│
│ of a ml of b % ethanol to c % is:  $V = [a \times (c - b)] / [100 - c]$.│
│                                                                       │
│ Step 4:          Repeat step 2, using 25 % ethanol for the resuspension.│
│                                                                       │
│ Steps 5 to n:    Carry out steps 3 and 4 repeatedly, increasing the ethanol│
│ concentration by a standard increment (e.g. 10 %) each time round.    │
│                                                                       │
│ Next step:       Analysis  of  the  polysaccharides  present  in  each  cut.  This│
│ is  conveniently  done  for  [³H]pentose-labelled  material  by  Driselase-│
│ digestion  (Section 4.2.5)  since  the  radioactive  products  of  arabinan,│
│ xyloglucan  and  xylan  are  [³H]arabinose, [³H]xylosyl-α-(1→6)-glucose and│
│ [³H]xylose respectively, which are readily resolved by paper chromatography│
│ in EtOAc/pyridine/$H_2O$ (8:2:1) (Panel 4.5.2 b).                    │
└─────────────────────────────────────────────────────────────────────┘
```

Trichloroacetic acid (TCA) at 10-25 % and 0°C precipitates proteins but not polysaccharides. Glycoproteins may or may not be precipitated, depending on the carbohydrate content and TCA concentration.

$(NH_4)_2SO_4$ is another protein precipitant; it is generally used at a final concentration of 20-80 % saturation [439]. Different proteins and glycoproteins are precipitated at different $(NH_4)_2SO_4$ concentrations. Polysaccharides are less well precipitated, but β-(1→3),(1→4)-glucans and acidic pectins may precipitate at certain $(NH_4)_2SO_4$ concentrations.

Calcium salts have occasionally been used to precipitate or gel those (highly acidic) pectins that can be cross-linked by Ca^{2+} bridges. Pectins that do not readily bind Ca^{2+} can sometimes be made to do so by treatment with pectin methylesterase [570].

Copper salts will form insoluble complexes with certain acidic poly-saccharides [29] and with some which contain mannose or xylose [271]. The polysaccharide is dissolved at ca. 1 % in water or NaOH, and to it is added Fehling's solution [= a freshly prepared 1:1 mixture of solution **A** (6.92 % $CuSO_4.5H_2O$ in 2.5 mM H_2SO_4) + solution **B** (12 % NaOH, 34.6 % potassium sodium tartrate tetrahydrate)] until a precipitate forms. If no precipitate forms, addition of a little ethanol can aid the process. After 4 h, the precipitate is centrifuged down, washed with water, and treated for 1 min with ethanol/conc HCl (20:1) at 0°C to release the Cu from the polysaccharide. The insoluble polysaccharide is then further washed with pure ethanol to remove the HCl. The polysaccharides that remained soluble in Fehling's solution can be freed from Cu by simple de-salting.

Barium hydroxide precipitates mannans and certain low-arabinose xylans [348]. The polysaccharide is dissolved at 1-5 % in water or 10 % NaOH and 0.18 M $Ba(OH)_2$ is added until a precipitate forms. [Note that open solutions of $Ba(OH)_2$ form a precipitate of $BaCO_3$ spontaneously by reaction with atmospheric CO_2.] The precipitate is centrifuged down, washed with $Ba(OH)_2$ solution, redissolved or resuspended in 0.1 M acetic acid, and de-salted. The polysaccharides not precipitated by $Ba(OH)_2$ are recovered from the supernatant by de-salting.

pH has a strong influence on the water-solubility of certain polysaccharides. For example, an NaOH-extract containing hemicelluloses can be fractionated into 'hemicellulose A' which precipitates upon neutralisation, and 'hemicellulose B' which does not [549].

Iodine has been suggested as a means of selectively precipitating linear polysaccharides, while leaving branched ones in solution [185].

Phase partitioning is possible when two wall polymers have very different physico-chemical properties. An example is the separation of pure lignin from lignin-carbohydrate complexes (LCCs). If a mixture of these two (see Section 3.4.3) is dissolved in pyridine/acetic acid/H_2O (9:1:4) and then shaken with chloroform, the lignin partitions into the lower ($CHCl_3$) phase while the LCCs remain in the upper (aqueous) phase owing to their higher polarity [328].

3.5.7: Purification of polymers by breakdown of contaminants

If two polymers in a mixture cannot be separated, but one of them can be destroyed, the other can be obtained pure. For instance, in a mixture of β-(1→3),(1→4)-glucan and arabinoxylan (grass hemicelluloses), the glucan can be selectively hydrolysed with β-(1$\rightarrow$3),(1$\rightarrow$4)-glucanase, and the hydrolysis-products removed by dialysis or GPC, leaving pure arabino-xylan [3]. Similarly, one method for the isolation of lignin depends on enzymic hydrolysis of the carbohydrate components of the wall (Section 3.4.3). It must be ascertained that the enzyme preparation does not contain any activities that would act on the polymer of interest. Unfortunately, very few pure endo-glycanases are available commercially; the development of genetic engineering techniques for the mass-production of desired enzymes should eventually circumvent this problem.

3.6: DETERMINATION OF PHYSICAL PROPERTIES OF WALL POLYMERS
3.6.1: Determination of molecular weight (M_r)

Gel-permeation chromatography (GPC, Section 3.5.2) can provide an estimate of M_r of soluble polymers if the gel is calibrated by reference to markers of known M_r. Errors arise from the tendency of certain polymers (especially those containing phenolics) to bind to the gel, and from the influence of molecular shape (globular molecules elute later than linear ones of the same M_r) [160]. Since many proteins (but not extensin) are globular, whereas many polysaccharides are elongated, the gel manufacturers quote the M_r fractionation range for these two groups separately. Suitable polysaccharide markers include commercially available dextrans [α-(1$\rightarrow$6)-glucans] of defined mean M_r; pure proteins, e.g. cytochrome c, ovalbumin, bovine serum albumin, myosin and thyroglobulin can be used as protein markers. Markers should be chosen that are chemically similar to the molecules of interest. An example of the use of GPC is to monitor changes in the M_r of wall-bound xyloglucan during auxin-induced growth [379].

Analytical GPC can also be performed by HPLC on columns such as 'Zorbax Bio Series' GF-250 or GF-450 [DuPont Co.]. HPLC has the advantage of speed (separation time ca. 10-30 min), but the resolution is little better than on low-pressure GPC columns (separation time ca. 2-10 h). Anomalies of elution may be accentuated on HPLC: e.g. monomeric extensin (M_r ca. 1′ x 10^5) co-chromatographs on Zorbax GF-250 with thyroglobulin (M_r ca. 6.7 x 10^5) [311].

SDS-gel electrophoresis can provide an estimate of the M_r of proteins [305], but the method is less reliable for glycoproteins. Extensins and AGPs behave anomalously on SDS gels because of their charges and high hydroxyproline and carbohydrate content. Acid-urea gels are useful for extensin [468], and SDS gels without the stacking gel for AGPs [287].

Centrifugation.   The ability of high **g** forces to sediment polymers at a rate related to molecular weight provides an independent estimate of M_r.   The sample is layered on to a pre-formed sucrose density gradient (made up in a buffer that suppresses polymer—polymer interactions), and ultra-centrifuged. For details, see [52].

Viscosity. Measurement of viscosity can provide an estimate of M_r if standards of known M_r are available or the relationship between viscosity and M_r has been published for the polymer of interest.   The method is appropriate for cellulose, which can be solubilised either by nitration [60] or by treatment with 'Cadoxen' [254].

3.6.2: Determination of charge

Ion-exchange chromatography and CTAB binding (Section 3.5.3) and, for proteins, 'chromatofocusing' (see Parmacia literature) can give information on charge. Another analytical method is <u>electrophoresis</u>, which depends on the ability of a charged polymer to move through a solution towards an electrode of opposite sign. Electrophoretic equipment is described in Section 4.5.4.   Methods for electrophoresis (including isoelectric focusing) of proteins in non-SDS polyacrylamide gels are well established [222]; the following account therefore refers mainly to polysaccharides.

The pH of the electrophoresis buffer determines the magnitude (and sometimes sign) of the polymer's net charge. A volatile buffer (Section 4.5.4) is usually chosen, e.g. pH 6.5 for electrophoresis of pectins [36, 481].   Glass-fibre paper is often used as the support for electrophoresis because many polysaccharides irreversibly bind to normal paper. Whatman GF/A glass fibre paper is marked out very lightly in dotted lines with a 2B pencil, and should not be bent. Untreated GF/A exhibits high electro-endo-osmosis (EEO), i.e. the tendency of neutral solutes to be swept towards the cathode by a mass flow of water through the paper; this can be minimised by treatment of the GF/A in 'Repelcote' or 2 % dichlorodimethylsilane in CCl_4 and drying before use [264]. Treated GF/A is difficult to wet, and should be soaked overnight in the electrophoresis buffer containing 0.2 % Tween-20; the samples (e.g. 2 μl) are applied to the paper while it is still wet. It is important to include a neutral marker (e.g. 2 μl of 2 %

dextran) on the electrophoretogram as a reference for EEO. A convenient negatively charged marker is picric acid, which is bright yellow; electrophoresis is stopped when the yellow spot approaches the end of the paper. The polymer's electrophoretic mobility is then quoted as $m_{picrate}$, defined as [(distance of polymer from dextran) / (distance of picrate from dextran)].

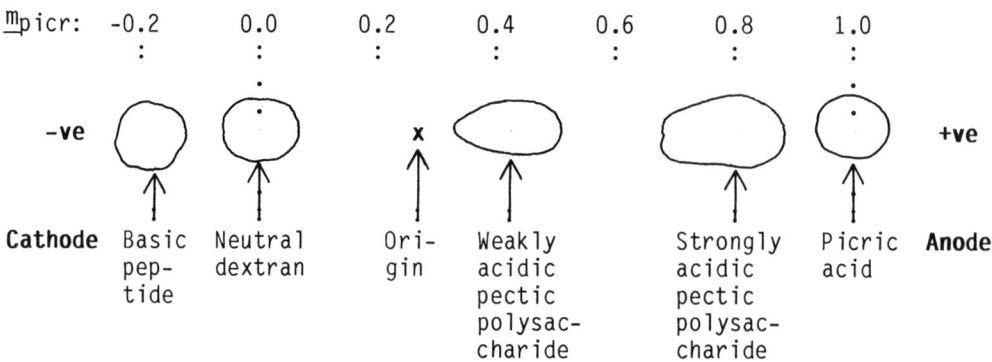

Owing to EEO, neutral polymers move slightly towards the cathode; they must move faster than dextran before being ascribed a positive charge. Acidic polymers with a moderate or high charge:mass ratio move towards the anode. Those with a very low charge:mass ratio may show little net movement, being balanced between the conflicting forces of electrophoresis and EEO. However, polymers remaining at the origin should be viewed with suspicion since immobility could be due to precipitation or irreversible binding to the paper: a simple test for this is to attempt to elute the material off the paper with electrophoresis buffer.

Radioactive polysaccharides on GF/A can be assayed directly by liquid scintillation counting of dried strips in non-Triton scintillant [481]; the counting efficiency is higher with GF/A than with ordinary filter paper, especially for ^3H. Non-radioactive polysaccharides can be stained on glass fibre paper by dipping in sulphonated naphthol solution (prepared by dissolving 5 g 1-naphthol in 25 ml conc. H_2SO_4, standing it at 25°C for 16 h, and then diluting it into 500 ml 96 % ethanol — the ethanolic solution can be stored) and heating at 100°C for a few minutes [36]. Papers stained in this way are corrosive and cannot conveniently be stored.

If a pectin exhibits a weak negative charge upon electrophoresis at pH 6.5, it could be a homogalacturonan with many of the —COOH groups

94

methyl-esterified, or a rhamnogalacturonan with many neutral side-chains but few methyl-ester groups, or a mixture. One simple way of distinguishing these possibilities is to electrophorese the sample before and after treatment with 0.5 M Na_2CO_3 at 0°C for 18 h. Any increase in electrophoretic mobility after treatment suggests the presence of methyl-ester groups.

3.7: ASSAYS FOR WALL COMPONENTS

It is often necessary to perform quantitative assays for the wall components under investigation: either total polymers, or the specific 'building blocks' thereof. A collection of useful assays is presented in Panel 3.7 a. Particularly convenient are assays 5 and 7-10, since they can detect the class sugar in question _free_ (e.g. glucose), as part as a _soluble polymer_ (e.g. starch) or as part of an _insoluble polymer_ (e.g. wall-bound polymers). This is possible because the treatment involves heating in quite concentrated acid, which quickly dissolves all polymers and breaks them down. The spectrophotometric assays are mostly designed for 1-ml cuvettes. The quantities can be scaled up or down to achieve other final volumes. Interference of common laboratory reagents in selected assays is described in Panel 3.7 b.

PANEL 3.7 a: Assays for wall components (for footnotes, see page 100)

1) TOTAL SOLUBLE POLYMERS [suitable standard: dextran]

    (i) Ethanol precipitation (qualitative): Mix 0.2 ml aqueous solution + 10 µl 10 % ammonium formate + 1 ml ethanol. Stand 1 h. Result = turbidity or precipitate. [See also Panel 3.7 b.]

    (ii) Refractive index (RI): Remove all low-M_r material (Sect 3.5.1). Measure RI with a refractometer. All polysaccharides and glycoproteins increase the RI of water, but different polymers to different extents.

2) TOTAL POLYMERS, SOLUBLE + INSOLUBLE[2]

    Gravimetric: Remove all low-M_r material (Section 3.5.1). Dry to constant weight in evacuated desiccator over P_2O_5 or anhydrous $CaCl_2$. Freeze-drying is a useful preliminary, but less effective at removing the last traces of H_2O if used alone. Use glass (not plastic) vessels to avoid static.

3) SOLUBLE PROTEIN [suitable standard: bovine serum albumin (BSA)]

(i) <u>TCA-precipitation</u> (qualitative): Mix 1 ml aqueous solution + 0.5 ml 75 % (w/v) trichloroacetic acid. Stand at 0°C for 1 h. Result = turbidity or precipitate. [See also Panel 3.7 b.]

(ii) <u>UV-absorbance</u> (quick): Read A_{280} in a quartz cuvette vs a blank containing same solvent/buffer. [Subject to interference by non-protein phenols, etc., but useful for rapid assay of column eluates.]

[See also Panel 3.7 b.]

(iii) <u>UV-absorbance</u> (rigorous): Perform test [3(i)]. Centrifuge down the precipitate in an Eppendorf tube. Reject supernatant. Resuspend pellet in 25 % TCA, and repeat centrifugation. Reject supernatant. Re-dissolve precipitate in 1 ml 1 M NaOH. Perform test [3(ii)]. [Not subject to interference by non-protein phenolics.]

(iv) <u>Coomassie Blue binding</u>: Reagent: dissolve 20 mg Coomassie Brilliant Blue G-250 in 20 ml 85 % H_3PO_4, add to 80 ml H_2O, filter through Whatman no.1 paper, and store in the dark at 25°C.
In tube **A** mix 0.5 ml aqueous sample (containing 1-30 µg protein) with 0.5 ml reagent. In tube **B** mix 0.5 ml H_2O with 0.5 ml reagent. Incubate at 25°C for 1 h. Read [(A_{595} minus A_{465}) of **A**] minus [(A_{595} minus A_{465}) of **B**] — this is proportional to protein concentration. [The following substances interfere: SDS, Triton X-100, alkali, salt (over 0.4 M) and guanidinium thiocyanate.] See also [412]

(v) <u>Lowry method</u>: Solutions required: **A** = 75 % w/v trichloroacetic acid, **B** = 1 M NaOH, **C** = 0.5 % $CuSO_4.5H_2O$ in 1 % sodium potassium tartrate [can be stored], **D** = 50 ml 2 % Na_2CO_3 + 1 ml **C** [use same day], **E** = Folin & Ciocalteu's phenol reagent (available commercially), diluted with H_2O to 1 N acidity, as determined by titration of a sample to the end-point of phenolphthalein with 1 M NaOH.
Perform test [3(i)]. Centrifuge down the precipitate in an Eppendorf tube. Reject supernatant. Resuspend pellet in 25 % TCA, and repeat centrifugation. Reject supernatant. [These steps are necessary for most plant protein sources, which tend to be contaminated with non-proteins phenolics that react in the Lowry assay.] Redissolve precipitate (5-100 µg protein) in 100 µl 1 M NaOH, with heating if necessary [if heating is used, the BSA standard should be subjected to similar heating]. Add 1 ml **D**, incubate at 25°C for at least 10 min, then add 100 µl **E**. After at least 30 min at 25°C, read A_{750}. For high protein concentrations, the reading can be kept on-scale by measuring A_{500} instead of A_{750}. [315]

(vi) ^{35}S-Labelling: Continuously label (Section 2.1.3) cells/plants with $Na_2{}^{35}SO_4$ as sole sulphur-source. High-M_r radioactive material is then presumed to be protein [not valid for algae which have sulphated polysaccharides]. After several cell cycles' labelling, specific radioactivity of Met & Cys residues = that of supplied $Na_2{}^{35}SO_4$. [Note: extensin contains little or no sulphur.]

(vii) <u>Biuret assay</u>: For an alternative, see reference [315].

4) INSOLUBLE[(2)] PROTEIN

Assays after hydrolysis [suitable standard: bovine serum albumin (BSA)]: Hydrolyse in 6 M HCl (Section 4.3.3 a); assay total or specific amino acids as in parts 14 & 15 of this Panel.

5) TOTAL CARBOHYDRATE, SOLUBLE + INSOLUBLE[(2)] [suitable standard: dextran]

Phenol/H_2SO_4: To 0.4 ml aqueous sample (2-15 µg carbohydrate) add 10 µl 80 % (w/w) phenol (available commercially — BDH), followed by 1 ml conc H_2SO_4 (which should be pipetted directly into the water so as to produce maximum mixing). Caution — becomes hot. Shake the solution and stand it on the bench for 10 min. Cool in a bath of tap-water for 10-20 min. Read A_{485}. All carbohydrates react in this test, but there is a ca. 3-fold difference in colour yield between the best (xylose) and worst (e.g. fucose).   [137]

6) CELLULOSE[(1,2)] [suitable standard: filter paper]

Updegraff method: Suspend the sample (containing 0.2-20 mg cellulose) in 3 ml 'acetic-nitric reagent' [$HOAc/H_2O/HNO_3$, 8:2:1], cover the tube with a glass marble, and heat in a boiling water bath for 30 min to hydrolyse non-cellulosic polysaccharides. Cool. Centrifuge (2,500 **g**, 5 min). Reject the supernatant. Resuspend the pellet in 10 ml H_2O. Re-centrifuge and reject the supernatant. Resuspend the pellet in 10 ml acetone. Re-centrifuge and reject the supernatant. Dry the pellet, then re-dissolve it in 1.0 ml of 67 % (v/v) H_2SO_4 with shaking at 25°C for 1 h. Transfer 1-100 µl of the solution (depending on the expected cellulose content, 20-0.2 mg) into a clean tube containing 0.5 ml H_2O, and add 1.0 ml 0.2 % anthrone in conc H_2SO_4. Proceed as in assay (7).   [510]

7) HEXOSE, FREE AND POLYMER-BOUND (SOLUBLE + INSOLUBLE[(1,2)]) [suitable standard: glucose, galactose or mannose]

Anthrone: To 0.5 ml aqueous solution or suspension (containing 5-50 µg hexose) add 1 ml 0.2 % anthrone in conc H_2SO_4. Mix (careful shaking is required because H_2SO_4 is much denser than H_2O. Caution: becomes hot). Incubate in boiling water bath for 5 min. Cool. Read A_{620}. Hexose (and 6-deoxyhexose) containing material goes blue; other sugar units give paler, greenish colours.   [135]

8) PENTOSE, FREE AND POLYMER-BOUND (SOLUBLE + INSOLUBLE[(1,2)]) [suitable standard: xylose or arabinose]

Orcinol: Solutions required: **A** = 6 % orcinol in ethanol, **B** = 0.1 % $FeCl_3.6H_2O$ in conc HCl.
To 0.5 ml of aqueous solution or suspension (containing 1-10 µg pentose) add 67 µl of **A** followed by 1.0 ml of **B**. Mix well. Incubate in a boiling water bath for 20 min. Cool. Mix again. Read A_{665}. Pentoses give a green to blue colour, depending on concentration.   [135]

9) URONIC ACID, FREE AND POLYMER-BOUND (SOLUBLE + INSOLUBLE[1,2]) [suitable standard: galacturonic acid or glucuronic acid]

m-Hydroxybiphenyl: Solutions required: **A** = 0.5 % borax ($Na_2B_4O_7$. 10 H_2O) in conc H_2SO_4, **B** = 0.15 % m-hydroxybiphenyl [also known as 3-hydroxydiphenyl or m-phenylphenol] in 1 M NaOH.

To 0.2 ml aqueous solution or suspension (containing 1-20 µg uronic acid) add 1 ml **A**. Mix (careful shaking is required because H_2SO_4 is much denser than H_2O. Caution: becomes hot). Incubate in boiling water bath for 5 min. Cool in a bath of tap water. Read A_{520}. Then add 20 µl **B**, mix thoroughly, incubate at 25°C for 5 min, and re-read A_{520}. Calculate increase in absorbance (pink colour), which indicates uronic acid content. (SDS interferes badly). [63]

10) 6-DEOXYHEXOSE, FREE AND POLYMER-BOUND (SOLUBLE + INSOLUBLE[1,2]) [suitable standard: fucose or rhamnose]

Cysteine/H_2SO_4:   Solutions required: **A** = 86 % (v/v) H_2SO_4, **B** = 3 % L-cysteine-hydrochloride monohydrate in water.

To 0.2 ml aqueous solution or suspension (containing 1-100 µg 6-deoxyhexose) add 1 ml ice-cold **A**. Mix well (difficult because H_2SO_4 is much denser than H_2O. Caution: becomes hot). Cool in a 25°C water bath. In-cubate in a boiling water bath for exactly 3 min. Cool in a 25°C water bath. Add 20 µl **B**. Mix well. Incubate at 25°C for 2 h. Read A_{380}, A_{396} and A_{427}. The difference ($A_{396} - A_{427}$) is proportional to 6-deoxyhexose content; ($A_{413} - A_{380}$) is an indication of hexose content (but the colour yield for glucose and fructose is about double that for galactose and mannose). If a large excess of hexose interferes with reliable estimation of 6-deoxyhexose, prolong the 100°C treatment to 10 min: this virtually abolishes colour production by hexoses. [135]

11) SOLUBLE PHENOLS [suitable standard: ferulic acid]

(i) Folin & Ciocalteu's reagent: Solutions required: **A** = half-strength Folin & Ciocalteu's phenol reagent (available commercially), **B** = saturated aqueous Na_2CO_3.

Mix 0.25 ml aqueous solution (containing 0.5-25 µg ferulic acid equivalents) with 0.25 ml of **A**. Incubate at 25°C for 3 min. Add 0.5 ml **B**. Incubate at 25°C for 1 h. Read A_{750} (or the highest wavelength offered by the spectrophotometer if lower). [162]

(ii) Bathochromic shift (difference spectrum):   Using a recording spectrophotometer, prepare an absorption spectrum with sample at pH 10 (adjusted with dilute ammonia) in the sample cuvette and an identical sample at pH 4 (adjusted with dilute HOAc) in the reference cuvette. A peak in the resulting difference spectrum, between 280 and 400 nm, is given by all phenols. The precise wavelength of the peak, and its height for a given concentration will vary with the type of phenol; but the method is useful for comparing like samples quantitatively.

12) LIGNIN + OTHER POLYMER-BOUND PHENOLS, SOLUBLE + INSOLUBLE[2]

(i) Acetyl bromide[1]: See Panel 3.4.3

(ii) [14C]Cinnamate labelling: Since cinnamate is the precursor of most major phenols (except gallic acid) in most plants (though grasses can by-pass cinnamate by the action of tyrosine ammonia-lyase), [14C]cinnamate is a valuable precursor to render phenolics easily detectable. However, the usual difficulties with precursor pools etc. (Chs 2 & 5) are met during attempts to obtain quantitative assays of phenolic synthesis by this method.

13) SILICA [suitable standard: amorphous silica]

Molybdenum blue: Solutions required: A = 2 % $(NH_4)_6Mo_7O_{24}.4H_2O$ in 0.6 M HCl. B = 2 % Metol [= 4-methylaminophenol sulphate] in 1.2 % Na_2SO_3 (discard when coloured). C = 10 % Oxalic acid. D = 25 % (v/v) H_2SO_4. E = 100 ml B + 60 ml C + 120 ml D + 20 ml H_2O (store at 4°C), F = 1 M NaOH. [Note: use metal and polythene laboratory ware to avoid contamination with silica from glassware.]
Heat the dried plant material (e.g. 25 mg dry weight containing 0.1-10 % Si) in a platinum crucible with 6 drops of conc H_2SO_4 until the acid stops fuming, and then ignite with a Fisher burner to yield a white ash. Add 0.4 g pure anhydrous Na_2CO_3, and heat to melting point; maintain it in the molten state for 20 min (with periodic swirling) to depolymerise the silicon compounds. Cool. Dissolve the products in 30 ml 0.3 M HCl, with warming if necessary. Shortly before the following step, bring the pH of the solution to the end-point of phenolphthalein by dropwise addition of 1 M NaOH.
To 0.5 ml of the sample solution, add 75 µl solution A. Incubate at 25°C for exactly 10 min. Add 375 µl soln E + 300 µl H_2O. Incubate at 25°C for 3 h. Read A_{810} (if this wavelength is not available on the spectro-photometer, use the highest that is available: the sensitivity will be rather lower).   [259,529]

14) AMINO ACIDS [suitable standards: leucine etc.]

Ninhydrin: Solutions required: A = 49 mg NaCN in 100 ml H_2O, B = 21 % HOAc, pH brought to 5.35 with 10 M NaOH, C = 2 ml A freshly mixed with 98 ml B, D = 3 % ninhydrin in 2-methoxyethanol, and E = 50 % propan-2-ol.
To 200 µl of aqueous solution (containing 0.4 - 8.0 µg amino acids) add 100 µl C followed by 100 µl D. Incubate in a boiling water bath for for 15 min. While the samples are still warm, add 1 ml of E and shake vigorously. Cool. Read A_{570} (or A_{440} for proline or hydroxyproline). If the absorbance is off scale, add more E; if this is done, the same should be done for the standard curve. All the amino acids (not Pro or Hyp) give similar colour yields per α-NH_2 group. [Proteins should be hydrolysed in 6 M HCl and dried (Panel 4.3.3 a) before the ninhydrin assay. Care is needed to avoid absorption by the HCl of NH_3 from laboratory air; protein-free blanks should be 'hydrolysed' in parallel. If uptake of NH_3 is a problem, alkaline hydrolysis (Panel 4.3.2) can be used as an alternative to acid hydrolysis.]   [429]

15) HYDROXYPROLINE

Kivirikko & Liesmaa method: Solutions required: **A** = 180 µl bromine dissolved in 50 ml of ice-cold 1.25 M NaOH and stored at 4°C for 3 days to 3 months, **B** = 16 % Na_2SO_3, **C** = 5 % p-dimethylaminobenzaldehyde in propan-1-ol, **D** = 6 M HCl.

To 300 µl of ice-cold aqueous sample (containing 0.1-2.0 µg hydroxyproline) add 300 µl of ice-cold **A**. Shake vigorously, and incubate at 0°C for 3-10 min. With the sample still in an ice bucket, add 15 µl **B**, shake, add 300 µl **C**, shake, add 150 µl **D**, and shake. Incubate in a 95°C water bath for 2½ min followed by a 20°C water bath for 10 min. Read A_{560}. [This method will detect free hydroxyproline and hydroxyproline-arabinosides but not peptide-bound hydroxyproline. Proteins should be hydrolysed (Panel 4.3.2) before assay.]   [292]

16) REDUCING SUGARS [suitable standards: glucose etc.]

(i) PAHBAH: Solutions required: **A** = 5 % p-hydroxybenzoic acid hydrazide in 0.5 M HCl, **B** = 10 ml **A** freshly mixed with 40 ml 0.5 M NaOH.

To 0.25 ml aqueous sample (containing 0.2-10 µg monosaccharide equivalents of reducing sugar) add 0.75 ml **B**. Incubate in a boiling water bath for 5 min. Cool. Read A_{410}.   [320]

(ii) Nelson—Somogyi method: For an alternative, see [463].

17) ESTERS [suitable standard: methyl p-hydroxybenzoate]

Hydroxylamine: To 50 mg cell walls add 2 ml 50 mM hydroxylamine hydrochloride in 95 % ethanol followed by 0.4 ml 6 M NaOH. Incubate on a boiling water bath for 5 min. Cool. Carefully add 4 ml 1 M HCl. Centrifuge (2,500 **g**, 5 min). To the clear supernatant, add 100 µl 5 % $FeCl_3$. Read A_{540}.   [333]

Wood & Siddiqui assay for methyl esters, e.g. of pectin: Solutions required: **A** = 2 % $KMnO_4$, **B** = 0.5 M sodium arsenite in 60 mM H_2SO_4, **C** = 20 mM pentane-2,4-dione / 2 M ammonium acetate / 50 mM acetic acid, **D** = 1.5 M NaOH, **E** = 2.75 M H_2SO_4.

Mix 250 µl of aqueous sample (solution or suspension), containing for example 25-100 µg pectin, with 125 µl **D**. Incubate at 25°C for ½ h to release MeOH from methyl esters. Add 125 µl **E**. Cool to 0°C. Add 100 µl **A** and incubate at 0°C for 15 min. Add 100 µl **B** + 300 µl H_2O, mix, and incubate at 25°C for 1 h. Add 1 ml of solution **C**, mix, close tube with a glass marble; incubate at 60°C for 15 min. Cool. Read A_{412}.   [558]

[1]These assays are extremely sensitive, and therefore susceptible to trace contaminants. A small piece of paper-dust, for example, would give a high reading in the anthrone assay owing to its cellulose content. Assays should therefore be performed in triplicate and any anomalously high readings discounted. An alternative is to pre-treat test tubes in a Bunsen flame or furnace to burn off any carbohydrate material prior to the assay.

[2]E.g. as part of a cell wall.

PANEL 3.7 b: Interference of common contaminants in polymer assays

No.	Contaminant	Concn	EtOH/[1] dextran	TCA/[2] BSA	UV [3]
1	NaCl	1.0 M			0.0
2	CaCl$_2$	0.5 M			0.0
3	LaCl$_3$	0.25 M			0.0
4	NaH$_2$PO$_4$	0.5 M			0.1
5	(NH$_4$)$_2$SO$_4$	1.5 M	**		0.0
6	NH$_4$HCO$_3$	0.5 M			0.0
7	Ammonium formate	0.5 M			0.2
8	NaOH	3.0 M	*	*	0.0
9	NH$_3$	1.0 M			0.0
10	Pyridine	10 % v/v			30
11	Imidazole	2.0 M			370
12	H$_3$BO$_3$	0.25 M			0.0
13	Acetic acid	10 % v/v			0.0
14	Trifluoroacetic acid	1.0 M			0.0
15	Trichloroacetic acid	12½% w/v		n/a	0.1
16	HCl	1.0 M			0.0
17	EDTA	0.25 M			1.9
18	CDTA	0.25 M	**		1.2
19	Urea	4.0 M			0.0
20	Guanidinium thiocyanate	2.25 M		**	4.0
21	SDS	1.5 % w/v			4.5
22	Triton X-100	1.5 % v/v		**	0.1
23	Ethanol	50 % v/v	n/a	*	0.0
24	Acetone	50 % v/v			11.5
25	DMSO	45 % v/v			0.1
26	Phenol/HOAc (2:1, w/v)	37½% v/v			27000
27	Chlorbutol	0.5 % w/v			2.9
28	Sucrose	1.0 M			0.1
29	Dextran	0.5 % w/v	n/a		0.1
30	Pectin	0.5 % w/v	**	**(4)	14
31	Bovine serum albumin	0.5 % w/v		n/a	290
32	Ferulic acid	25 mM [Na$^+$ salt]		**	20500
33	Tannic acid	0.5 % w/v			35000
34	Culture medium[5]	½-strength			0.2

* = Formation of a precipitate by dextran[1] or BSA[2] was prevented.

** = Contaminant itself formed precipitate on addition of EtOH[1] or TCA[2].

Blank = The contaminant did not interfere

[1] The effect of each contaminant was tested on the ability of dextran to form a precipitate with ethanol in the qualitative total soluble polymer assay — see Panel 3.7 a, test [1(i)]. n/a = not applicable.

[2] The effect of each contaminant was tested on the ability of bovine serum albumin to form a precipitate with trichloroacetic acid in the soluble protein assay — see Panel 3.7 a, test [3(i)]. n/a = not applicable.

[3] Approximate UV-absorbance of the contaminant at 280 nm.

[4] Possibly due to presence of protein in commercial sample of pectin.

[5] The culture medium recommended in Panel 2.7.1 a.

4 Wall polymers: chemical characterisation

4.1: SPECIFIC CHEMISTRY OF KNOWN PRIMARY CELL WALL POLYMERS

4.1.1: Classes of polysaccharide

The polysaccharides of the primary cell wall are classified as pectic, hemicellulosic and cellulosic (Panel 4.1.1). The definitions adopted here are: **Cellulose** = β-(1$\rightarrow$4)-D-glucan. **Pectins** = polysaccharides rich in D-galacturonic acid, the side-chains of such polysaccharides, and chemically similar polysaccharides. [An alternative definition of 'pectin' is 'chelating-agent-extractable polysaccharide plus chemically similar inextractable polysaccharides'. In practice the two definitions encompass a similar set of polysaccharides.] **Hemicelluloses** = Non-cellulosic wall polysaccharides other than pectins. Although this is rather a negative definition, hemicelluloses so defined generally share a number of common properties, e.g. they are extractable by NaOH but not by cold chelating agents and many can hydrogen-bond to cellulose. The three classes of wall polysaccharide differ greatly in sugar composition, further supporting the validity of the classification. The major sugars are:

Cellulose	Glc
Pectins	GalA, Ara, Gal, Rha
Hemicellulose	Glc, Xyl, Ara, GlcA

4.1.2: Cellulose

Cellulose is an unbranched polymer of D-glucopyranose residues joined by β-(1$\rightarrow$4) linkages:

$$\xrightarrow{\rightarrow 4}\text{Glc}p\xrightarrow[\beta]{1\rightarrow4}\text{Glc}p\xrightarrow[\beta]{1\rightarrow4}\text{Glc}p\xrightarrow[\beta]{1\rightarrow4}\text{Glc}p\xrightarrow[\beta]{1\rightarrow4}\text{Glc}p\xrightarrow[\beta]{1\rightarrow4}\text{Glc}p\xrightarrow[\beta]{1\rightarrow4}\text{Glc}p\xrightarrow[\beta]{1\rightarrow4}\text{Glc}p\xrightarrow[\beta]{1\rightarrow4}\text{Glc}p\xrightarrow[\beta]{1\rightarrow}.$$

Cellulose (α-cellulose), obtained as the insoluble residue after extraction of pectin and hemicellulose, often contains traces of Man and Gal [257] but these may arise from contaminant hemicelluloses. Cellulose is usually ca. 20-30 % of the dry weight of the primary wall, although examples are known where the figure is much lower, especially in tissue recently regenerated from isolated protoplasts [58]. Cellulose accounts for 40-90% of the secondary cell wall and is the world's most abundant organic compound.

PANEL 4.1.1: Chemistry of the main polysaccharides of the primary cell wall

Polysaccharide	Major monomers[1]	Typical sequences & linkages
α-CELLULOSE	β-Glc$\underline{p}$ (β-Man$\underline{p}$?)	Glc-1→4-Glc... Unbranched. Mannose possibly a minor impurity?
HEMICELLULOSES		
Xyloglucan	β-Glc$\underline{p}$, α-Xyl$\underline{p}$, β-Gal$\underline{p}$, α-Fuc$\underline{p}$, Ara$\underline{f}$	Backbone as in cellulose. Side-chains: Fuc-1→2-Gal-1→2-Xyl-1→6-Glc, Gal-1→2-Xyl-1→6-Glc, Xyl-1→6-Glc & others Gal partially acetylated.
Xylans	β-Xyl$\underline{p}$, α-Ara$\underline{f}$, α-Glc$\underline{p}$A	Backbone: Xyl-1→4-Xyl... Side-chains: Ara & GlcA as mono & short oligosacchs. Some GlcA as 4-O-methyl ether. Some of Xyl acetylated. Some of Ara feruloylated in Monocots.
Callose	β-Glc$\underline{p}$	Glc-1→3-Glc... Unbranched?
β-(1→3),(1→4) -glucan	β-Glc$\underline{p}$	Glc-1→4-Glc-1→3-Glc-1→4-Glc-1→4-Glc-1→3-Glc-1→4-Glc-1→4-Glc-1→3-Glc-1→4-Glc.... Unbranched. A few stretches with up to 10 contiguous 1→4 bonds. Contiguous 1→3 bonds rare or absent.
PECTINS		
Homogalacturonan	α-Gal$\underline{p}$A, α-Rha$\underline{p}$	GalA-1→4-GalA-1→2-Rha-1→4-GalA-1→4-GalA-1→4-GalA-1→4-GalA-1→4-GalA... Unbranched. Some GalA as methyl ester.
RG-I	α-Gal$\underline{p}$A, α-Rha$\underline{p}$, β-Gal$\underline{p}$, α-Ara$\underline{f}$, Fuc, $\overline{X}$yl	Backbone: GalA-1→2-Rha-1→4-GalA-1→2-Rha-1→4-GalA... Side-chains rich in Ara &/or Gal attached to O-4 of Rha. Some GalA as methyl ester. Some acetylation (of GalA?).
RG-II	α-Gal$\underline{p}$A, β-Rha$\underline{p}$, α-Gal$\underline{p}$, α-Fuc$\underline{p}$, α-Ara$\underline{p}$, Ara$\underline{f}$, β-Gal$\underline{p}$A, α-$\overline{R}$ha$\underline{p}$, Api$\underline{f}$, $\overline{\beta}$-Glc$\underline{p}$A, $\overline{K}$DO, Ace$\overline{f}$A, Xyl$\underline{p}$, Glc	Two different complex heptasaccharides and several smaller oligosaccharides as repeating units attached to a GalA-rich core. Some Fuc and all Xyl as 2-O-methyl ethers.
Apiogalact- uronans	α-Gal$\underline{p}$A, Api$\underline{f}$	Backbone: GalA-1→4-GalA... Side-chains Api and Api$_2$. Restricted to growing fronds of certain water plants.

[1]In order of decreasing abundance.

The DP of cellulose can be estimated by GPC or viscometry (Ch 3) [76, 336]. In secondary cell walls the DP is typically about 14 000 (M_r 2.3×10^6); in primary walls, on the other hand, it tends to be lower, and also biphasic with about half the cellulose of DP less than 500 and the other half of DP 2500-4500 [60,336]. This is of unknown significance.

The biological function of cellulose is presumed to be skeletal, providing shape and strength to the cell wall. When cellulose synthesis is inhibited by addition of 2,6-dichlorobenzonitrile, algal cells burst under their own turgor pressure [420].

4.1.3: Pectins

Pectins have been described as 'block' polymers [263]. For example, they contain 'smooth' blocks (**homogalacturonan**) consisting mainly of contiguous unbranched α-D-galacturonic acid [pyranose] residues,

$$\xrightarrow{4}\text{Galp}A\xrightarrow[\alpha]{1\to4}\text{Galp}A\xrightarrow[\alpha]{1\to4}\text{Galp}A\xrightarrow[\alpha]{1\to4}\text{Galp}A\xrightarrow[\alpha]{1\to2}\text{Rhap}\xrightarrow[\alpha]{1\to4}\text{Galp}A\xrightarrow[\alpha]{1\to4}\text{Galp}A\xrightarrow[\alpha]{1\to4}\text{Galp}A\xrightarrow[\alpha]{1\to4}\text{Galp}A\xrightarrow[\alpha]{1\to}$$

and, apparently elsewhere in the same molecule, 'hairy' blocks containing numerous other sugars (especially Rha, Gal and Ara) [130]. The hairy blocks are recognised by their resistance to pectinase (Fig. 3.5.2 b) [118, 261,343,490].

Within homogalacturonan blocks, there are occasional Rha residues. The spacing of these may be regular (about every 25 GalA units), as hinted by the release of discretely-sized oligosaccharides upon selective hydrolysis of Rha bonds by dilute acid [406]; but this has been questioned [345]. It has been suggested that the Rha residues act as molecular 'punctuation marks', with the intervening homo-GalA block being in some cases fully methylesterified and in others not esterified at all [263,508]:

UUUUUR Û R UUUUUUUUUUUUUUUUUUUUUUUUUUR Û Û Û Û Û Û Û Û Û Û Û Û Û Û Û Û Û

      U = Galacturonic acid
      Û = Galacturonic acid methyl ester
      R = Rhamnose

However, other workers favour a more random distribution of methyl ester groups [131]. Blocks with about a dozen or more consecutive un-esterified GalA residues can become cross-linked via Ca^{2+} bridges (Ch 7), whereas esterified and partially esterified blocks cannot [570]. A block-wise distribution of ester groups would thus enable local Ca^{2+}-bridging. The

Rha residues may also act as anchorage points for neutral Ara- and/or Gal-
rich (arabinan, galactan or arabinogalactan) side-chains [263]:

$$\xrightarrow{4}\underline{GalpA}\xrightarrow[\alpha]{1\to4}\underline{GalpA}\xrightarrow[\alpha]{1\to4}\underline{GalpA}\xrightarrow[\alpha]{1\to4}\underline{GalpA}\xrightarrow[\alpha]{1\to2}\underline{Rhap}\xrightarrow[\alpha]{1\to4}\underline{GalpA}\xrightarrow[\alpha]{1\to4}\underline{GalpA}\xrightarrow[\alpha]{1\to4}\underline{GalpA}\xrightarrow[\alpha]{1\to4}\underline{GalpA}\xrightarrow[\alpha]{1\to}.$$

$$\beta\Big|1\to4$$

$$\underline{Galp}\xrightarrow[\beta]{1\to4}\underline{Galp}\xrightarrow[\beta]{1\to4}\underline{Galp}\xrightarrow[\beta]{1\to4}\underline{Galp}\xrightarrow[\beta]{1\to4}\underline{Galp}$$

Pectinase-resistant 'hairy' blocks include the rhamnogalacturonans
(RG-I and RG-II), which have been best characterised in the walls of
cultured sycamore cells. **RG-II**, which has a DP of about 60, contains 12
dif-ferent sugars, namely GalA (including the unusual β-linked form), Rha,
Ara (including the unusual pyranose form), Gal, Fuc, MeFuc, MeXyl, Api,
GlcA, AceA (unique to RG-II), KDO (previously known only from prokaryotes)
and Glc [118]. It seems to be built up of a GalA-rich core bearing as
side-chains several very precise sub-blocks (the 'hairs') in the form of
two different heptasaccharides and a number of smaller units [352,465].

A: $\underline{Rhap}\xrightarrow[\alpha]{1\to2}\underline{Arap}\xrightarrow[\alpha]{1\to4}\underline{Galp}\xrightarrow[\alpha]{1\to2}\underline{AcefA}\xrightarrow[\beta]{1\to3}\underline{Rhap}\xrightarrow[\beta]{1\to3'}\underline{Api}\xrightarrow[?]{1\to}$ $\boxed{\begin{array}{c}\text{RG-II}\\\text{CORE}\end{array}}$

$$\alpha\Big|1\to2$$

$$\textbf{Me-Fuc}\underline{\textbf{p}}$$

B:

$$\textbf{GalpA}$$
$$\beta\Big|1\to3$$
$\underline{Galp}\xrightarrow[\alpha]{1\to2}\underline{GlcpA}\xrightarrow[\beta]{1\to4}\underline{Fucp}\xrightarrow[\alpha]{1\to4}\underline{Rhap}\xrightarrow[\beta]{1\to3'}\underline{Api}\xrightarrow[?]{1\to}$ $\boxed{\begin{array}{c}\text{RG-II}\\\text{CORE}\end{array}}$
$$\alpha\Big|1\to2$$
$$\textbf{GalpA}$$

C: $\underline{Rhap}\xrightarrow[\alpha]{1\to5}KDO\xrightarrow{2\to}$ $\boxed{\begin{array}{c}\text{RG-II}\\\text{CORE}\end{array}}$

It was possible to isolate these side chains by very mild acid hydrolysis
because they are linked to the core via particularly acid-labile glycosidic
linkages (Api and KDO — cf. Section 4.2.2) [465]. There appears to be a
defined number of copies of each side chain per RG-II molecule, e.g. 4 of
heptasaccharide **A**, 1 of heptasaccharide **B**, 1 of hexasaccharide (**B** minus the

β-GalA residue), etc [352]. It is not yet known whether the oligosaccharide side-chains are attached to the backbone in any particular sequence.

RG-I, which has a DP of about 1000, has as the major part of its
backbone a repeating disaccharide unit [GalA-Rha],

$$\xrightarrow{4}Ga\underline{l}pA\xrightarrow[\alpha]{1\to2}Rha\underline{p}\xrightarrow[\alpha]{1\to4}Ga\underline{l}pA\xrightarrow[\alpha]{1\to2}Rha\underline{p}\xrightarrow[\alpha]{1\to4}Ga\underline{l}pA\xrightarrow[\alpha]{1\to2}Rha\underline{p}\xrightarrow[\alpha]{1\to4}Ga\underline{l}pA\xrightarrow[\alpha]{1\to2}Rha\underline{p}\xrightarrow[\alpha]{1\to4}Ga\underline{l}pA\xrightarrow[\alpha]{1\to},$$

about half the copies of which bear Ara- and/or Gal-rich side-chains
('hairs') attached to the 4-position of the Rha [343]: these side-chains
can also be regarded as blocks. It is not yet clear whether they comprise
a large but finite number of precisely specified structures comparable with
the side-chains of RG-II, or whether RG-I has a more random repertoire of
side chains, whose structures are governed statistically. Such randomness
would not preclude the existence of preferred structures — for instance:

> there is a high probability that the sugar residue directly linked to
> the backbone Rha unit will be D-Gal rather then L-Ara;

> the next few sugars are also frequently Gal [mainly linked via pyran-
> ose β-(1→4) bonds, or less often β-(1→6) bonds, or more rarely
> still β-(1→3) or β-(1→2) bonds], and more rarely L-Ara;

> the most distal residues tend to be L-Ara [mostly furanose and linked
> via α-(1→5) bonds; some acting as branch points, with an additional
> sugar attached to O-2 or O-3 [cf. 343]].

In cultured sycamore cells the mean DP of an RG-I side chain is about 7
(range = 1 to at least 15), and at least 30 different side chains have been
detected. These side chains were isolated by degradation of the backbone
GalA residues with lithium/ethylenediamine [364] followed by GPC [314].
RG-I has been localised by antibody-labelling to the outer surface of the
primary cell wall plus the middle lamella [360].

Other Ara- and/or Gal-rich polysaccharides, which may well arise from
hairy blocks of pectin, are the arabinans, galactans and arabinogalactans.
Although having a similar overall sugar composition to the side chains of
RG-I, some are of considerably higher M_r. Some of these polysaccharides
are present in solution in the apoplast (extractable in cold aqueous buffer
or present in the culture medium of cell suspensions), but most can only be
extracted by harsher methods, usually involving heating. Many of them are
electrophoretically neutral, and, after extraction, remain highly water-
soluble. It is possible that some or all of the neutral Ara- and/or Gal-

rich polysaccharides had originally been associated with acidic pectic backbones, but been freed either in vivo by wall-bound enzymes or in vitro by the extraction conditions. The Gal residues tend to be concentrated towards the reducing terminus, and the Ara residues towards the non-reducing termini [24,26]. Tha arabinogalactan domains of AGPs (see Section 4.1.6) differ from those of pectins in that the Gal residues of the former are predominantly $(1 \to 3)$ and $(1 \to 6)$ linked [157].

In some plants, especially the Chenopodiaceae (e.g. spinach, beet), ferulate and p-coumarate residues are attached to non-reducing terminal Ara and Gal residues. The feruloylated polysaccharides cannot be extracted from the cell wall by cold water, but can be extracted by degradative heat treatment as a mixture of acidic and neutral species [170,428]. By digestion with Driselase, the phenolic acids can be solubilised from cell walls or polysaccharides still attached to the sugar residue to which they were originally bound. The products from spinach include [168]:

3-O-Feruloyl-L-Arap-α-(1→3)-L-Ara
and   6-O̅-Feruloyl-D-Galp̲-β-(1→4)-D-Gal,

together with similar products tentatively identified as

3-O-p-Coumaroyl-L-Arap-α-(1→3)-L-Ara
and   6-O̅-p̲-Coumaroyl-D-Galp̲-β-(1→4)-D-Gal.

Several water plants, especially Lemna and Spirodela, possess apio-galacturonan in their cell walls — a pectin in which the GalA-rich backbone bears many Api and Api$_2$ side chains [207]:

$$\overset{\to 4}{}\text{Ga}\underline{\text{lp}}\text{A}\overset{1\to 4}{\underset{\alpha}{}}\text{Ga}\underline{\text{lp}}\text{A}\overset{1\to 4}{\underset{\alpha}{}}\text{Ga}\underline{\text{lp}}\text{A}\overset{1\to 4}{\underset{\alpha}{}}\text{Ga}\underline{\text{lp}}\text{A}\overset{1\to 4}{\underset{\alpha}{}}\text{Ga}\underline{\text{lp}}\text{A}\overset{1\to 4}{\underset{\alpha}{}}\text{Ga}\underline{\text{lp}}\text{A}\overset{1\to 4}{\underset{\alpha}{}}\text{Ga}\underline{\text{lp}}\text{A}\overset{1\to 4}{\underset{\alpha}{}}\text{Ga}\underline{\text{lp}}\text{A}\overset{1\to 4}{\underset{\alpha}{}}\text{Ga}\underline{\text{lp}}\text{A}\overset{1\to}{\underset{\alpha}{}}$$

1→3	1→2		1→3	1→3	1→3		1→2
Apif	**Apif**		**Apif**	**Apif**	**Apif**		**Apif**
1→3'					1→3'		
Apif					**Apif**		

In addition, the pectins of some plants contain small amounts of xylose (e.g. apple fruit but not lemon peel [132]). It is not known whether the Xyl and Api are arranged block-wise. Pectins, including RG-I, may also be acetylated, and the acetyl groups may occur on O-2 or O-3 of the GalA residues [298,364]. There is also a suggestion that silanolate groups (C—O—Si) are covalently linked to pectins [438].

Commercial pectin is extracted under acidic conditions, and is likely to have lost some side-chains and undergone a small amount of backbone breakdown, but it maintains a high degree of methylesterification. Commercial polygalacturonic acid is similar but it has been de-esterified by alkali. RG-I and II and apiogalacturonan are not available commercially.

It is impossible to give a reliable estimate for the M_r of pectins as they occur in the cell wall. Since they are inefficiently extracted at $25°C$, only a small proportion of the total wall pectin wall can be studied in intact form. Estimates of DP made on total pectin extracted by harsher treatments, or on commercial pectin, are of dubious significance since the polymer will have undergone some breakdown. Estimates of the DP of the deliberately produced fragment RG-I are about 2000 [345]. One possible approach to the estimation of the M_r of intact pectins is to study pectins produced early in the process of wall regeneration by isolated protoplasts: these cells release a large proportion of their newly synthesised pectin into the medium until they have deposited a cellulosic wall framework [225]. However, isolated protoplasts have altered physiology and may produce pectin with an unusual M_r, and the suggested approach ignores the possibility that pectins are built up into larger structures in the cell wall by transglycosylation (Ch 6).

The precise functions of pectins are unclear. Unfortunately there are no wall mutants with which hypotheses could be tested. Pectins are present in much larger amounts in primary walls than in secondary walls, suggesting a rôle in growth: whether this rôle is enabling, restraining, or both, is unknown. They are highly hydrophilic polysaccharides, and the water that they introduce into the matrix may loosen the wall [416], enabling the skeletal cellulose microfibrils to separate — necessary for wall expansion. On the other hand, pectins can form cross-links via Ca^{2+} bridges and probably also via covalent bonds (Ch 7): they may therefore serve the opposite function of resisting the expansion of the wall. However, any rôle found for pectins in Dicots will have to be largely served by other polysaccharides in grasses, which often contain only traces of pectin. Pectins are present in high concentration in the middle lamella [261,360], where they presumably serve the function of cementing adjacent cells together: evidence for this is the rapid release of single cells from plant tissues by pectinolytic enzymes and (in a few selected tissues)

108

by chelating agents. Finally, pectins are probably a source of biologically active oligosaccharides (Ch 9).

4.1.4: Hemicelluloses

Hemicelluloses, like pectins, are polysaccharides built up of a
variety of different sugars. They are not extractable in cold water, but
can be effectively solubilised in alkali (Section 3.4.1). The principal
hemicelluloses are the following.

Xylans typically make up roughly 5 % of the primary cell wall in
Dicots, 20 % of the primary walls of grasses, and 20 % of the secondary
walls in both Dicots and grasses [345]. They have a backbone of β-(1→4)-
linked D-Xylp residues, some of which carry single α-L-Araf and/or α-D-
GlcpA residues attached to the 2- and/or 3-0-position:

$$\overset{\rightarrow 4}{-}\text{Xylp}\underset{\beta}{\overset{1\rightarrow 4}{-}}\text{Xylp}\underset{\beta}{\overset{1\rightarrow 4}{-}}\text{Xylp}\underset{\beta}{\overset{1\rightarrow 4}{-}}\text{Xylp}\underset{\beta}{\overset{1\rightarrow 4}{-}}\text{Xylp}\underset{\beta}{\overset{1\rightarrow 4}{-}}\text{Xylp}\underset{\beta}{\overset{1\rightarrow 4}{-}}\text{Xylp}\underset{\beta}{\overset{1\rightarrow 4}{-}}\text{Xylp}\underset{\beta}{\overset{1\rightarrow 4}{-}}\text{Xylp}\underset{\beta}{\overset{1\rightarrow}{-}}\text{Xylp}\underset{\beta}{\overset{1\rightarrow}{-}}$$

α|1→3	α|1→2		α|1→3	α|1→3	α|1→2	α|1→3	α|1→3		α|1→3
Araf	Araf		GlcpA	Araf	Araf	Araf	Araf		GlcpA

The presence of GlcA residues gives these polysaccharides a net negative
charge, which can be exploited in anion-exchange chromatography [138].
Other reported side chains are short oligosaccharides of Ara, and the
following specific oligosaccharides:

D-Galp-β-(1→5)-L-Araf-	[24],
D-Xylp-β-(1→2)-L-Araf-	[24],
D-Xylp-α-(1→3)-L-Araf-	[200]
and D-Galp-β-(1→4)-D-Xylp-β-(1→2)-L-Arap-	[24].

Modifications occur on at least 3 of the sugars: many of the Xyl residues
carry 2- or 3-0-acetyl ester groups, about half the GlcA residues carry 4-
0-methyl ether groups, and (in grasses) a small proportion of the terminal
Ara residues carry 5-0-feruloyl ester groups [282].   p-Coumaroyl,
diferuloyl and p-hydroxybenzoyl ester groups also occur [231], but their
mode of linkage in unknown.

The proportion of the Xyl residues that bear carbohydrate side chains
varies greatly between different xylans: heavily arabinosylated and/or
glucuronosylated xylans tend to be more water-soluble (because they cannot
self-aggregate by H-bonds [16,341] and less able to H-bond to cellulose.
They can therefore be fractionated by affinity chromatography on a

cellulose column (Section 3.5.5), and eluted by a gradual increase in NaOH concentration. Some heavily acetylated xylans can be extracted intact from the cell wall in DMSO [69].

Xyloglucans typically make up about 20 % of the primary cell walls of Dicots, and 1-5 % of the primary walls of grasses. They appear to be absent from most secondary walls, but are the major components of the thick 'storage' walls of some seeds, e.g. nasturtium and tamarind [418]. Xyloglucans are neutral polysaccharides and have a backbone identical with cellulose: a linear polymer of β-(1→4)-linked D-Glcp residues. Attached to the 6-O-position of about 70-80 % (fewer in grasses) of the Glc residues are side chains e.g. [345]:

$$D-Xylp-\alpha-(1\to6)-backbone$$
$$D-Galp-\beta-(1\to2)-D-Xylp-\alpha-(1\to6)-backbone$$
$$L-Fucp-\alpha-(1\to2)-D-Galp-\beta-(1\to2)-D-Xylp-\alpha-(1\to6)-backbone$$
and $$L-Araf-?-(1\to2)-D-Xylp-\alpha-(1\to6)-backbone$$

These side chains occur in different ratios in different xyloglucans, but typically the first accounts for about 80-90 % of all side chains. Other side-chains of unknown structure also occur, but the sugars and linkages shown above can be regarded as typical. Seed storage xyloglucans do not possess Fuc residues.

The side chains tend to be arranged in a definite order, often forming a block of three consecutive substituted Glc residues followed by one unsubstituted Glc. Most cellulases, e.g. that of Trichderma viride, specifically chop the backbone at unsubstituted Glc residues, yielding isolated blocks. Two blocks (a nonasaccharide and a heptasaccharide) from the walls of cultured sycamore cells account between them for over half the initial weight of xyloglucan polysaccharide [512].

$$\xrightarrow{\to4}Glcp\xrightarrow[\beta]{1\to4}Glcp\xrightarrow[\beta]{1\to4}Glcp\xrightarrow[\beta]{1\to4}Glcp\xrightarrow[\beta]{1\to4}Glcp\xrightarrow[\beta]{1\to4}Glcp\xrightarrow[\beta]{1\to4}Glcp\xrightarrow[\beta]{1\to4}Glcp\xrightarrow[\beta]{1\to4}Glcp\xrightarrow[\beta]{1\to4}Glcp\xrightarrow[\beta]{1\to}$$

with Xylp side chains (α, 1→6) on residues 1, 2, 3 and 5, 6, 7 and 9; and on the residue 7 Xylp a Galp (β, 1→2) to which a Fucp (α, 1→2) is attached.

It is unclear whether these two blocks alternate along the polysaccharide chain [240] or occur more randomly. Xyloglucans contain substantial numbers of acetyl ester groups, apparently on the galactose residues [576]. The DP of xyloglucan has been studied in view of the interest in auxin-stimulated xyloglucan turnover (Ch 8); a typical value is 600-700 [379].

β-(1→3),(1→4)-Glucans are abundant neutral polysaccharides of the cell walls of grasses. Their occurrence in Dicots is controversial [195], but they are certainly not major components. They appear to be composed of an unbranched chain of β-D-Glcp residues. About 30 % of the linkages are (1→3), the others being (1→4). The distribution of (1→3) and (1→4) linkages can be investigated by Smith degradation (Section 4.2.7) and by use of a specific endoglucanase (EC 3.2.1.73) which cleaves any (1→4) bond that is immediately preceded by a (1→3) bond, thereby generating analytically useful oligosaccharides [565].

$$\overset{\rightarrow 3}{-}\text{Glc}\underline{p}\overset{1\rightarrow 4}{\underset{\beta}{-}}\text{Glc}\underline{p}\overset{1\rightarrow 4}{\underset{\beta}{-}}\text{Glc}\underline{p}\overset{1\rightarrow 3}{\underset{\beta}{-}}\text{Glc}\underline{p}\overset{1\rightarrow 4}{\underset{\beta}{-}}\text{Glc}\underline{p}\overset{1\rightarrow 4}{\underset{\beta}{-}}\text{Glc}\underline{p}\overset{1\rightarrow 4}{\underset{\beta}{-}}\text{Glc}\underline{p}\overset{1\rightarrow 3}{\underset{\beta}{-}}\text{Glc}\underline{p}\overset{1\rightarrow 4}{\underset{\beta}{-}}\text{Glc}\underline{p}\overset{1\rightarrow 4}{\underset{\beta}{-}}\text{Glc}\underline{p}\overset{1\rightarrow}{\underset{\beta}{-}}$$
$$\quad * \qquad\qquad\qquad\qquad\quad * \qquad\qquad\qquad\qquad\quad *$$

* = bonds cleaved by β-(1→3),(1→4)-endo-glucanase (EC 3.2.1.73)

The most common blocks are [(1→4)-(1→4)-(1→3)] and [(1→4)-(1→4)-(1→4)-(1→3)]; these two blocks probably occur at random [86]. A few considerably longer stretches of (1→4) linkages also occur, e.g. the block [(1→4)$_{10}$-(1→3)]; such blocks can probably hydrogen-bond to cellulose, and there is evidence that they may occur relatively regularly — once every ca. 50-70 residues [235]. The occurrence of the sequence [(1→3)-(1→3)] is more controversial (see [235]).

Callose is another polymer of β-D-Glcp residues, but essentially all the linkages are (1→3):

$$\overset{\rightarrow 3}{-}\text{Glc}\underline{p}\overset{1\rightarrow 3}{\underset{\beta}{-}}\text{Glc}\underline{p}\overset{1\rightarrow 3}{\underset{\beta}{-}}\text{Glc}\underline{p}\overset{1\rightarrow 3}{\underset{\beta}{-}}\text{Glc}\underline{p}\overset{1\rightarrow 3}{\underset{\beta}{-}}\text{Glc}\underline{p}\overset{1\rightarrow 3}{\underset{\beta}{-}}\text{Glc}\underline{p}\overset{1\rightarrow 3}{\underset{\beta}{-}}\text{Glc}\underline{p}\overset{1\rightarrow 3}{\underset{\beta}{-}}\text{Glc}\underline{p}\overset{1\rightarrow 3}{\underset{\beta}{-}}\text{Glc}\underline{p}\overset{1\rightarrow 3}{\underset{\beta}{-}}\text{Glc}\underline{p}\overset{1\rightarrow}{\underset{\beta}{-}}$$

It normally occurs in small amounts in healthy cell walls, where it has been suggested to form a thin, rapidly turning-over layer between the plasma membrane and the wall proper [349]. However, upon injury it often accumulates rapidly and in large quantities. It also occurs in certain specialised cell walls [541], e.g. at phloem sieve plates.

β-Mannans occur in moderate amounts in certain secondary cell walls. The presence of small amounts of mannose suggests that primary walls may also contain such polysaccharides (but α-Man is a major component of some glycoproteins, and it has not been established whether the primary wall mannose is α- or β-linked). Mannans from secondary cell walls possess a backbone of β-(1→4)-linked D-Manp residues, interrupted by a smaller proportion (25-35 %) of β-D-Glcp residues. In Gymnosperms D-Galp residues are attached as side-chains to the 6-position of certain Man residues [24].

$$\xrightarrow{4}\text{Man}\underline{p}\xrightarrow[\beta]{1\to4}\text{Man}\underline{p}\xrightarrow[\beta]{1\to4}\text{Glc}\underline{p}\xrightarrow[\beta]{1\to4}\text{Man}\underline{p}\xrightarrow[\beta]{1\to4}\text{Man}\underline{p}\xrightarrow[\beta]{1\to4}\text{Man}\underline{p}\xrightarrow[\beta]{1\to4}\text{Glc}\underline{p}\xrightarrow[\beta]{1\to4}\text{Man}\underline{p}\xrightarrow[\beta]{1\to4}\text{Glc}\underline{p}\xrightarrow[\beta]{1\to4}\text{Man}\underline{p}\xrightarrow[\beta]{1\to}$$

$$\alpha\Big|1\to6$$
$$\text{Gal}\underline{p}$$

4.1.5: Extensins

Extensins are present in the primary cell walls of Dicots in widely varying quantities, making up 1-10 % of the wall [308]. They are particularly abundant in the walls of cultured cells and perhaps also in epidermal cell walls (cf. Section 10.1.5). They are present in the primary walls of grasses in considerably smaller amounts [345], and are apparently absent from secondary walls. They are strongly basic glycoproteins with a polypeptide backbone of unusual amino acid composition [97,458]. It is extremely rich in hydroxyproline (ca. 30-40 mol %) [cf. proline ca. 1-10 mol %], and also contains much serine (ca. 12 mol %), lysine (ca. 10 mol %), valine (ca. 6 mol%) and tyrosine (ca. 10 mol %). Histidine is abundant (ca. 10 mol %) in some extensins, and virtally absent in others. These six amino acids thus account for most of the polypeptide; others occur in much smaller amounts. Partial proteolysis of de-glycosylated tomato extensins followed by amino acid sequencing has shown that certain oligopeptide sequences are repeated many times per polypeptide. These include:

    Ser-Hyp-Hyp-Hyp-Hyp-Lys   [307]
    Ser-Hyp-Hyp-Hyp-Hyp-Val-Tyr-Lys-Tyr-Lys   [307]
    Ser-Hyp-Hyp-Hyp-Hyp-Val-Lys-Pro-Tyr-His-Pro-Thr-Hyp-Val-Thr-Lys [459]

The nucleotide sequence of the gene encoding a carrot extensin has been worked out [97], and from it the amino acid sequence predicted (Fig. 4.1.5); this confirms the remarkable repetition that typifies extensin, but does not distinguish betwen Hyp and Pro residues since Hyp is formed as a post-translational modification of Pro.

```
N-Lys-Tyr-Thr-Tyr-Ser-
  Ser-Hyp-Hyp-Hyp-Hyp-Glu-His-
  Ser-Hyp-Hyp-Hyp-Hyp-Glu-His-
  Ser-Hyp-Hyp-Hyp-Hyp-Tyr-His-Tyr-Glu-
  Ser-Hyp-Hyp-Hyp-Hyp-Lys-His-
  Ser-Hyp-Hyp-Hyp-Hyp-Thr-Hyp-Val-Tyr-Lys-Tyr-Lys-
  Ser-Hyp-Hyp-Hyp-Hyp-Met-His-
  Ser-Hyp-Hyp-Hyp-Hyp-Tyr-His-Phe-Glu-
  Ser-Hyp-Hyp-Hyp-Hyp-Lys-His-
  Ser-Hyp-Hyp-Hyp-Hyp-Thr-Hyp-Val-Tyr-Lys-Tyr-Lys-
  Ser-Hyp-Hyp-Hyp-Hyp-Lys-His-Ser-Hyp-Ala-Hyp-Val-His-His-Tyr-Lys-Tyr-Lys-
  Ser-Hyp-Hyp-Hyp-Hyp-Thr-Hyp-Val-Tyr-Lys-Tyr-Lys-
  Ser-Hyp-Hyp-Hyp-Hyp-Lys-His-Ser-Hyp-Ala-Hyp-Glu-His-His-Tyr-Lys-Tyr-Lys-
  Ser-Hyp-Hyp-Hyp-Hyp-Lys-His-Phe-Hyp-Ala-Hyp-Glu-His-His-Tyr-Lys-Tyr-Lys-Tyr-Lys-
  Ser-Hyp-Hyp-Hyp-Hyp-Thr-Hyp-Val-Tyr-Lys-Tyr-Lys-
  Ser-Hyp-Hyp-Hyp-Hyp-Thr-Hyp-Val-Tyr-Lys-Tyr-Lys-
  Ser-Hyp-Hyp-Hyp-Hyp-Lys-His-Ser-Hyp-Ala-Hyp-Val-His-His-Tyr-Lys-Tyr-Lys-
  Ser-Hyp-Hyp-Hyp-Hyp-Thr-Hyp-Val-Tyr-Lys-
  Ser-Hyp-Hyp-Hyp-Hyp-Glu-His-
  Ser-Hyp-Hyp-Hyp-Hyp-Thr-Hyp-Val-Tyr-Lys-Tyr-Lys-
  Ser-Hyp-Hyp-Hyp-Hyp-Met-His-
  Ser-Hyp-Hyp-Hyp-Hyp-Thr-Hyp-Val-Tyr-Lys-Tyr-Lys-
  Ser-Hyp-Hyp-Hyp-Hyp-Met-His-
  Ser-Hyp-Hyp-Hyp-Hyp-Val-Tyr-
  Ser-Hyp-Hyp-Hyp-Hyp-Lys-His-His-Tyr-Ser-Tyr-Thr-
  Ser-Hyp-Hyp-Hyp-Hyp-His-His-Tyr-C.
```

Fig. 4.1.5: Predicted sequence of an extensin coded for by the genome of carrot [97]. The Figure arbitrarily shows all the Pro codons as Hyp; in fact a small proportion remain as Pro, i.e. are not post-translationally modified. Basic amino acids are underlined; tyrosine is **emboldened** to draw attention to potential sites of isodityrosine formation.

A number of different extensins can be separated by cation-exchange chromatography; different tissues (or species) have different extensin patterns [459,470].

Attached to the polypeptide backbone are carbohydrate side chains, which together make up ca. 50-60 % of the weight of extensin. Most of the Hyp residues bear short oligosaccharide side chains, principally the tri- and tetrasaccharides shown on the next page [5,20]. These units can be isolated intact (still attached to the Hyp group) by alkaline hydrolysis (Section 4.3.2). Extensins also contain single α-D-Galp residues glycosidically linked to some of the Ser residues [308]; these linkages are alkali-labile, although the reaction with alkali is anomalously slow (Section 4.3.1). Some analyses have also shown the presence of traces of other sugars e.g. Glc; these possibly arise from contaminating polysaccharides.

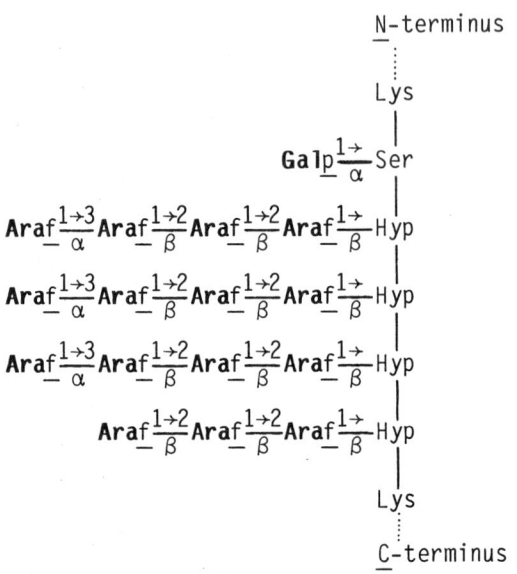

Newly synthesised extensin molecules are readily isolated from the wall with salt (Section 6.4.2). The bulk of the extensin is covalently bound to the wall and cannot be extracted even by such powerful solvents as boiling SDS, PAW or anhydrous HF. This is attributed to inter-polypeptide cross-linking via isodityrosine bridges [169] (Ch 7).

The suggested biological rôles of extensin include the skeletal construction of the cell wall, the organisation of cellulose microfibrils, the restriction of growth, and the exclusion of invading pathogens. Extensin levels vary between different cells by as much as 100-fold, suggesting a fifth possibility: no rôle at all. However, this seems improbable because extensin is produced in very large amounts in certain tissues, accounting for up to 20 % of the total cellular protein synthesising activity in cultured sycamore cells. This is particularly high in view of the fact that sycamore cultures are normally grown under nitrogen-limitation, which would impose selective pressure not to make wasteful proteins.

4.1.6: Arabinogalactan-proteins (AGPs)

AGPs are slimy and/or water-soluble glycoproteins found in the apoplastic fluid of many plant tissues, including suspension culture media, xylem sap, the stylar canal, and probably also the wall water of most tissues [157]. The polypeptide backbone accounts for 2-10 % of the weight.

The polypeptide chain is rich in Hyp and Ser, but differs from that of extensin in being rich in Ala. Some of the Hyp residues may or may not [32] bear short oligo-Ara side chains like those of extensin. Most Hyp residues carry long polysaccharide-like side chains attached via a β-D-Galp linkage [483]. The large side chains are rich in L-Ara and D-Gal and it is likely that many of the arabinogalactans described as polysaccharides are in fact AGPs in which the protein moiety has been overlooked. The long side chains are based on a highly branched core of β-D-Galp residues linked by (1→3) and (1→6) bonds; to this are linked L-Araf, smaller amounts of L-Arap D-GlcpA and D-GalpA, and traces of L-Rha and D-Man residues. AGPs will bind to certain aromatic β-glycosides known as Yariv's antigens — a property very useful in AGP purification [15]. The biological rôle of AGPs is unclear: suggestions include lubrication, and cell/cell recognition.

4.1.7: Enzymic glycoproteins and lectins

These proteins tend to be present in the wall in much smaller amounts than extensins and AGPs. They are also highly diverse, so it is less easy to pick out structural generalisations. One probable wall enzyme that has received detailed investigation is the major basic peroxidase of horseradish and turnip: the complete amino acid sequence has been defined, and the carbohydrate side chains have been preliminarily investigated [188,544]. In the case of horseradish, there are 8 oligosaccharide side chains, each attached to the Asn residue of a different Ser-X-Asn or Thr-X-Asn sequence in the polypeptide backbone (X represents an unspecified amino acid). The oligosaccharide groups are rich in Man, GlcNAc, Xyl and Fuc and may thus be similar to those of many animal glycoproteins, but Ara residues are also sometimes present. Acid phosphatase has also been studied [111].

4.1.8: Cutin, suberin, lignin and tannins

(i) Cutin is a hydrophobic and relatively indigestible polymer found in the outer epidermal walls of stems, leaves, and possibly also roots, including root hairs [223]. It is often concentrated in an outer cuticle, but cutin also extends into the polysaccharide-rich 'cell wall proper' [252]. Cutin is primarily a polyester of hydroxy-fatty acids e.g:

$$CH_2\text{-}CH_2\text{-}CH_2\text{-}CH_2\text{-}CH_2\text{-}CH_2\text{-}CH_2\text{-}CH_2\text{-}CH_2\text{-}CH_2\text{-}CH_2\text{-}CH_2\text{-}CH_2\text{-}CH_2\text{-}CH_2\text{-}COOH,$$
$$|$$
$$OH$$

$$CH_2-CH_2-CH_2-CH_2-CH_2-CH_2-CH_2-CH-CH_2-CH_2-CH_2-CH_2-CH_2-CH_2-CH_2-COOH,$$
$$\quad | \qquad\qquad\qquad\qquad\qquad\qquad\; |$$
$$\quad OH \qquad\qquad\qquad\qquad\qquad\qquad OH$$

$$CH_2-CH_2-CH_2-CH_2-CH_2-CH_2-CH-CH_2-CH_2-CH_2-CH_2-CH_2-CH_2-CH_2-CH_2-COOH,$$
$$\quad | \qquad\qquad\qquad\qquad\qquad |$$
$$\quad OH \qquad\qquad\qquad\qquad\;\; OH$$

$$CH_2-CH_2-CH_2-CH_2-CH_2-CH_2-CH_2-CH_2-CH=CH-CH_2-CH_2-CH_2-CH_2-CH_2-CH_2-CH_2-COOH \quad \text{and}$$
$$\quad |$$
$$\quad OH$$

$$CH_2-CH_2-CH_2-CH_2-CH_2-CH_2-CH_2-CH_2-CH-CH-CH_2-CH_2-CH_2-CH_2-CH_2-CH_2-CH_2-COOH.$$
$$\quad | \qquad\qquad\qquad\qquad\qquad\quad \backslash\; /$$
$$\quad OH \qquad\qquad\qquad\qquad\qquad\quad\; O$$

The —COOH group of one acid is esterified with an —OH group of another [253,296]. Since some of the acids have two or even three —OH groups, the polymer may be a branched structure. The ratio of the different acids varies with species, tissue and environment; cutins from growing and non-growing tissues tend to be rich in C_{16} and C_{18} acids respectively [297].

Cutinised walls and cuticles also contain phenolic material, including esterified ferulic acid, but it is unclear whether the feruloyl groups are attached to cutin itself or to the polysaccharides that occur in association with it.

(ii) Suberin is another polyester, found in certain specialised cell walls e.g. of cork, seed coats and the Casparian strip of the endodermis. The major monomers are unsaturated, omega-hydroxy-C_{16} to -C_{24} fatty acids, as well as α,omega-dicarboxylic acids and C_{20} to C_{30} acids and alcohols. It contains more phenolic material than cutin, but the chemistry of the phenolics and their cross-links are very poorly understood [296,297].

(iii) Lignin (see Section 3.3) is the second most abundant organic substance on Earth, making up 20-30 % of the dry weight of wood. It is laid down initially in the middle lamellae and primary walls of certain cells which have developed secondary walls, principally vessel elements, tracheids, fibres and sclereids. It later also accumulates in the secondary walls of these cells [436]. Other cells can also lay down lignin-like material rapidly, for instance as a response to fungal infection [421]. Lignin is a hydrophobic filler, replacing the wall's water and permanently preventing further growth. Its hydrophobicity,

116

plasticity and indigestibility allow it to serve several biological rôles, e.g. waterproofing the water-conducting system of the xylem, exclusion of pathogens, and provision of physical strength.

Since lignin is formed by random polymerisation of three main monomers [p-coumaryl (= p-hydroxycinnamyl) (H), coniferyl (C) and sinapyl (S) alcohols — Fig. 3.3.1 a-c], using whichever of these the cytoplasm can provide (Section 3.3), variation between lignins is expected to be continuous rather than of the discontinuous type seen for instance in the hemicelluloses between xylans and xyloglucans. Continuous variation is observed in monomer ratios between taxa, between tissues within one plant, between different layers of a single wall, and between defined sub-classes of lignin [358]. Gymnosperms have C but [except in the Gnetales] no S units, whereas Angiosperms have C and S units. Grasses have C, S and H units. The lignin of poplar (a Dicot) has a different S/C ratio in different tis-sues, e.g. cork (S/C = 0.2), primary xylem (S/C less than 0.2), young secondary xylem (S/C = 1.1) and old secondary xylem (S/C = 2.0) [522]. The lignin S/C ratio increases from tip to base in bamboo shoots and decreases in response to light in poplar stems. Lignin from primary walls and middle lamellae of birch xylem has a lower S/C ratio than that from secondary walls of the same cells. Polysaccharide-bound poplar lignin has a higher S/C ratio than unattached lignin from the same source [358].

The three monomers are linked together in a large number of ways (Fig. 3.3.2 c). It seems that variation also occurs between lignins in the ratio of abundance of these different linkages. A convenient method for distinguishing some of the different modes of linkage is to feed tissues with [5-^{3}H]ferulic acid, which, via, coniferyl alcohol, is incorporated into lignin: if coupling occurs through position 5, the ^{3}H is displaced, but if the coupling is via other positions, it remains and the lignin is labelled with ^{3}H (Fig. 4.1.8). Controls are run with [ring-U-^{14}C]ferulic acid, which labels lignin whatever the inter-unit linkages. By this method, auxin was shown to increase the proportion of linkages formed via ring carbon atom 5 of the newly synthesised lignin in pine twigs [504].

Extracted lignins show a great spread in apparent M_r, but it is impossible to obtain a valid estimate of the M_r of lignin as it occurs in the cell wall because lignin can only be solubilised by degradation.

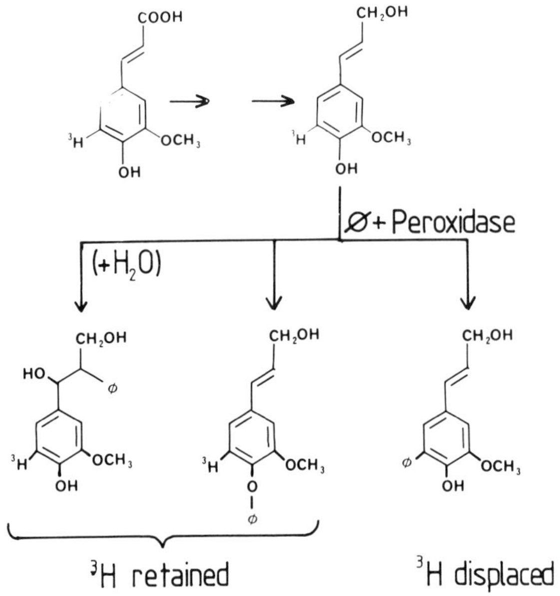

Fig. 4.1.8: A radiochemical method for monitoring the occurrence in lignin of inter-unit bonds involving the 5-position.
Ø is a non-radioactive aromatic monomer with which the [5-³H]coniferyl alcohol molecule is able to couple.

(iv) Tannins of the proanthocyanidin type seem to be linked covalently to polysaccharides [234]. These polysaccharides probably include those of the wall, but this is not firmly established, and it is not known if the walls of living cells contain tannins. Some proanthocyanidin can be extracted from plant tissue in aqueous or organic solvents, but 10-20-fold more cannot. Inextractable material can only be studied by characterisation of its breakdown products (see Section 4.4.5). Extractable tannin can be analysed by GPC, which shows M_r values ranging from about 2 000 to 20 000 (occasionally up to 200 000) [551]. There is variation in the building blocks (catechin vs. epicatechin; number of —OH groups) and in the inter-unit bonds (C-4—C-6 vs. C-4—C-8).

4.2: ANALYTICAL DEGRADATION OF CARBOHYDRATE MOIETIES

The basic strategy for determining the structure of a complex polymer is to break it down into workable (small) fragments, and then to separate these by chromatography and investigate their chemical composition.

Polysaccharide chains and the carbohydrate moieties of glycoproteins can be degraded by the following treatments [23,323,395; see also 364 for some novel alternatives], yielding useful products which can be separated and structurally elucidated by the methods described in Section 4.5.

4.2.1: Acid-catalysed hydrolysis ('complete')

The simplest and most informative single way to characterise carbohydrate polymers is to hydrolyse them to their constituent monosaccharides. These are then analysed by chromatography. Hydrolysis can be effected by acids or enzymes [alkalies are slow to hydrolyse glycosidic bonds and quickly break down any free monosaccharides formed]. Hot acids are used at various temperature/time combinations which are a compromise imposed by **(a)** the resistance of glycosidic bonds and **(b)** the instability of free monosaccharides in hot acid. Resistance of glycosidic bonds to hydrolysis depends mainly on the nature of the glycosyl residue rather than that of the aglycone (see Section 3.1.2) to which it is attached. Conditions for hydrolysis of some representative bonds are:

Class	Specific examples	Conditions recommended
Very labile	Apif, KDOp	0.1 M TFA, 50°C, 24 h[3]
Other neutral furanosyl	Araf	0.1 M TFA, 100°C, 1 h
6-Deoxyhexopyranosyl	Fucp, Rhap	1.0 M TFA, 100°C, 1 h
Other neutral pyranosyl	Glcp[1], Galp, Manp, Xylp, Arap, GlcpNAc[2], GalpNAc	2.0 M TFA, 120°C, 1 h
Acidic pyranosyl	GalpA, GlcpA	[2.0 M TFA, 120°C, 1 h][4]
2-Aminoglycosyl	GlcpN, GalpN	[2.0 M TFA, 120°C, 1 h][4]

[1]Except in cellulose owing to its crystallinity.

[2]Except in chitin owing to its crystallinity. Also, the removal of N-acetyl groups will be a side reaction, impeding hydrolysis of the glycosidic bond; the resulting free base (GlcN residues) is much more resistant to acid hydrolysis [see footnote [4]]. If necessary, the partially hydrolysed material can be dried in vacuo, re-acetylated with acetic anhydride/pyridine (1:1 at 100°C for 1 h), re-dried, and the hydrolysis then repeated for a more complete yield of monosaccharide.

[3]With very low acid concentrations, impurities in the sample can easily buffer the pH, preventing hydrolysis; hence moderate acid concentrations are preferred, but at low temperatures.

[4] Incomplete hydrolysis achieved.

119

A recommended set of conditions for all-round use is **2 M TFA at 120°C for 1 h**, which gives efficient release of neutral, non-cellulosic sugar residues, with little decomposition of the free monosaccharides (Panel 4.2. 1 a). TFA is the preferred acid because it is volatile.

Uronic acid glycosyl linkages are relatively acid-resistant, and 2 M TFA at 120°C for 1 h gives a low yield of monosaccharide. More severe conditions cannot be used because free uronic acids are readily decarboxylated in hot acid. One approach is hydrolysis for 4-24 h at 100°C in 70-90 % formic acid, in which free uronic acids are relatively stable. The products include formyl esters, which are subsequently hydrolysed with 2 M TFA at 100°C for 1 h [25]. Neither TFA nor formic acid gives a quantitative yield of uronic acids, although they can provide a good qualitative picture of the uronic acids present, which can be backed up by the quantitative m-hydroxybiphenyl assay for total uronic acids (Panel 3.7 a). If a uronic acid is glycosidically linked to a neutral sugar [e.g. in

...D-Ga lpA-α-(1→2)-L-Rha...

of RG-I (Section 4.1.3)], then inefficient hydrolysis will also lower the yield of the latter (Section 4.2.2).

Special conditions are required for hydrolysis of cellulose, which exists in semi-crystalline microfibrils not readily penetrated by 2 M acid. For total acid hydrolysis (Panel 4.2.1 b), cellulose is dissolved in cold

---

PANEL 4.2.1 a: Hydrolysis of non-cellulosic polysaccharides in TFA

Sample: 10 mg of dried cell walls (see Panel 1.2.1 a).

Step 1:   Suspend sample (10 mg) in 1 ml 2 M trifluoroacetic acid (TFA) (= 0.85 ml H_2O + 0.15 ml pure TFA) in a tube.

Step 2:   Seal the tube. This can be done by melting the neck of the tube (use soda glass) in a flame and drawing it out slowly with forceps. More conveniently, use Pyrex screw-necked culture tubes with tightly-fitting caps lined with Teflon pads.

Step 3:   Heat at 120°C for 1 h. This can be done in an autoclave (15 psi), oven or block heater.

Step 4:   Cool. Centrifuge (5 min, 2,000 **g**). Harvest supernatant.

Next steps: Either apply 10-100 μl of hydrolysate directly on to Whatman no. 1 paper for chromatography (Penels 4.5.2); Or remove TFA in vacuo (see Section 4.5.1).

---

72 % (w/w) H_2SO_4, which swells the microfibrils, and then hydrolysed in hot 3 % acid [1].

4.2.2: Acid-catalysed hydrolysis (partial)

Additional information about the composition and structure of poly-saccharides can be obtained from partial hydrolysis. Here the goal is to halt hydrolysis at the oligosaccharide stage, rather than aiming for 100 % yield of monosaccharides. Since oligosaccharides are easier to purify than polysaccharides, they can give clear-cut evidence that, for instance, two different monosaccharides were part of the same polysaccharide molecule. Also, the structure of an oligosaccharide can indicate the type(s) of sugar—sugar linkage present in the polysaccharide: for example, maltose [D-Glcp-α-(1→4)-D-Glc] is diagnostic of starch, and cellobiose [D-Glcp-β-(1→4)-D-Glc] of cellulose. Particularly high yields of a given oligosac-charide are possible when neighbouring glycosyl bonds differ greatly in the

---

Panel 4.2.1 b: Hydrolysis of total wall polysaccharides in H_2SO_4[1]

Sample: 25 mg of dried cell walls (see Panel 1.2.1 a).

Step 1:   Suspend sample (25 mg) in 0.5 ml 72 % (w/w) [ca. 58 % v/v or ca. 11 M] H_2SO_4 at 25°C. Shake or stir until dissolved (ca. 1 h). [Lignin & protein do not dissolve well.]

Step 2:   Add solution to 10.5 ml H_2O. Mix well. Heat at 120°C for 1 h (for this size of sample an autoclave is convenient).

Step 3:   Cool. Add about 25 μl of 1 % bromophenol blue. Partially neutralise[1] the acid by slow addition, with rapid stirring, of 25 ml of 0.18 M $Ba(OH)_2$ (= enough to neutralise, and precipitate as $BaSO_4$, about 80 % of the H_2SO_4; the indicator should not go blue).

Step 4:   Complete the neutralisation by stirring with 1-2 g of finely divided $BaCO_3$ powder. Both the base ($BaCO_3$) and the resulting salt ($BaSO_4$) are insoluble, so a slight excess of base does not produce an alkaline solution (which would degrade the sugars). The reaction with $BaCO_3$ is rather slow, so warming and continued stirring is recommended until the indicator goes (and remains on a few minutes' standing) blue.

Step 5:   Remove $BaSO_4$ + remaining $BaCO_3$ by centrifugation (15 min, 3,000 **g**). Collect supernatant. To remove further traces of Ba^{2+} from solution, freeze and thaw the solution, and repeat centrifugation.

Next steps: Chromatography of supernatant, after concentration by (freeze) drying.

[1]For an H_2SO_4 method that greatly simplifies neutralisation, see [55a].

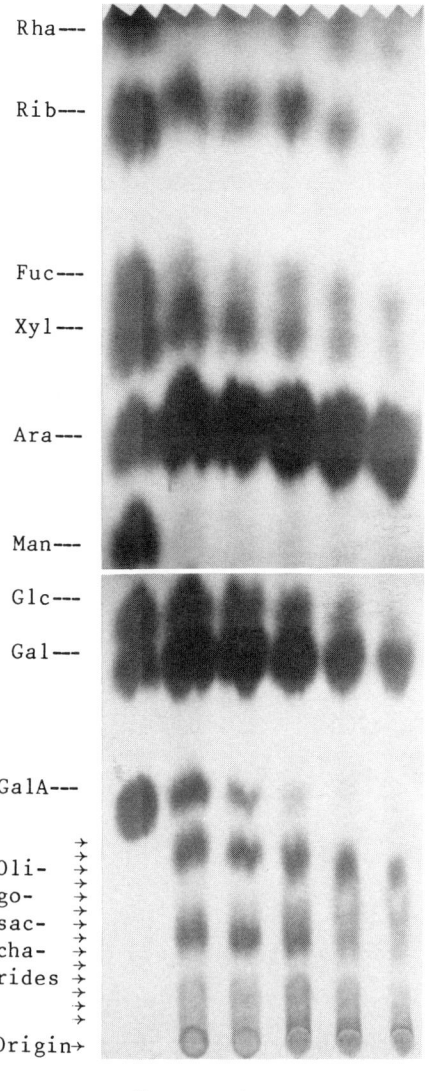

Rha---
Rib---
Fuc---
Xyl---
Ara---
Man---
Glc---
Gal---
GalA---
Oli-
go-
sac-
cha-
rides
Origin→

M   a   b   c   d   e

Fig. 4.2.2 a: Paper chromatography of sugars released from rose cell walls by graded acid-hydrolysis. Walls were heated at 120°C for 1 h in **(a)** 2 M, **(b)** 1 M, **(c)** 0.5 M, **(d)** 0.25 M, or **(e)** 0.125 M TFA, and the products were separated on Whatman no. 1 paper in BuOH/HOAc/H$_2$O (12:3:5) followed in the same dimension by EtOAc/pyridine/H$_2$O (8:2:1). Stain: aniline hydrogen phthalate. **M:** 50 µg of each named sugar.

ease with which they are hydrolysed by acid. Thus, linkages involving apiose or KDO (which occur at occasional sites within RG-II — see Section 4.1.3) are 35 % cleaved by 0.1 M TFA at 40°C for 24 h, liberating specific oligosaccharides (Fig. 4.2.2 b); other glycosidic bonds are relatively unaffected [465].

Differences in acid-lability are illustrated in Fig. 4.2.2 a. Keeping the hydrolysis conditions constant at 120°C for 1 h, a reduction in acid concentration from 2 M to 0.125 M had little effect on the yield of monomeric arabinose and fucose, while reducing that of monomeric xylose, glucose and galactose, and almost totally preventing the release of any monomeric galacturonic acid.

Arabinofuranosyl linkages are cleaved by 0.1 M TFA at 100°C for 1 h; sequences of the type (Sugar)p-Araf-... are released as intact units. For instance, in spinach cell walls, the rare Arap residues can be isolated in good yield from the polysaccharide sequence

Arap-Araf-Araf-Araf-Araf-...

as the disaccharide Arap-α-(1→3)-Ara by hydrolysis in 0.1 M TFA at 100°C for 1 h. The great majority of the Ara units (which were present as Araf residues) are recovered as free monosaccharide.

In wall polymers, most Araf residues are side chains [in RG-I, extensin, arabinoxylan and arabinogalactan (Fig. 4.2.2 b)], as shown by the fact that mild acid hydrolysis yields free arabinose while the rest of the polysaccharide remains polymeric. For instance, mild acid converts arabinoxylan to free (dialysable) arabinose plus polymeric (non-dialysable) xylan: this result could only be obtained from a structure like

```
Xyl-Xyl-Xyl-Xyl-Xyl-Xyl-Xyl-Xyl-Xyl-Xyl-Xyl-Xyl-Xyl-Xyl-Xyl-Xyl-Xyl-...,
 |       |   |       |       |   |   |           |               |
Ara     Ara Ara     Ara     Ara Ara Ara         Ara             Ara
```

where the Xylp backbone is uninterrupted by labile Araf units, and not from

```
Xyl-Ara-Ara-Xyl-Xyl-Xyl-Xyl-Ara-Xyl-Xyl-Xyl-Ara-Xyl-Xyl-Xyl-Xyl-Ara-...,
 |       |   |       |       |   |   |           |               |
Ara     Ara Ara     Ara     Ara Ara Ara         Ara             Ara
```

where the backbone would be labile to mild acid, yielding dialysable oligosaccharides e.g. Xyl-Xyl-Xyl-Ara.

Pyranose likages involving 6-deoxyhexose residues (rhamnose or fucose) are more labile to mild acid than those involving otherwise similar hexose residues. Fucose residues often occur as non-reducing termini (e.g. in xyloglucan) and the comments in the previous paragraph apply (Fig. 4.2.2 b). Rhamnose residues occur mid-chain in the backbone of homogalacturonan and RG-I. Thus mild acid hydrolysis of RG-I, can generate oligosaccharides like D-GalpA-α-(1$\rightarrow$2)-L-Rha (Fig. 4.2.2 b). According to some workers, selective hydrolysis of Rha residues in homogalacturonan gives oligosaccharides with presumed structures of the type (GalA)$_n$-Rha where n is about 25 [406].

Neutral pyranose residues show some variation in acid-lability, e.g. (1$\rightarrow$6) bonds are more resistant than (1$\rightarrow$4) bonds. However, they do not exhibit such large differences as are seen between p and f bonds. A complex polysaccharide composed of neutral pyranoses might therefore yield many oligosaccharides, only the smallest of which are likely to be produced in large enough quantities for analysis. Thus it is recommended in partial hydrolysis to aim for a maximum yield of di- or trisaccharides [17,200]: theoretically, this will be obtained when about 67 % or 50 % respectively of the sugar—sugar bonds in the polysaccharide have been hydrolysed [23].

The di- and trisaccharides are then isolated and separated chromato-graphically, and investigated structurally. Overlapping structures may allow the reconstruction of any repeating unit that might occur in the polysaccharide (Fig. 4.2.2 b [344,512]).

Glycosidic linkages involving uronic acid residues are more stable in hot 2 M TFA than are those from neutral sugars. Thus, disaccharides of the type Uronic acid—Neutral sugar (= aldobiouronic acids), are relatively easy to isolate, e.g. GlcpA-(1→2)-Xyl from glucuronoxylan or GalpA-(1→2)-Rha from rhamnogalacturonan-I [298].

A problem that arises during partial hydrolysis is that glycosidic linkages may be formed in hot acid as well as degraded. Thus, totally spurious oligosaccharides may turn up. This 'acid reversion' can be minimised by use of low polysaccharide concentrations during hydrolysis (less than 10 mg/ml). If an oligosaccharide is suspected of being an acid reversion product, the hydrolysis is repeated at a lower acid concentration: a reversion-product would diminish relative to a true hydrolysis-product.

Mild acid hydrolysis can also be used to generate large fragments of polysaccharides. These may serve as useful soluble models of an insoluble polysaccharide. The formation of large fragments is achieved with very mild acid, and this has some special properties: for example, when poly-saccharides are heated at pH 3-4, uronic acid linkages are hydrolysed more rapidly than are neutral sugar pyranose residues, although both at a rate much lower than in 2 M TFA. This reversal of the rank order of acid lability is considered to be due to hydrolysis catalysed intramolecularly by the uronic acid residues themselves [455]. Neutral sugar residues adjacent to uronic acid residues within a single polysaccharide molecule may also be hydrolysed. This may explain the release of traces of monosac-charide Gal and Ara as well as GalA when pectins are heated at pH 3.5; Fuc, Xyl and Glc are not released from xyloglucan when heated in the same solution, indicating that pectin cannot catalyse intermolecular hydrolysis.

4.2.3: Acetolysis

Acetolysis complements partial acid-hydrolysis [323]. Thus, in acetolysis, (1→6) linkages are the most labile in contrast to the situation in acid hydrolysis. 6-Deoxyhexose residues are relatively resistant to acetolysis. Thus, oligosaccharides that would be difficult to isolate in good yield by acid hydrolysis can be isolated by acetolysis (Panel 4.2.3; Fig. 4.2.3).

PANEL 4.2.3: **Isolation of short repeating oligosaccharides from xyloglucan by acetolysis.** [see also 444]

Sample: Xyloglucan, purified from culture filtrate of spinach cells by differential precipitation with ethanol (Panel 3.5.6).

Step 1:   Stir the dry polysaccharide (10 mg) at 0°C in 1 ml of acetic anhydride containing 4 % of concentrated H_2SO_4.

Step 2:   When the polysaccharide has dissolved (which occurs readily, even with water-insoluble types, owing to acetylation of the polysaccharide), stirring can be discontinued.

Step 3:   Continue incubation, in a sealed vessel, for 1-10 days at 25°C. Linkages are selectively broken, forming glycosyl-acetates.

Step 4:   Stop reaction at required degree of de-polymerisation by pouring the mixture into stirred water (10 ml) at 0°C, and adjust pH to 3-4 with $NaHCO_3$.

Step 5:   Shake thoroughly with $CHCl_3$ (10 ml), and separate two phases; inorganic salts are left in (upper) aqueous phase, while oligosaccharide-acetates are recovered in the (lower) organic phase.

Step 6:   Dry the $CHCl_3$ solution in vacuo and de-acetylate the resulting solid by treatment with 1 ml of 0.1 M methanolic barium methoxide [prepared by cautious addition of Ba metal to anhydrous methanol] for 1 h.

Next steps: Removal of $Ba(OMe)_2$ by addition of CO_2; chromatography to isolate the expected oligosaccharides, e.g. Fuc-Gal-Xyl, Gal-Xyl and Glc-Glc [see structure — Fig. 4.2.3].

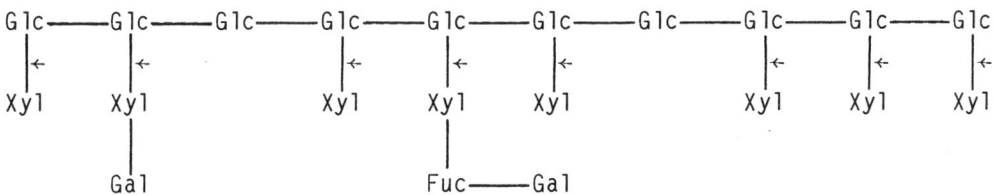

Fig. 4.2.3: Leitmotiv of xyloglucan, showing bonds (←) particularly labile to acetolysis.

125

4.2.4: Alkaline degradation

Monosaccharides are degraded in alkali, especially on warming in the presence of O_2. The products are organic acids. Even under very mildly basic conditions, monosaccharides are interconverted <u>via</u> **ketoses** (e.g. glucose $\leftrightarrow$ fructose $\leftrightarrow$ mannose). Such conditions can occur in warm pyridine and on anion-exchange resins e.g. Dowex-1. Reducing termini of oligo- and polysaccharides undergo similar reactions.

Polysaccharides undergo alkali-catalysed β-elimination reactions, which can be put to good analytical use [548]. These reactions occur in structures where a sugar is glycosidically linked to an oxygen atom which is β to a neutral C=O group:

$$\text{Glycosyl—O—C—C—C=O}$$

This general structure occurs in three particular sites (Fig. 4.2.4).

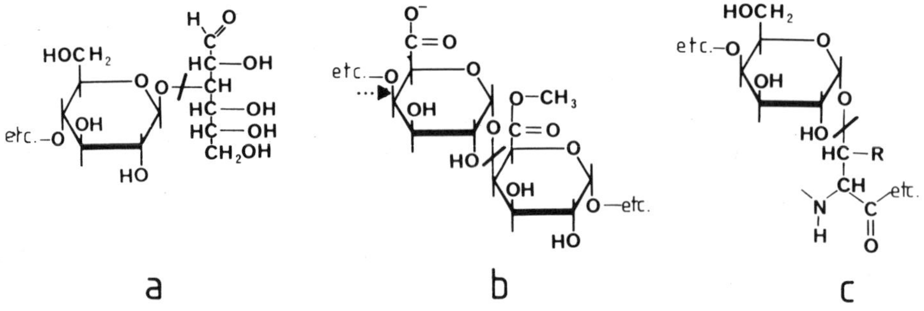

```
     a              b              c
```

Fig. 4.2.4: **Three common situations in which sugar bonds can be broken by alkali-catalysed β-elimination reactions (/).**
   (a) the reducing end of a polysaccharide e.g. β-(1→3),(1→4)-glucan;
   (b) a mid-chain linkage in a (partially) methyl-esterified pectin [the bond marked ···→ is stable to alkali because the neighbouring C=O group is acidic];
   (c) a sugar attached to serine (R = —H) or threonine (R = —CH₃).

126

The first (Fig. 4.2.4 a) is at the reducing terminus of a polysaccharide, where C-1 (of an aldose) or C-2 (of a ketose) can form a C=O group (Section 3.1.1) and thereby favour the elimination of any sugar residue attached to C-3 or C-4 respectively. β-Elimination splits off the reducing terminal sugar, in the form of a saccharinic acid, exposing the next sugar unit as the new reducing terminus. Alkaline degradation allows the examination of a polysaccharide from its reducing end, complementing the ability to work from the non-reducing end(s) by enzymic hydrolysis (Section 4.2.5). The breakdown products released by alkali at 25°C depend on the glycosidic linkage of the penultimate sugar ('Q') to the reducing terminus ('R') [23]:

in **Q**-(1→2)-**R**, **R** is stable;

in **Q**-(1→3)-**R**, **R** is released as a metasaccharinic acid;

in **Q**-(1→4)-**R**, **R** is first slowly converted to the corresponding ketose (see above) and this is then rapidly released as isosaccharinic acid;

in **Q**-(1→6)-**R**, **R** is (slowly) converted to lactic acid.

When two or more sugars ('P', 'Q') are attached to one reducing terminus ('R'), more complex degradation products may result. Not all even-tualities have been studied, but, for example,

in $\begin{array}{c} \text{P-}(1\to6) \\ \\ \text{Q-}(1\to4) \end{array}\Big\rangle\text{R},$ **R** is released as **P**-(1→6)-metasaccharinic acid.

Since upon β-elimination the 'Q' unit becomes the new reducing terminus it may itself be attacked. By repetition of this, polysaccharides are progressively 'peeled' from the reducing terminus, releasing analytically useful saccharinic acid fragments. The peeling runs smoothly in (1→3)-linked polysaccharides; (1→4)-linked polysaccharides eventually succumb to a competing side reaction that stops further peeling.

The second site that fulfills the requirements for β-elimination (Fig. 4.2.4 b) is where a sugar is glycosidically linked to O-4 of an esterified uronic acid unit, a situation that occurs frequently in pectins. β-Elimination cleaves the polysaccharide chain, leaving the former uronic acid unit with a C=C group. Alkali also hydrolyses methyl ester groups from pectin, forming acidic C=O groups (—COOH) that do not promote elimination. Ester-hydrolysis and β-elimination are thus competing reactions. Conditions favouring elimination are mild alkali (a pH as low as 6.8 will suffice at 100°C) and heating (since the Q_{10} of elimination is high). Elimination can also be favoured by use of hot methanolic sodium methoxide as base, as the methyl esters will then constantly be reformed by transesterification. A useful application of this type of elimination has been the generation of pectic polysaccharide fragments suitable for electrophoresis [480].

The third example of β-elimination is described in Section 4.3.1.

Alkali also strips other ester-linked groups from polysaccharides, e.g. acetyl and feruloyl groups. Suitable conditions for de-acylation are 0.5 M NaOH at 25°C for 1-5 h, under N_2. The progress of hydrolysis is followed by back-titration of a portion of the NaOH with standard HCl to the end point of phenolphthalein, since for every mole of ester hydrolysed one mole of NaOH is neutralised. However this does not distinguish between acetyl, feruloyl and uronoyl esters, and is not very sensitive. Different esters can be distinguished by chromatography (especially HPLC) of the hydrolysis products. They can also be distinguished by simple volatility tests, which are readily applied to [U-^{14}C]-labelled material: ferulic acid and related compounds are not volatile, acetic acid is volatile at low pH but not (e.g. as sodium acetate) at high pH, and methanol is volatile regardless of pH.

4.2.5: Enzymic hydrolysis

Hydrolysis with pure enzymes is the ideal way to degrade polysaccharides because it can be extremely specific, and is carried out under very mild conditions. The main problem is the lack of commercially-available pure enzymes. Impure enzyme preparations are often used, but in these cases the contaminating enzyme activities must be identified if valid conclusions about polysaccharide structure are to be drawn.

The main sources of enzymes are saprophytes. Soil organisms probably produce enzymes that cleave all naturally-occurring sugar—sugar bonds. Purification of these enzymes is difficult because most saprophytes produce not only several different activities (cellulase, pectinase, xylanase etc.) but also many isozymes of each of these [561]. Potential solutions to the problem are: **(a)** Production of a given activity is favoured if the micro-organism is grown on the target polysaccharide as sole carbon source [80]. **(b)** Purification should if possible include an affinity adsorption step, as this separates on the basis of biological function rather than an arbitrary physical constant e.g. M_r or pI [419]. **(c)** Unwanted activities can be inhibited. Many enzymes that hydrolyse glycosyl bonds are inhibited by the mono- or disaccharide end-product (e.g. cellulase by glucose or cello-biose): specific bonds in a polysaccharide can be protected from contaminant enzymes by inclusion of 50 mM of appropriate sugar(s) in the hydrolysis reaction mixture [170]. **(d)** Ideally, the gene coding for the desired enzyme is transferred to and expressed in an organism (e.g. E. coli) that produces few or no enzymes active against the plant cell wall. This has been achieved for endo-β-(1→3),(1→4)-glucanase [Bass Brewing, plc, personal communication], pectinase [579], cellulase [516], pectate lyase [516] and pectin methylesterase [404].

Polysaccharide-hydrolysing enzymes fall into two categories: endo- and exo-hydrolases. Exo-enzymes (glycosidases) attack the non-reducing end(s), cleaving off sugars one (or in a few enzymes two) at a time. They are specific for the glycosyl residue, e.g. β-glucosidase will attack non-reducing terminal β-D-Glcp residues (but not α-D-Glcp, β-L-Glcp, β-D-Galp or β-D-Glcf); they will not usually attack if any substituent is present on the residue e.g. an acetyl or feruloyl group or another glycosyl residue. They tend to be much less specific for the aglycone: e.g. they will often attack a residue equally well whether the aglycone is a polysac-charide, oligosaccharide or monosaccharide, or (in model substrates) methanol or 4-nitrophenol (Section 6.5). However, there are exceptions to this rule: e.g. a plant α-xylosidase will attack xylosyl termini of xyloglucan and its repeating heptasaccharide, but not the disaccharide D-Xylp-α-(1→6)-D-Glc or the model substrate 4-nitrophenyl-α-D-xylopyranoside [140]. Commercially available glycosidases of reasonable purity include β-D-galactosidase (from E. coli), α-D-mannosidase (from jack bean) and α-L-fucosidase (from mammalian epididymis) [359]. Many other glycosidase preparations are sold which, although containing the activity stated on the bottle, also contain large amounts of other enzymes. This is a serious problem with enzymes from saprophytes; it is less serious with enzymes of animal origin (e.g. α-fucosidase from bovine epididymis) since here the contaminants are less likely to be enzymes that degrade plant cell walls.

Endo-enzymes cleave appropriate sugar—sugar bonds mid-chain in the polysaccharide. They are often named by adding '-ase' to the name of the substrate (e.g. pectinase,· xylanase and mannanase). They are inhibited by the wrong kind of substitution (e.g. acetyl or feruloyl groups) on the target glycosyl residue; but, unlike the exo-enzymes, they often positively require another glycosyl substituent on the glycosyl residue to be hydrolysed. These requirements for enzyme action can give valuable information on the structure of a polysaccharide. No polysaccharide endo-hydrolase active on the plant cell wall is available commercially in pure form except endo-$(1\rightarrow3),(1\rightarrow4)$-glucanase [from Biosupplies Australia Pty, Parkville, Victoria]. Some are available in impure form, and others can be extracted and purified by published techniques.

A useful enzyme source is **'Driselase'**, a mixture of endo- and exo-enzymes from the fungus <u>Irpex</u> <u>lacteus</u> (=<u>Polyporus</u> <u>tulipiferae</u>). It is highly active on plant cell walls, as seen by protoplast yield (Section 2.7.6). Activities present include:

<u>Exo-hydrolases</u>: α-D-galactopyranosidase, β-D-galactopyranosidase, β-D-glucopyranosidase, α-D-mannopyranosidase, β-D-mannopyranosidase, α-L-arabinofuranosidase, β-D-xylopyranosidase, α-L-fucopyranosidase, cellulose-cellobiohydrolase.

<u>Endo-hydrolases</u>: cellulase [β-$(1\rightarrow4)$-D-glucanase], β-D-galactanase, α-L-arabinanase, pectinase [α-$(1\rightarrow4)$-D-galacturonanase], β-D-mannanase, xylanase [β-$(1\rightarrow4)$-D-xylanase].

Driselase also contains an **endo-$(1\rightarrow4)$-β-glucanase** that attacks all the β-glucose bonds of xyloglucan whether or not they are substituted by α-xylose residues [cf. most cellulases can only attack xyloglucan at the <u>ca.</u> 1 in 4 glucose residues that do not carry xylose side-chains — see Section 4.1.4]. Since Driselase lacks α-xylosidase, but is rich in α-L-fucosidase, β-D-galactosidase and α-L-arabinosidase, it completely digests xyloglucan to the disaccharide D-Xyl$p$-α-$(1\rightarrow6)$-D-Glc [= isoprimeverose], a useful diagnostic breakdown product. The use of Driselase is described in Panel 4.4.2 a and illustrated in Fig. 4.2.5.

Driselase often completely dissolves non-lignified walls, yielding monosaccharides plus Xyl-α-$(1\rightarrow6)$-Glc [177]. Exhaustive Driselase digestion is a useful alternative to acid for the hydrolysis of non-ligni-fied walls. Since it works at 25°C, any unexpected labile components are retained. It also has the advantage of distinguishing between α- and β-xylose residues, which are released as disaccharide (see above) and mono-saccharide respectively. Acetylated β-xylose residues are resistant to Driselase, and should be de-acylated with alkali (Panel 4.4.1) for maximum yield of monosaccharide.

130

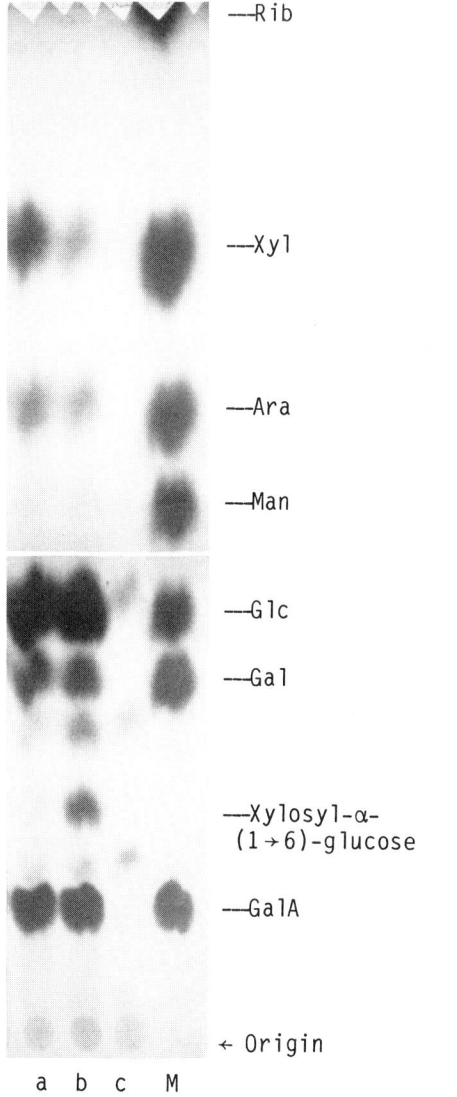

		—Rib
		—Xyl
		—Ara
		—Man
		—Glc
		—Gal
		—Xylosyl-α-(1→6)-glucose
		—GalA
		← Origin

a   b   c   M

Driselase is a useful starting material from which to purify individual enzymes. Its endo-**pectinase** has a particularly high pI and is readily purified by 'Chromatofocusing'. A rapid method for the purification of an endo-**xylanase** from Driselase has been reported [79]; for an alternative source of xylanase, see [281].

A useful enzyme for analysis of β-(1→3),(1→4)-D-glucans is a specific **endo-β-(1→3),(1→4)-glucanase** (EC 3.2.1.73) that attacks the second bond in the sequence

...Glc-(1→3)-Glc-(1→4)-Glc-....

It thus cuts the chain at the (1→4) bond next to each (1→3) bond, releasing the analytically useful oligosaccharides shown on the following page,

Fig. 4.2.4: Paper chromatography of sugars released from rose cell walls by Driselase digestion. The walls (1.25 mg) were treated with 125 μl 0.25 % Driselase (de-salted) at pH 4.7 and 25°C for 16 h and the products chromatographed on Whatman 3MM paper in BuOH/HOAc/H$_2$O (12:3:5) for 8 h followed in the same direction by EtOAc/pyridine/H$_2$O (8:2:1) for 30 h and stained with aniline hydrogen phthalate (**lane b**). Little free xylose has been released, but there is a major spot of xylosyl-α-(1→6)-glucose, which arises from xyloglucan. In **lane a**, the walls were pre-treated with 2 M TFA at 120°C for 1 h prior to Driselase: the xylosyl-α-(1→6)-glucose is replaced by xylose + glucose. **Lane c** shows a control from which the cell walls were omitted, and indicates the low degree of contamination by Driselase autolysis products. **Lane M:** marker mixture.

...Glc-(1→3)-Glc-(1→4)-Glc-(1→4)-Glc-(1→3)-Glc-(1→4)-Glc-(1→4)-...

→   ...Glc-(1→3)-Glc + Glc-(1→4)-Glc-(1→3)-Glc + Glc-(1→4)-...,

giving information on the number of consecutive (1→4)-linked units that intervene between each pair of (1→3)-linked units. Methods have been published for the purification of this enzyme [564], and, as already mentioned, it is available commercially.

Pectinases, which release the hairy domains (rhamnogalacturonans) of pectin, can be purified from ripening fruit, from culture filtrates of Colletotrichum lindemuthianum [144], from Driselase, or from commercial pectinase preparations. Some pectinases act on methyl-esterified galact-uronate residues, others on acidic galacturonate residues. The action of the latter type is augmented by de-esterification of the pectin (e.g. in 0.5 M Na_2CO_3 at 0°C for 16 h followed by 25°C for 3 h). Pectinases generally require a stretch of about 3 contiguous galacturonate residues (in the correct esterification state) for action, which explains why they will not attack the backbone of RG-I and RG-II. Other pectin-attacking enzymes include pectin- and pectate-lyases (eliminases) [121,122], which catalyse the non-hydrolytic cleavage (Section 4.2.4) of pectin chains at neutral and acidic galacturonate residues respectively. Pectin methylesterase, which hydrolyses off the methyl-ester groups, is available commercially from citrus peel but such preparations are contaminated with glycosidases. The arabinan domains of pectins are cleaved by an endo-arabinanase, which can be purified from Bacillus subtilis [543].

Cellulases can be isolated from plants (avocado is a rich source [48]), or from fungal enzyme preparations. A preparation from Trichoderma viride can be used, without further purification, to cleave xyloglucan into hepta- and nonasaccharides (Panel 9.2.2) [177]. α-Xylosidase is absent, and contamination with α-fucosidase and is low enough that good yields of nonasaccharide can be obtained with less than 10 % loss of fucose residues. Cellulase will not necessarily attack cellulose if the latter is crystalline and thus able to exclude the enzyme; efficient hydrolysis of cellulose requires a second enzyme, cellobiohydrolase [562], of which Driselase is a rich source.

132

4.2.6: Borohydride reduction

Reduction is a straightforward chemical reaction (Panel 2.4.6) by which it is possible to identify the **reducing terminus** of a poly- or oligosaccharide. All the sugars in a polysaccharide are protected from $NaBH_4$ except the single reducing terminus, which gets reduced to the corresponding alditol (e.g. mannose → mannitol):

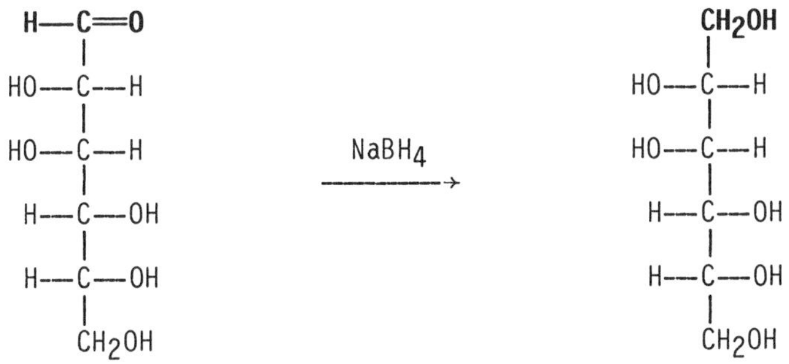

```
  H——C══0                                      CH2OH
      |                                          |
HO——C———H                                 HO——C———H
      |                                          |
HO——C———H            NaBH4                 HO——C———H
      |         ------------→                    |
  H——C——OH                                   H——C——OH
      |                                          |
  H——C——OH                                   H——C——OH
      |                                          |
     CH2OH                                      CH2OH
```

The alditol can be identified after hydrolysis by chromatography [107]. The method can be made very sensitive by use of NaB^3H_4 since the product is then a [^{3}H]alditol. In addition, $NaBH_4$-reduction can indicate **(a)** the DP of the polysaccharide (deduced from the monosaccharide : alditol ratio after hydrolysis as shown below for the pentadecasaccharide, Glc_{15})

Glc-Glc-Glc-Glc-Glc-Glc-Glc-Glc-Glc-Glc-Glc-Glc-Glc-Glc-Glc $\xrightarrow{\text{NaBH}_4}$

Glc-Glc-Glc-Glc-Glc-Glc-Glc-Glc-Glc-Glc-Glc-Glc-Glc-Glc-Glucitol $\xrightarrow{\text{TFA}}$

14 Glucose + 1 Glucitol

and, in the case of small oligosaccharides, **(b)** the linkage [(1→2), (1→3) etc.] of the penultimate sugar residue to the reducing terminus [deduced by electrophoresis in molybdate buffer (Section 4.5.4)].

Polysaccharides extracted in NaOH/NaBH₄ lack reducing termini and do not react with NaB^3H_4. In addition, some polysaccharides as they occur in the intact cell wall may lack (accessible) reducing termini. However, an oligosaccharide cleaved from such a polymer by partial hydrolysis <u>will</u> have an intact, accessible reducing terminus.

```
┌─────────────────────────────────────────────────────────────────────────┐
│ PANEL 4.2.6:   NaBH₄-Reduction of an oligosaccharide                      │
│                                                                           │
│ Sample:          0.1 mg of xyloglucan nonasaccharide (see Panel 9.2.2).   │
│                                                                           │
│ Step 1:          Incubate sample in 0.2 ml of 0.5 M NaBH₄ dissolved in 1 M│
│ NH₃ at 25°C for 4 h.                                                      │
│                                                                           │
│ Step 2:          Destroy excess NaBH₄ by addition of 30 μl acetic acid    │
│ (bringing the pH to ca. 4.8).  [H₂ gas is evolved at this stage, and this │
│ will be ³H₂ if NaB³H₄ was used:  use a fume cupboard.]                     │
│                                                                           │
│ Step 3:          Pass solution through a 1.5-ml bed volume column of Dowex-│
│ 50 (previously washed in 1 M HCl followed by plentiful H₂O) in a Pasteur  │
│ pipette, and elute with 3 ml H₂O.  This step removes NH₄⁺ and yields the  │
│ reduced sample in acetic acid/H₃BO₃.                                      │
│                                                                           │
│ Step 4:          Dry in vacuo.  [Conveniently performed in a 'Speed-Vac'.]│
│                                                                           │
│ Step 5:          Re-suspend the sample in 0.1 ml methanol/acetic acid (10:1)│
│ and re-dry in vacuo.  Repeat this step several times (e.g. 6 times) to    │
│ remove H₃BO₃ as its volatile methyl-ester.  Re-dissolve in H₂O (0.2 ml).  │
│                                                                           │
│ Step 6:          If it is important to ascertain complete removal of the  │
│ H₃BO₃ (e.g. if the sample is subsequently to be analysed by PE in molybdate│
│ or acetate buffer), take a 20-μl sample of the aqueous solution for assay │
│ of borate [71].*                                                          │
│                                                                           │
│     Next steps:  either hydrolysis in 2 M TFA followed by chromatography  │
│ [expected products:  glucose, xylose, galactose, fucose and glucitol (=   │
│ sorbitol) in the molar ratio 3:3:1:1:1];  or paper electrophoresis in     │
│ molybdate buffer (Panel 4.5.4 d).                                         │
│                                                                           │
│     *To 10 μl, add 90 μl 0.3 M HCl, cool in an ice-bath, and add 0.5 ml   │
│ of pre-cooled (0°C) conc H₂SO₄.  Mix carefully and re-cool to 0°C.  Add 0.5│
│ ml of 0.092 % carmine N.F.40 (dissolved in conc H₂SO₄).  Mix.  Incubate at│
│ 25°C for 45 min.  Read A₅₈₅ (the value of which is proportional to the    │
│ amount of boron in the range 1-100 nmol).  Treat another 10-μl sample in  │
│ the same way but using pure H₂SO₄ instead of carmine solution [this allows│
│ correction for any discoloration of carbohydrates in the acid treatment]. │
└─────────────────────────────────────────────────────────────────────────┘
```

   Treatment of **esters** with $NaBH_4$ (or, more effectively, $LiBH_4$) reduces
them to two alcohols, e.g.

$$\text{Galacturonic acid methyl-ester} \xrightarrow{\ \ LiBH_4\ \ } \text{methanol + Galactose.}$$

This reaction converts uronic acid residues to the more easily analysed

neutral hexose residues. Since $LiBH_4$ reduces esters but not free acids

[475], the material may need to be esterified in vitro, e.g. with diazo-

methane [22] or activated with a carbodi-imide [28]. Alternatively, $LiBH_4$-

reducibility gives a way of identifying which uronic acid residues in a

polysaccharide are naturally esterified: in this application, Gal derived

from in-vivo esterified GalA residues must be distinguished from Gal derived from Gal residues [475]. This can be done by subtraction [(Gal obtained after LiBH4 reduction) <u>minus</u> (Gal obtained without reduction)] or by working with polysaccharide labelled <u>in vivo</u> with a [^{3}H]- or [^{14}C]uronic acid (Panel 2.1.1), which is not a precursor for Gal residues.

4.2.7: Periodate oxidation

Periodate oxidation allows the identification of certain sugar—sugar linkages in poly- and oligosaccharides [64,238]. NaIO4 oxidises compounds in which —OH groups are attached to two or more adjacent carbon atoms, and breaks the C—C bond:

$$R\!-\!\!\underset{\underset{OH}{|}}{\overset{\overset{H}{|}}{C}}\!-\!\!\underset{\underset{OH}{|}}{\overset{\overset{H}{|}}{C}}\!-\!R' \quad\rightarrow\quad R\!-\!\!\overset{\overset{H}{|}}{C}\!\!=\!\!O \;+\; O\!\!=\!\!\overset{\overset{H}{|}}{C}\!-\!R'$$

(aldehyde oxidation-products)

This grouping occurs repeatedly on monosaccharides, which are thus extensively oxidised, but in polysaccharides certain —OH groups are protected:

(a) <u>in all sugar units except the reducing terminus</u> there is no free —OH group on C-1 and either C-4 (in furanose rings) or C-5 (in pyranose rings);

(b) <u>in all sugar units except non-reducing termini</u> there is no free —OH group at the point of attachment of another sugar residue to the sugar in question;

(c) <u>in the case of acylated polysaccharides</u> there is no free —OH group at the point of attachment of an ester group (e.g. acetate or ferulate).

Such protection restricts oxidation to certain C atoms, or prevents it altogether, and analysis of the oxidation-products and/or intact sugars obtained after NaIO4-treatment gives information about the polysaccharide's structure. Since the aldehydes produced by NaIO4 are unstable, it is usual to reduce them to the corresponding alcohols with NaBH4 (Section 4.2.6) prior to hydrolysis with trifluoroacetic acid. The products expected of NaIO4 — NaBH4 — TFA treatments are listed in Panels 4.2.7 a and b. Two of the products (glycollaldehyde and glyceraldehyde) are very labile in hot acid; if it is important to assay these, the TFA solution is supplemented with bromine water, which oxidises the aldehydes (including free monosaccharides) to the corresponding acids (glycollic acid, glyceric acid, gluconic acid, etc.), which are much more stable.

PANEL 4.2.7 a: Reaction products expected from the major <u>pyranose</u> residues upon treatment with $NaIO_4$ — $NaBH_4$ — complete hydrolysis.

Sugar	Substi-tuted on position	Products (listed in order, starting with products derived from C-1 of the sugar residue)
D-Gal	0[*]	Glycollaldehyde + Formic acid + Glycerol
	2	D-Glyceraldehyde + Glycerol
	3	D-Galactose
	4	Glycollaldehyde + D-Threitol
D-Glc	0[*]	Glycollaldehyde + Formic acid + Glycerol
	2	D-Glyceraldehyde + Glycerol
	3	D-Glucose
	4	Glycollaldehyde + Erythritol
D-Man	0[*]	Glycollaldehyde + Formic acid + Glycerol
	2	L-Glyceraldehyde + Glycerol
	3	D-Mannose
	4	Glycollaldehyde + Erythritol
L-Gal	0[*]	Glycollaldehyde + Formic acid + Glycerol
	2	L-Glyceraldehyde + Glycerol
	3	L-Galactose
	4	Glycollaldehyde + L-Threitol
D-GalA	0[*]	Glycollaldehyde + Formic acid + L-Glyceric acid
	2	D-Glyceraldehyde + L-Glyceric acid
	3	D-Galacturonic acid
	4	Glycollaldehyde + D-Threonic acid
D-GlcA	0[*]	Glycollaldehyde + Formic acid + L-Glyceric acid
	2	D-Glyceraldehyde + L-Glyceric acid
	3	D-Glucuronic acid
	4	Glycollaldehyde + L-Erythronic acid
L-Ara<u>p</u>	0[*]	Glycollaldehyde + Formic acid + Ethane-1,2-diol
	2	D-Glyceraldehyde + Ethane-1,2-diol
	3	L-Arabinose
	4	Glycollaldehyde + Glycerol
D-Xyl	0[*]	Glycollaldehyde + Formic acid + Ethane-1,2-diol
	2	D-Glyceraldehyde + Ethane-1,2-diol
	3	D-Xylose
	4	Glycollaldehyde + Glycerol
L-Rha	0[*]	Glycollaldehyde + Formic acid + propane-1,2-diol
	2	D-Glyceraldehyde + propan-1,2-diol
	3	L-Rhamnose
	4	Glycollaldehyde + 1-Deoxy-D-erythritol
L-Fuc	0[*]	Glycollaldehyde + Formic acid + propane-1,2-diol
	2	L-Glyceraldehyde + propan-1,2-diol
	3	L-Fucose
	4	Glycollaldehyde + 1-Deoxy-L-threitol

Note that the products listed for pyranose sugars substituted on position no. are also expected from sugars substituted on position(s):

0[*]	6
2	2+6
3	2+3; 2+4; 3+4; 3+6; 2+3+4; 2+3+6; 2+4+6; 3+4+6; 2+3+4+6
4	4+6

[*] Meaning not substituted at all; i.e. a non-reducing terminus.

An interesting potential use of $NaIO_4$-oxidation is to study the structure of rhamnogalacturonan-I: Rha residues bearing Ara- or Gal-rich side chains would be recovered as free rhamnose, and those lacking such side chains would yield propan-1,2-diol (Panel 4.2.7 a). The ratio of these two products would indicate the degree of branching in RG-I.

An extension of $NaIO_4$-treatment known as **Smith degradation** [196] allows study of the distribution of particular bonds within a polysaccharide molecule (Panel 4.2.7 c). Polysaccharides remain of high M_r during $NaIO_4$-oxidation, but linkages between the oxidised residues are exceedingly acid-labile, and can be deliberately broken without affecting the glycosidic linkages of unoxidised sugar residues. Stretches of contiguous $NaIO_4$-resistant residues will be released as intact oligosaccharides (with the reducing terminus glycosidically linked to a single oxidation-product) during the sequence $NaIO_4$ — $NaBH_4$ — very mild acid hydrolysis. Examples of Smith degradation are:

(a) Determination of whether β-(1→3),(1→4)-glucans contain <u>contiguous</u> (1→3)-links. Since attachment of another sugar to C-3 renders glucopyranose residues completely resistant to $NaIO_4$ (Panel 4.2.7 a), any contiguous (1→3)-linked units would survive as an intact oligosaccharide, which could be isolated by chromatography (Sections 4.5.2 and 4.5.6):

..Glc<u>p</u>-(1→4)-Glc<u>p</u>-(1→3)- **Glc<u>p</u>-(1→3)-Glc<u>p</u>-(1→4)**-Glc<u>p</u>-(1→4)-Glc-...

→   **Glc<u>p</u>-(1→3)-Glc<u>p</u>-(1→2)**-<u>Erythritol</u>

                         +   smaller degradation products.

(b) Determination of the number of contiguous branch-points in xylan. In an unbranched xylan [...Xyl<u>p</u>-(1→4)-Xyl<u>p</u>-(1→4)-Xyl<u>p</u>-(1→4)-...] each mid-

---

PANEL 4.2.7 b: Reaction products expected from the major <u>furanose</u> residues upon treatment with $NaIO_4$ — $NaBH_4$ — complete hydrolysis.

Sugar	Substituted on position	Products (listed in order starting with products derived from C-1 of sugar)
L-Ara<u>f</u>	0* or 5 only	Glycollaldehyde + Glycerol
	(2 &/or 3) ± 5	L-Arabinose
D-Gal<u>f</u>	0*	Glycollaldehyde + Glycerol + Formaldehyde
	2 &/or 3	L-Arabinose + Formaldehyde
	5 &/or 6	Glycollaldehyde + D-Threitol
	(2 &/or 3) **and** (5 &/or 6)	D-Galactose

* Meaning not substituted at all; i.e. a non-reducing terminus.

PANEL 4.2.7 c: NaIO4-Oxidation (Smith degradation) of [3H]arabinoxylan to determine the chain-length of contiguous branched xylose residues.[1]

Sample: [3H]Arabinoxylan,[2] e.g. prepared by growth of spinach cells in the presence of [3H]arabinose (Section 2.2; Fig. 2.5), and purified from the medium by differential ethanol precipitation (Panel 3.5.6).

Step 1: Suspend sample (less than 0.2 mg total polysaccharide) in 0.1 ml of reagent [50 mM NaIO4 (sodium metaperiodate) made up in 0.25 M formic acid, pH adjusted to 3.7 with 1 M NaOH]. If sample contains more than 0.2 mg polysaccharide, increase volume of reagent in proportion [also in steps 3-5].

Step 2: Incubate in a capped tube at about 4°C in the dark for 6 days. Shake at intervals if the polysaccharide tends to precipitate.

Step 3: Add 20 μl ethane-1,2-diol, and incubate for a further 1 h to destroy excess NaIO4.

Step 4: Add 0.2 ml 0.5 M NaBH4 / 1 M NaOH; incubate 4-24 h at 25°C.

Step 5: Cool to 0°C. Add 30 μl HOAc to destroy excess NaBH4.

Step 6: De-salt, e.g. by dialysis against H_2O (Panel 3.5.1 a).

Step 7: Add TFA to the high-M_r material to give a final acid concentration of 0.01 M, check the pH is ca. 2.0, and heat at 100°C for 10 min to cleave the linkages involving non-sugar fragments. Dry *in vacuo*.

Controls: Complete acid-hydrolysis (Panel 4.2.1 a) of samples identical to those used in steps 1 and 7. Comparison [e.g. by PC in EtOAc/Py/H_2O (8:2:1)] of the [3H]monosaccharides obtained will indicate whether the NaIO4 gave complete oxidation of Ara residues and what proportion of Xyl residues were protected.

Next steps: Separate the oligomers [(Xyl)$_n$-glycerol] either by PC in EtOAc/HOAc/H_2O (10:5:6) or by GPC on Bio-Gel P-2. Relate the distribution of radioactivity[3] to the chromatographic behaviour of markers [preferably oligoxylans prepared from xylan by partial acid hydrolysis (Panel 9.2.1); alternatively commercially available malto-oligosaccharides]. (Xyl)$_n$-glycerol will generally run between (Xyl)$_n$ and (Xyl)$_{n+1}$.

[1]This method does not distinguish between xylose residues that bear acetyl side chains and those that bear sugar side chains; to investigate specifically those that bear sugar side chains (and hence by difference those that only bear acetyl side chains), the sample is de-acylated with NaOH (Section 4.2.4) before NaIO4 treatment. Hemicellulose samples extracted in NaOH will already have been de-acylated.

[2]The same method can be used for non-radioactive polysaccharide samples, the volumes of reagents being stepped up if necessary (see step 1).

[3]If the sample was non-radioactive, stain the PC with AgNO3 (Panel 4.5.2 f) or assay GPC fractions by the orcinol test (Panel 3.7 a).

chain Xyl residue yields glycollaldehyde + glycerol upon Smith degradation (Panel 4.2.7 a). The presence of a substituent, e.g. Ara, GlcA or acetate, on position 2 and/or 3 of a given Xyl makes that residue NaIO$_4$-resistant. Contiguous substituted Xyl residues are recovered as an oligosaccharide, whose DP (estimated by chromatography) indicates the number of contiguous substituents:

```
...Xyl-Xyl-Xyl-Xyl-Xyl-Xyl-Xyl-Xyl-...
          |   |   |   |
         Ara Ara Ara Ara
```

→   **Xyl-Xyl-Xyl-Xyl**-Glycerol   + smaller degradation products.

4.2.8: Methylation analysis

This is the most powerful technique for investigation of sugar—sugar bonds in non-radioactive polysaccharides. Specialised gas-chromatography/mass-spectrometry equipment is required to derive full benefit from the method, which will therefore not be discussed in detail. For more information, see references [56,342,512,533]. In outline, all free —OH groups in a polysaccharide are methylated (Panel 4.2.8); the methyl groups stay attached during acid hydrolysis, and the resulting methyl-sugars are analysed. A given '—OH' group in a sugar may be protected from methylation by involvement in a pyranose or furanose ring, or by having another sugar glycosidically linked to it. (Methylation removes any acyl groups unless special techniques are used [540].) For instance, cellulose [(1→4)-glucopyranan] yields 2,3,6-tri-O-methyl-glucose from each mid-chain Glc residue, whereas callose [(1→3)-glucopyranan] yields 2,4,6-tri-O-methyl-glucose.

It is not immediately obvious from methylation analysis whether a sugar is in the pyranose or furanose form: for instance 2,3-di-O-methyl-arabinose could arise from either ...(1→4)-Arap-... or ...(1→5)-Araf-.... These two possibilities can be distinguished by mild acid hydrolysis, which preferentially hydrolyses furanosyl linkages (Section 4.2.2). Methylation analysis has the advantage over NaIO$_4$-oxidation that it can look at single —OH groups, whereas NaIO$_4$ can only view them in pairs because it reacts specifically with diol groups.

Radioactive methyl-sugars are more efficiently detected on TLC than on GC, but with poorer resolution; it is best to work with polysaccharides in which only one sugar (say, mannose) is labelled. The TLC system is then 'only' required to separate the different methyl-mannoses, rather than these plus the methyl-glucoses, methyl-galactoses, methyl-xyloses etc. If the polysaccharide of interest has many different residues, **specific** in vivo labelling (Section 2.1.2) will be necessary. Few authentic methyl-sugars suitable for use as markers are available commercially; they generally have to be prepared by methylation of polysaccharides and other sugar derivatives of known structure.

4.2.9: Determination of optical isomerism of a monosaccharide

Enzymic degradation (Section 4.2.5) will indicate whether the sugar units of a polysaccharide are the D- or the L-enantiomer. For example, if pure β-galactosidase releases galactose, this is **D**-galactose because the enzyme is specific for the D-isomer.

Acid hydrolysis and other chemical techniques do not discriminate between D- and L-isomers, which are chromatographically inseparable. To identify a free monosaccharide as D- or L-, it is treated with an enzyme for which it is the specific substrate. Thus if free glucose is oxidisable to gluconate by glucose oxidase, it was **D**-glucose. Oxidation is monitored by a change in stainability (gluconate is undetectable with the aniline hydrogen phthalate stain for sugars: Panel 4.5.2 f) or by a shift in R_F (see Section 4.5). One source of useful enzymes is the mung bean seedling, which contains a variety of sugar kinases. A crude extract (Panel 5.4.1) of the seedlings is mixed with ATP and the unknown sugar, and the products are examined by chromatography or electrophoresis. A substrate sugar will be converted to the corresponding sugar-phosphate, which has a very different R_F or electrophoretic mobility. Mung bean extract [155] contains active kinases that phosphorylate:

D-glucose,	but not L-;
D-galactose,	but not L-;
D-glucuronic acid,	but not L-;
D-galacturonic acid,	but not L-;
L-arabinose	but not D-.

Other suitable enzymes include the sugar dehydrogenases [42]. Many enzymes active on monosaccharides are available commercially.

A non-enzymic alternative is to convert the free monosaccharide into its D-(+)-2-octyl-glycosides [192]. [Monosaccharide (e.g. 1 mg) is treated with 0.5 ml D-(+)-octan-2-ol (plus a drop of neat TFA) at 130°C in a sealed

PANEL 4.2.8: Methylation of polysaccharides[1]
           [Care! wear gloves and work in a fume cupboard.]

Sample: 0.5 mg of polysaccharide, reduced with $NaBH_4$[2] and freed
from cations on acid-washed Dowex-50 (see Panel 4.2.6).

Step 1:   [Prep of methylsulphinyl carbanion] Place 1 g NaH (bought
as a suspension in oil) in a dry, 200-ml round-bottomed flask and, while
flushing the flask with dry N_2, wash the NaH 3 times in pentane. Continue
flushing to evaporate the pentane.   Still flushing, add 10 ml **dry** DMSO
('silylation grade', or previously dried by distillation from CaH_2; stored
over 0.4-nm molecular sieves) and stir at 50°C for about 45 min [until H_2
evolution stops: $(CH_3)_2SO + NaH \rightarrow CH_3SOCH_2^-Na^+ + H_2$]. Estimate carbanion
concentration (should be ca. 4 M) by diluting 0.5 ml into 5 ml water and
titrating to pH 7 with 0.1 M HCl. Store the rest of the carbanion solution
under N_2, frozen in small, tightly-sealed containers at about 4°C.

Step 2:   Dry the polysaccharide in a screw-necked tube (containing a
magnetic stirrer bar) in vacuo over P_2O_5 at 40°C overnight.

Step 3:   Add 1 ml **dry** DMSO (see step 1), flush the tube with dry N_2,
screw down the cap tightly, and stir to dissolve the polysaccharide. If
necessary, place the tube in a sonicating water bath to speed dispersion.

Step 4:   With a dry syringe, add enough carbanion solution (ca. 45
µl) to make 0.17 M. Flush with N_2, cap tightly and stir for 2 h at 25°C.

Step 5:   Transfer 5 µl of the solution on to a crystal of triphenyl-
methane: a red colour indicates that unreacted carbanion remains, and it
is safe to continue. [If not, repeat step 4.]

Step 6:   Add 12 µl of methyl iodide[3] (MeI) [slowly, taking 10-15
s, and with cooling]. Stir for 1 h at 25°C.

Step 7:   Add 0.35 mmol carbanion and stir 2 h.

Step 8:   Add 25 µl MeI, and stir 1 h.

Step 9:   Add 0.52 mmol carbanion and stir 2 h.

Step 10:   Add 250 µl MeI and stir 1-16 h.

Step 11:   De-salt the methylated polysaccharide on a 15-ml bed volume
column of Sephadex LH-20 with $CHCl_3$/methanol (1:1) as eluent, and collect
the carbohydrate-rich material [e.g. detected by the anthrone reaction
(Panel 3.7 a) or detection of radioactivity] eluting at V_0. Dry in vacuo.

Next steps: Acid hydrolysis (e.g. in 2 M trifluoroacetic acid at 120°C
for 1 h); chromatographic analysis of the methyl-sugars by TLC or (after
reduction and acetylation: see Panel 4.5.8) by GC.

[1]See [344] for more details and [533] for optimisation for micro-analysis.

[2]Note: if the polysaccharide contains uronic acids which are of interest,
these should be de-esterified by treatment with Na_2CO_3 (0.5 M at 0°C for 16
h, followed by 25°C for 3 h).   Also, replace steps 6-10 by a single
addition of 100 µl MeI followed by stirring for 2-16 h.   This is to
minimise elimination reactions (Section 4.2.4).

[3]The concentration of pure MeI is about 16 M.

tube for 16 h; the solution is then dried. Four products are obtained
(α/β, pyranose/furanose), but often one or two of these predominate(s).]
Since the aglycone itself is optically-active, the major glycoside obtained
from a D-sugar is <u>not</u> the mirror image of that obtained from the
corresponding L-sugar; they are different compounds [e.g. D-(+)-2-octyl-D-
galactoside and D-(+)-2-octyl-L-galactoside], which are separable chromato-
graphically. Separation has generally been by GC [192], but the use of TLC
[e.g. on silica-gel in EtOAc/acetone (2:1)] or HPLC are feasible
alternatives. Authentic standards are synthesised by treatment of
commercial D- and L-sugars with D-(+)-octan-2-ol under identical conditions
and are run as markers, preferably internal (see Section 4.5.2) if the
sample is radioactive.

4.3: ANALYTICAL DEGRADATION OF GLYCOPROTEINS

The carbohydrate moieties of glycoproteins can be studied as for
polysaccharides (Section 4.2), except that, since linkage to the polypeptide backbone blocks the reducing terminus of the carbohydrate, no reaction
occurs with $NaBH_4$ (Section 4.2.6) and alkaline 'peeling' (Section 4.2.4)
cannot take place unless the alkali splits the sugar—amino acid bond.

Chopping of glycoproteins into workable fragments can be approached
in two ways: (a) splitting the sugar—amino acid bond so as to release
pure carbohydrate, and (b) splitting the polypeptide backbone under
conditions which do not affect glycosidic bonds. The various bonds have
the following properties (+ = susceptible; - = resistant) [308]:

| Type of bond | Aqueous alkalies | | Hot aqueous acids | | Hydra- | Protease |
	cool	hot	2 M TFA	6 M HCl	zinolysis	
Sugar—Sugar	-	-	+	+(*)	-	-
Sugar—Hyp	-	-	+	+(*)	-	-
Sugar—Asn	-	+	+	+(*)	+	-
Sugar—Ser	+	+	+	+(*)	+	-
Sugar—Thr	+	+	+	+(*)	+	-
Amino acid— Amino acid	-	+	-/+	+	+	+

(*With degradation of the sugar.)

4.3.1: Splitting the sugar—amino acid bond

Carbohydrate chains attached to **serine** or **threonine** residues can be
removed by alkali-catalysed β-elimination because the amino acid has a

142

neutral C=O group β to the glycosyl linkage (see Section 4.2.4). The products are a free carbohydrate and an unsaturated amino acid residue. Carbohydrates are alkali-labile, and unsaturated amino acids are difficult to analyse: both problems are solved by inclusion of NaBH$_4$ in the reaction mixture since the reducing terminal sugar is then converted to a stable alditol (Section 4.2.6) and the unsaturated amino acid is reduced to alanine (from Ser) or 2-amino-butyric acid (from Thr). Typical reaction conditions are 50 mM NaOH / 1 M NaBH$_4$ at 45°C for 16 h [158]. Reduction of the unsaturated amino acids is aided by addition of PdCl$_2$ as catalyst [201]. β-Elimination occurs at different rates in different glycoproteins; the Gal—Ser bond of extensin is particularly unreactive [308].

Sugars N-glycosidically attached to **asparagine** are removed by hydrolysis in hot alkali (e.g. 1 M NaOH / 1 M NaBH$_4$ at 100°C for 10 h [317] or by hydrazinolysis [359,487]. The GlcNAc—Asn bond is specifically hydrolysed by certain endo-N-acetylglucosaminidases [359]. There is no known way of splitting sugar—**hydroxyproline** bonds without also breaking glycosidic linkages within the carbohydrate moieties themselves.

Complete de-glycosylation of glycoproteins, leaving a simple poly-peptide backbone suitable for peptide mapping, can be achieved with anhydrous HF or HF in pyridine [435] (caution — HF dissolves glass) or with trifluoromethanesulphonic acid [129,462].

4.3.2: Splitting the polypeptide backbone

Glycopeptide fragments can be generated by three methods. [Acid hydrolysis is little use because peptide bonds are much more stable to acid than are glycosidic bonds.]

(a) Hot alkali splits peptide bonds, but neither sugar—sugar bonds nor sugar—hydroxyproline bonds. It can therefore yield intact fragments of the type (sugar)$_n$—Hyp, the best known example being [309]:

L-Araf-α-(1→3)-L-Araf-β-(1→2)-L-Araf-β-(1→2)-L-Araf-β-(1→4)-L-Hyp,

which is obtained from extensin (Panel 4.3.2). Another example is in the demonstration of a sugar—amino acid linkage in an arabinogalactan-protein (AGP). This was achieved by 'pruning' the polysaccharide by mild acid hydrolysis, and then subjecting the residual core glycoprotein to hot alkali: the products included D-Galp-α-(1→4)-L-Hyp [483].

(b) Hydrazinolysis [359,487] cleaves peptide bonds much more rapidly than sugar—sugar bonds. Specificity may be improved by addition of hydrazine sulphate [183], but this has been questioned [487]. Hydrazinolysis preserves sugar—Hyp linkages, but the sugar—Asn bond is cleaved, and sugar—Ser or sugar—Thr bonds may be β-eliminated.

(c) Enzymic hydrolysis can be absolutely specific for the polypeptide backbone. Commercial enzymes that have been used include trypsin, chymotrypsin, pepsin, collagenase, papain, subtilisin and, especially, **Pronase** [467,502]. Samples of 'Pronase E' from Sigma, analysed in the author's laboratory, had slight arabinosidase activity as a contaminant; such non-protease activities can be minimised by pre-incubation of the Pronase alone, before use. Each batch of protease used on a glycoprotein substrate should be checked to see if it is releasing carbohydrate units.

PANEL 4.3.2: **Isolation of hydroxyproline oligoarabinosides by alkaline hydrolysis** [see 309]

Sample: 10 mg cell walls from suspension-cultured sycamore cells.

Step 1: Suspend the walls (10 mg) in 1 ml of saturated (ca. 0.18 M) $Ba(OH)_2$. [$Ba(OH)_2$ must be stored in an airtight bottle to exclude CO_2.]

Step 2: Flush the tube with N_2 (if possible), and seal e.g. by use of a tightly-fitting, Teflon-lined, screw cap.

Step 3: Incubate in an oven at 110°C for 5-10 h.

Step 4: Cool. Add 25 µl 1% bromothymol blue. Transfer the suspension into a small beaker [to give a shallow layer], and place this in a desiccator containing CO_2 [e.g. formed from a mixture of Na_2CO_3 and HCl]. Swirl the desiccator gently until bromothymol blue goes (and stays) yellow.

Step 5: Freeze. Thaw.

Step 6: Centrifuge (10 min at 2,500 **g**) to pellet the cellulosic residue + $BaCO_3$. Harvest the clear supernatant.

Step 7: Resuspend the pellet in 1 ml H_2O, and repeat the centrifugation. Remove the new supernatant and add to the old.

Step 8a either: Dry in vacuo and redissolve residue in 0.1 ml H_2O.
Step 8b[1] or: Acidify the solution with 0.1 ml 0.1 M H_2SO_4, and pass through a 1-ml bed volume column of acid-washed, water-rinsed, Dowex-50 (in a Pasteur pipette plugged with glass wool). Wash with 3 x 1 ml H_2O, rejecting the washings, and then elute the amino acids with 2 ml of 2 M NH_3. Then proceed as in step 8a.

Next steps: Analysis, e.g. by electrophoresis at pH 2 (Section 4.5.4).

[1]If the amount of Hyp-Ara_n is very small, step 8b is recommended as it removes carbohydrates (and their degradation products) and allows larger amounts of amino acids and to be applied to the electrophoretogram.

Extensin is exceptionally resistant to proteases, but can be rendered more susceptible by prior removal of the Araf residues by mild acid hydrolysis [308]; the Galp residues remain attached and could presumably be located on specific oligopeptides.

4.3.3: Analysis of the amino acids

The amino acid composition of a glycoprotein can be determined after complete hydrolysis (Panel 4.3.3 a). In accurate work, allowance is made for degradation of amino acids (especially serine and threonine) during hydrolysis. Tryptophan is lost during acid hydrolysis, and Asn and Gln are hydrolysed to Asp and Glu respectively. Sugars are largely destroyed by 6 M HCl and thus cannot be analysed in the same sample. Acid hydrolysis of glycoproteins in the presence of a large excess of polysaccharide results in a brown or black product. This is partly caused by reactions between sugars and amino acids and could be a cause of low recovery of the latter. Recovery can be estimated by addition of traces of ^{14}C-labelled amino acids as internal standards; however, it is generally above 90 % even with hydrolysates of whole cell walls which undergo severe browning.

A second method is base hydrolysis, as described in Panel 4.3.2. $Ba(OH)_2$ is a useful base for this purpose as it is easily removed prior to

---

PANEL 4.3.3 a: Acid hydrolysis of wall protein for amino acid analysis

Sample: Capsicum fruit cell walls, washed in PAW (Panel 1.2.1 c) [1]

Step 1:   Suspend the dry walls at not more than 4 mg/ml in 6 M HCl containing 10 mM phenol. Seal the tube (preferably under N_2), either by melting and drawing out the neck or by means of a well-fitting Teflon-lined screw cap.

Step 2:   Incubate at 110°C for ca. 20 h.

Step 3:   Cool. Open the tube and centrifuge down the insoluble black material (2,500 g, 5 min). Dry the supernatant in vacuo, preferably with a KOH trap. Re-dissolve in H_2O (0.1 ml) and re-dry.

Step 4:   Re-dissolve the dried material. This is sometimes difficult, but can usually be achieved with 1 M NH_3.

Next step: Separate the amino acids by chromatography (Sections 4.5.2, 4.5.3, 4.5.7; see also Panel 4.3.3 b) and/or electrophoresis (Section 4.5.4). Electrophoresis is particularly suitable for hydroxyproline.

[1]The same technique is suitable for purified proteins; they should be thoroughly de-salted before hydrolysis.

chromatography. Base hydrolysis gives better recoveries of tryptophan, but Asn and Gln are still converted to Asp and Glu respectively.

Amino acids can be analysed by paper electrophoresis, and various forms of chromatography (PC, TLC, HPLC) — see Section 4.5. For HPLC, the amino acids are usually derivatised prior to analysis: this leads to better resolution, and allows the incorporation of a chromophore into the molecules. A simple and effective derivatisation reagent is phenylisothiocyanate (PITC) (Panel 4.3.3 b).

PANEL 4.3.3 b: Derivatisation of amino acids with PITC [106]

Sample: A hydrolysate of glycoprotein or cell wall (e.g. Panel 4.3.3 a, omitting step 4).

Step 1: Dry the sample thoroughly. Re-dry in vacuo from 10 μl of soln **A** [EtOH/H_2O/triethylamine, 2:2:1].

Step 2: Add 20 μl of freshly prepared soln **B** [EtOH/H_2O/triethylamine/phenylisothiocyanate (7:1:1:1)]. Mix; incubate at 25°C for 20 min.

Step 3: Dry in vacuo. Store samples at -20°C.

Next step: HPLC (see Section 4.5.7).

4.4: ANALYTICAL DEGRADATION OF WALL PHENOLICS

4.4.1: Alkaline hydrolysis of esters of phenolic acids

The ester linkage between ferulic acid (or a related phenolic acid) and a polysaccharide is readily hydrolysed by cold NaOH, releasing the sodium salt, e.g. sodium ferulate [168,170]. If necessary, the ferulate can be separated from the Na^+ by acidification (forming uncharged ferulic acid) and partitioning into an immiscible organic solvent e.g. butan-1-ol (Panel 4.4.1). However, this is a potential point at which losses in yield may be incurred, and it may be unnecessary if the volume of NaOH can be kept low: up to 25 μl of 0.1 M NaOH, after acidification with 1 μl acetic acid, can be loaded in toto as a spot on a standard thickness silica-gel TLC plate and run successfully in benzene/acetic acid (9:1). Although the isolation of a phenolic compound by NaOH-treatment is generally taken as evidence for an ester bond, care is required since certain ether bonds (e.g. Fig. 3.3.2 b) are unusually labile both to alkali and to hot acid [146,540].

146

```
┌─────────────────────────────────────────────────────────────────────────┐
│ PANEL 4.4.1:   Release of phenolic acids from walls by alkaline hydrolysis│
│                                                                           │
│ Sample:         Cell walls (10 mg dry weight) isolated from wheat         │
│                 coleoptiles (see Panel 1.2.1 a).                          │
│                                                                           │
│ Step 1:         To 10 mg cell walls, add 1 ml of 0.1 M NaOH.  Cap the tube,│
│ preferably under N₂, and incubate at 25°C in the dark for 1 h.            │
│                                                                           │
│ Step 2:         Add 0.1 ml 2 M trifluoroacetic acid (TFA) (brings the pH  │
│ down to about 1) followed by 1 ml butan-1-ol or ethyl acetate.(1)         │
│                                                                           │
│ Step 3:         Cap the tube and shake vigorously.  Separate the two phases│
│ by centrifugation (10 min at 2,500 g).   Withdraw and keep the upper      │
│ (organic) phase, taking great care not to get any of the lower phase.     │
│                                                                           │
│ Step 4:         To the aqueous phase, add a further 1 ml BuOH or EtOAc.    │
│                                                                           │
│ Step 5:         Repeat step 3.                                            │
│                                                                           │
│ Step 6:         Combine the two BuOH/EtOAc solutions;  add 1 ml 0.01 M TFA.│
│                                                                           │
│ Step 7:         Repeat step 3.  Read the A₃₂₀ for a rough estimate of the │
│ ferulic acid content of the sample.                                       │
│                                                                           │
│ Step 8:         Dry BuOH or EtOAc in vacuo or under a stream of N₂.  Re-  │
│ dissolve resiude in acetone and store cold and dark.                      │
│                                                                           │
│ Next steps:     TLC on silica-gel in Bz/HOAc (9:1) (Section 4.5.3) or HPLC│
│ on C₈ or C₁₈ reversed-phase silica (Section 4.5.6) to resolve the phenolic│
│ acids (ferulic, p-coumaric, p-hydroxybenzoic, diferulic and unknowns).    │
│                                                                           │
│ (1)Ethyl acetate evaporates quicker than butanol in step 8, but is not such│
│ a good solvent for diferulic acid.                                        │
└─────────────────────────────────────────────────────────────────────────┘
```

Step 1 restated with LaTeX: incubate at $25°C$; Step 2 footnote $^{(1)}$; Step 7 Read the A_{320}; Next steps C_8 or C_{18}.

4.4.2: Isolation of feruloyl-oligosaccharides from feruloyl-polysaccharides

Characterisation of the ferulate—polysaccharide linkage requires isolation of a small fragment still bearing this linkage intact. The most generally useful method of obtaining such fragments is enzymic hydrolysis with a mixture of polysaccharide-hydrolases (e.g. Driselase) which lacks esterase activity (Panel 4.4.2 a). The various endo- and exo-hydrolases of Driselase will cleave wall polysaccharides to monosaccharides [and one major disaccharide (Section 4.2.5)], but if a sugar residue is feruloylated its glycosidic linkage is protected from Driselase [168]. Thus

sugar-sugar-sugar-sugar-**sugar-sugar**-sugar-sugar-sugar-sugar-sugar-...
                          |
                      ferulate

can be converted by Driselase to

                    sugar-sugar
                        | + n free sugars.
                    ferulate

The glycosyl linkage to the 'right' of the ferulate group cannot be hydrolysed by Driselase, and a fluorescent feruloyl-disaccharide is thus
recovered which is easily separated from the other sugars by chromatography. In grass arabinoxylans, a single feruloyl group can protect <u>two</u>
glycosyl linkages [3,214,282]:

```
Xyl-Xyl-Xyl-Xyl-Xyl-Xyl-Xyl              5 Xyl  +  2 Ara  +  Xyl-Xyl
 |   |       |                                                 |
Ara Ara     Ara                 →                             Ara
            |                                                  |
          Ferulate                                          Ferulate.
```

[The reason for this is probably as follows. Driselase first removes non-
feruloylated Ara side chains by arabinosidase action, and then cleaves the
xylan backbone by xylanase and β-xylosidase. Arabinosidase cannot act on
feruloylated Ara residues, and the xylanase and xylosidase cannot act on
arabinosylated xylose residues.]

The structures of feruloyl-oligosaccharides give information not only
on the ferulate—sugar linkage, but also on the adjacent sugar—sugar
linkage, helping to establish the nature of the polysaccharide involved.
If it is desired to investigate the sugar residues located further away
from the ferulate, a method is required of causing the Driselase to 'stall'
at linkages which it would normally attack. One approach, used on
feruloyl-pectin from spinach, is to include an enzyme-inhibitor: 40 mM D-
galactose inhibits the galactanase activity of Driselase and favours the
isolation of relatively long feruloyl-oligosaccharides which include part
of the GalA-rich backbone rather than just the Gal-rich side chains [170],
as shown in the following interpretative diagram [→,←,↓ = glycosidic bonds].

Ferulate-Gal↓Gal (+ non-feruloylated products)

Driselase alone

```
                            GalA←GalA←GalA←GalA←GalA←GalA←GalA←GalA←...
                              ↓
Ferulate-Gal↓Gal↓Gal→Gal→Gal→Rha
                              ↓
                            GalA→GalA→GalA→GalA→GalA→GalA→GalA→GalA→-...
```

Driselase +
galactose

Ferulate-Gal↓Gal↓Gal→Gal→Gal→Rha   (+ non-feruloylated products)
```
                              ↓
                            GalA
```


148

```
┌─────────────────────────────────────────────────────────────────────────┐
│ PANEL 4.4.2 a:   Enzymic isolation of feruloyl-disaccharides from spinach │
│ cell walls                                                                │
│                                                                           │
│ Sample:        Cell walls (20 mg) of spinach callus (Panel 1.2.1 c)[1]    │
│                                                                           │
│       (A) Partial purification of Driselase [168][2]                      │
│ Step 1:         Stir 5 g Driselase for 15 min in 50 ml of buffer A (50 mM │
│ acetic acid, adjusted to pH 5 with 1 M NaOH) to dissolve the enzyme.      │
│ Step 2:         Centrifuge (10 min, 2,500 g) and collect the supernatant. │
│ Step 3:         Add 26 g solid (NH4)2SO4 per 50 ml of supernatant, with   │
│ constant stirring.  When crystals have dissolved, stand at 0°C for 15 min.│
│ Step 4:         Centrifuge (10 min at 2,500 g) and reject the supernatant.│
│ Step 5:         Resuspend pellet in 50 ml of fresh 52 % (w/v) (NH4)2SO4,  │
│ and repeat step 4.                                                        │
│ Step 6:         Dissolve pellet in 5 ml H2O, and de-salt on Sephadex G-25 │
│ (Panel 3.5.1 b).  Collect the protein (fastest-eluting brown material), and│
│ freeze-dry [yield ca. 0.5 g].  Store the dry powder below 0°C.            │
│                                                                           │
│       (B) Isolation of feruloyl-oligosaccharides                          │
│ Step 7:         To 20 mg cell walls, add 1 ml 0.5 % purified Driselase in │
│ buffer B (1 % pyridine, 1 % acetic acid, 0.05 % chlorbutol, pH 4.7).  Incu-│
│ bate 1-40 h at 25°C, preferably with shaking.  Fer-Ara2 is released in 1 h;│
│ Fer-Gal2 takes longer.  Run an enzyme-only control without the cell walls.│
│ Step 8:         Stop the digestion either by drying the sample on to      │
│ chromatography paper or by freezing.                                      │
│                                                                           │
│ Next steps:     Paper chromatography in BuOH/HOAc/H2O (12:3:5) (Panel 4.5.2│
│ a) to separate the feruloyl-oligosaccharides.  [For preparative work, use │
│ column chromatography on Bio-Gel P-2 in Buffer B:  Fer-Ara2 elutes at kav.│
│ = 1.4, i.e. later than monosaccharides, and is thereby greatly purified.] │
│                                                                           │
│ [1]Similar results are obtained with walls from spinach and beet plants.  │
│ With grass cell walls, the products are Fer-Ara-Xyl and Fer-Ara-Xyl2 [3]. │
│ [2]Driselase is incompletely soluble and contains phenolic contaminants.  │
│ Purification can be omitted, but 5 times the enzyme concentration is then  │
│ required, and it is imperative to run Driselase-only controls so as to     │
│ distinguish between Driselase- and spinach-derived phenolics.              │
└─────────────────────────────────────────────────────────────────────────┘
```

An alternative to enzymic methods for the preparation of feruloyl-oligosaccharides is the use of mild acid hydrolysis. Feruloyl—sugar bonds have an acid-stability intermediate between pyranose and furanose sugar—sugar linkages (Sections 4.2.1-2). Thus treatment with mild acid will hydrolyse most of the arabinofuranose linkages but few feruloyl ester bonds and very few sugar pyranose bonds. In the case of spinach feruloyl-pectin the major reaction [170] is:

$$\text{Feruloyl-Arap-Araf-}\underline{\text{Araf}}\text{-Araf-Araf-Araf-}\ldots$$
$$\rightarrow \quad \text{Feruloyl-Arap-Ara} \quad + \quad \underline{n}\ \text{Ara.}$$

If still milder acid is used (Panel 4.4.2 b), even furanose bonds are only partially hydrolysed, and a series of larger feruloyl-oligosaccharides is obtained:

Feruloyl-Arap-Ara,

Feruloyl-Arap-Araf-Ara,

Feruloyl-Arap-Araf-Araf-Ara,

Feruloyl-Arap-Araf-Araf-Araf-Ara,

Feruloyl-Arap-Araf-Araf-Araf-Araf-Ara   etc.,

which can be separated by paper chromatography [170]. In grass feruloyl-arabinoxylans, mild acid releases a feruloyl-monosaccharide [3]:

```
Xylp-Xylp-Xylp-Xylp-Xylp-Xylp-Xylp-...
 |         |         |
Araf      Araf      Araf
                     |
                   Ferulate
```

$$\rightarrow \quad \text{Xylp-Xylp-Xylp-Xylp-Xylp-Xylp-Xylp-}\ldots \quad + \quad 2\ \text{Ara} \quad + \quad \begin{array}{c}\textbf{Ara}\\|\\\textbf{Ferulate.}\end{array}$$

PANEL 4.4.2 b: **Preparation of feruloyl-oligosaccharides by partial acid hydrolysis of feruloyl-polysaccharides**

Sample: Cell walls of spinach, beet, or grasses (Panel 1.2.1 a).

Step 1:   To three replicate 10-mg samples of dried cell walls, add 1 ml of oxalic acid solution of 8, 16 and 32 mM.

Step 2:   Cap or seal the tubes and incubate at 100°C for 1.5 h.

Step 3:   Cool. Add about 50 mg finely powdered $BaCO_3$ and shake till neutral. Freeze. Thaw. Centrifuge (10 min at 2,500 **g**).

Step 4:   Dry the supernatant (e.g. in a Speed-Vac), re-dissolve in 50 µl H_2O, and load as a spot on to Whatman no. 1 chromatography paper. Dry. Run overnight in BuOH/HOAc/H_2O (12:3:5) (Panel 4.5.2 a).

Step 5:   Dry the paper and examine under UV light, in the presence of NH_3 vapour. For typical results see Fig. 4.4.2.

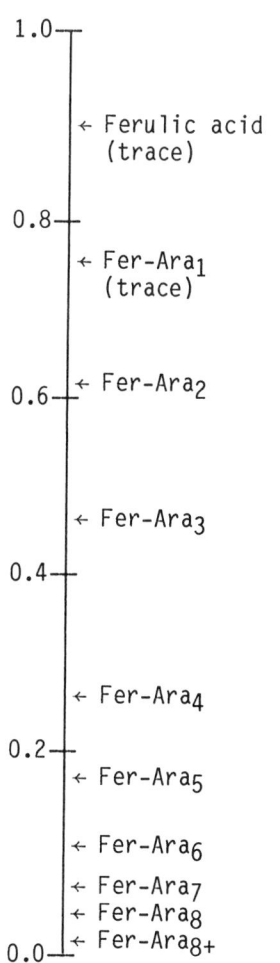

R_F

1.0 —

← Ferulic acid (trace)

0.8 —

← Fer-Ara$_1$ (trace)

0.6 — ← Fer-Ara$_2$

← Fer-Ara$_3$

0.4 —

← Fer-Ara$_4$

0.2 —

← Fer-Ara$_5$

← Fer-Ara$_6$

← Fer-Ara$_7$

← Fer-Ara$_8$

0.0 — ← Fer-Ara$_{8+}$

FIG. 4.4.2: Fluorescent products of the partial acid hydrolysis of feruloyl polysaccharides from spinach. The products were separated by paper chromatography in BuOH/HOAc/H$_2$O (12:3:5), and examined under long-wavelength UV light while the paper was being fumed with NH$_3$ vapour [170].

4.4.3: Analysis of lignin

The traditional way to characterise a lignin is by oxidation in alkaline nitrobenzene [95,452,482]. This converts some of the aromatic units of lignin into 4-hydroxybenzaldehyde, 4-hydroxy-3-methoxybenzaldehyde and 4-hydroxy-3,5-dimethoxybenzaldehyde, which can be separated as their 2,4-dinitrophenylhydrazones by TLC on silica-gel in EtOAc/ligroin (boiling point 75°-120°) (1:2) [431] and quantified spectrophotometrically after elution from the silica-gel. The ratio of these three compounds is a guide to the relative contribution to lignin synthesis of the three major precursors (Sections 3.3.1 and 4.1.8). The method has two main limitations: **(a)** the same three aldehydes are obtained from phenolics other than lignin, e.g. ferulic acid, cutin and suberin; and **(b)** the aldehydes arise only from the 'non-condensed' units of lignin, i.e. those in which the benzene ring is not directly linked to another.

Other methods of lignin analysis have also been proposed, including thioacidolysis (which is more specific for lignin, but still does not detect 'condensed' units) [312] and permanganate oxidation (which may be difficult to perform but does allow the investigation of 'condensed' units) [148].

4.4.4: Analysis of cutin and suberin

Methods potentially suitable for the little-known aromatic units of cutin and suberin are outlined in Sections 4.4.1 and 4.4.3, though the hydrophobicity of cutin and suberin makes methanolic NaOMe (methoxide) or Ba(OMe)$_2$ preferable to aqueous NaOH for splitting ester bonds.

The aliphatic units can be de-polymerised by methanolysis with NaOMe, or by hydrogenolysis with $LiAlH_4$ — in both cases after exhaustive extraction of material soluble in $CHCl_3$, methanol and H_2O (used sequentially, and preferably under reflux). For methanolysis, the polymer is heated in 0.5 M NaOMe in dry methanol containing 6 % pure, dry methyl acetate in a sealed tube at 80°C for 18 h. The reaction is:

$$R-\!\!-C\!\!=\!\!O \ + \ CH_3OH \ \rightarrow \ R-\!\!-C\!\!=\!\!O \ + \ HO-\!\!-R'.$$
$$\quad\quad | \quad\quad\quad\quad\quad\quad\quad\quad\quad | $$
$$\quad\quad O-\!\!-R' \quad\quad\quad\quad\quad\quad O-\!\!-CH_3$$

The solution is then neutralised with the calculated quantity of acetic acid, and the hydroxy-fatty acid methyl esters are extracted into diethyl ether [251], and separated either by silica-gel TLC (Panel 4.4.4) or by GC.

4.4.5: Analysis of tannins [234]

Hydrolysable tannins can be hydrolysed with dilute NaOH (Panel 4.4.1) to yield gallic and/or ellagic acids, which are separated by chromatography (e.g. Panel 4.5.2 d). As these acids are labile in alkali, some loss is inevitable, but to minimise this the hydrolysis is performed under N_2. Ellagic acid has a low solubility, but can be dissolved in pyridine.

Proanthocyanidins can be degraded to anthocyanins by a treatment with

---

PANEL 4.4.4: R_F-Values of methyl esters of common cutin acids on silica-gel TLC in $CHCl_3$/EtOAc (7:3) [from ref. 253]

	R_F
Hexadecanoic acid (= Palmitic acid)	0.70
16-Hydroxy-hexadecanoic acid	0.45
7-Hydroxy-hexadecane-1,16-dioic acid	0.50
8-Hydroxy-hexadecane-1,16-dioic acid	0.50
16-Hydroxy-9-oxo-hexadecanoic acid	0.34
16-Hydroxy-10-oxo-hexadecanoic acid	0.34
9,16-Dihydroxyhexadecanoic acid	0.15
10,16-Dihydroxyhexadecanoic acid	0.15
18-Hydroxy-octadec-9-enoic acid	0.47
9,18-Dihydroxy-octadecanoic acid	0.24
10,18-Dihydroxy-octadecanoic acid	0.24
9,18-Dihydroxy-10-methoxy-octadecanoic acid*	0.19
10,18-Dihydroxy-9-methoxy-octadecanoic acid*	0.19
9,18-Dihydroxy-10-methoxy-octadec-12-enoic acid*	0.19
10,18-Dihydroxy-9-methoxy-octadec-12-enoic acid*	0.19
9,10,18-Trihydroxy-octadecanoic acid	0.07
9,10,18-Trihydroxy-octadec-12-enoic acid	0.07

*The 9-hydroxy-10-methoxy and 9-methoxy-10-hydroxy groupings arise from methanolysis of a 9,10-epoxide (see Section 4.1.8).

butan-1-ol/conc HCl (19:1) at 95°C for 1 h in the dark in the presence of O_2 (= in an open tube). The anthocyanins are assayed by their colour (e.g. $\underline{A}_{550}$); the value is corrected for the background brown colour which often contributes to $\underline{A}_{550}$ in samples containing much non-tannin material.

4.5: SEPARATION AND IDENTIFICATION OF DEGRADATION PRODUCTS

4.5.1: Removal of degradation reagents

The techniques of Sections 4.2-4 convert complex polymers into degradation-products small enough for detailed characterisation. These products are often contaminated with the reagent used for degradation, and are too small to be separated from this reagent by dialysis. The present section outlines methods for their removal, which may be necessary for subsequent analysis.

Alternatives to the methods given below include GPC on Bio-Gel P-2, which is suitable for all but the smallest degradation-products, and the simple option of not removing the reagent. The latter approach is possible with very small (e.g. radioactive) samples where reagent volumes involved are very small, and when the analytical technique to be used will accommodate relatively large amounts of impurities in the sample (e.g. paper chromatography).

(a) Volatile substances are removed by drying, either in vacuo or under a stream of N_2 or filtered air. Numerous small samples (less than 5 ml) are conveniently dried in a Savant 'Speed-Vac', or similar device, which centrifuges the samples at the same time as submitting them to a vacuum (with optional heating): solvent rapidly boils away, but centrifugation stops the solution splashing out of the tube. A 'Speed-Vac' can dry forty 1-ml samples at a time (more smaller samples, fewer larger), taking ca. 2 h for aqueous samples or ½ h for ethanolic samples. This type of drying has the advantage over rotary evaporation and freeze-drying that the sample is concentrated to a small spot at the bottom of the tube.

Substances readily removed by drying include:
trifluoroacetic acid, acetic acid, formic acid, hydrochloric acid, acetic anhydride,
pyridine, ammonia, hydrazine (with re-drying from toluene),
pyridinium acetate or formate (but not pyridinium trifluoroacetate),
ethyl acetate, methanol, ethanol, butanol, phenol (slowly).

(b) Non-volatile alkalies. Barium methoxide and $Ba(OH)_2$ are removed from solution by addition of CO_2 (Panel 4.3.2), forming insoluble $BaCO_3$. It should be established that the compounds of interest do not co-precipitate with the $BaCO_3$. Sodium methoxide, NaOH and Na_2CO_3 cannot easily be precipitated. They are removed from solution by passage through a short column of Dowex-50 (pre-washed with 1 M HCl followed by copious volumes of methanol or water; allow about 2 ml of Dowex per mmol of alkali). This method cannot be used for purification of compounds that themselves have a positive charge (e.g. amino acids), as these also bind to the column. It should be established that the compounds of interest do not precipitate upon neutralisation, as they may then also be lost on the column.

(c) Borates and borohydrides (after conversion of the latter to borates by mild acidification with acetic acid) are freed from Na^+ and other cations on Dowex-50 (Panel 4.2.6), and the resulting boric acid is then removed as its volatile methyl ester by repeated drying from methanol/acetic acid (9:1).

(d) Non-volatile acids can be rapidly removed by passage through a short column of Dowex-1 (previously washed in 2 M NaOH followed by large volumes of water; allow 2 ml of Dowex per mmol of H^+). However, this method is not recommended for reducing sugars e.g. monosaccharides because these can be epimerised on anion-exchange columns. A better method, suitable for oxalic acid and H_2SO_4, is neutralisation with $Ba(OH)_2$, the barium salts of both these acids being insoluble (Panel 4.2.1 b).

(e) Periodate, if it cannot be removed by dialysis because the M_r of the compound of interest is too low, can be removed by precipitation as its barium salt. Alternatively, if the periodate-oxidised sample has been reduced with alkaline $NaBH_4$, and the borohydride removed as in (c), it will be safe to remove periodate by passage through a short column of Dowex-1 as in (d) because no reducing sugars will remain.

(f) Enzymes can often be removed from low-M_r degradation-products by precipitation of the former with ethanol or acetone.

4.5.2: Paper chromatography (PC) [219,456]

Although one of the oldest chromatographic methods, PC is very effective and is strongly recommended for many applications described in this book. Since PC is cheap, little money is spent advertising it but it is nevertheless valuable. Some of its advantages are: good 1-dimensional

154

resolution of 10 common monosaccharides without derivatisation (GalA, Gal, Glc, Man, Ara, Xyl, Fuc, Rib, Rha, MeXyl: better than any HPLC system currently available); good sensitivity even with non-radioactive samples (0.1 µg of arabinose by staining); excellent tolerance of impurities (salts, proteins etc.), simplifying sample-preparation and thus minimising losses; up to 100 samples can be run simultaneously per tank; convenient working times (typically 16 h: i.e. samples left running overnight and ready for further analysis next morning); cheap, simple to operate and little equipment to go wrong; convenient storage of compounds after chromatography (flat sheets, not test tubes); ideal for detection of radioactivity (areas of special interest can later be re-analysed by the cutting of smaller fraction sizes; the whole sample can be scintillation-counted and still easily recovered for further analysis — Section 4.6); it is certain that any non-volatile compound loaded on the paper will be present, after chromatography, somewhere between the origin and the solvent front [whereas a compound applied to a chromatography column may bind irreversibly and never elute]; easily scaled up to the preparative level.

The stationary phase is a sheet of paper (e.g. Whatman no. 1 for small samples, or Whatman 3MM for large samples), which is hung from a glass trough containing a solvent (mobile phase) (Fig. 4.5.2 a). The whole assembly is housed in a large glass tank and closed with a greased, well-fitting, glass lid. The sheets of paper are normally 46 x 57 cm [or 23 x 57 cm for narrow tanks], and are hung (long edges vertical) with the samples dried on to the paper 9 cm from a short edge (Fig. 4.5.2 b). The solvent seeps down the paper by capillary action, and the different compounds in the sample are swept along at characteristic speeds relative to the solvent front (R_F). If the compounds of interest all have low R_Fs, the solvent is allowed to drip off the end of the paper for a suitable time to improve resolution: in this case, the bottom short edge is serrated so that the solvent drips off evenly (Fig. 4.5.2 b).

The samples are loaded as either spots or streaks (the latter parallel with the short edge — see Fig. 4.5.2 b). Spot loadings should normally be about 1-1.5 cm in diameter; smaller spots do not improve the resolution, and may worsen it if they contain so much solute per mm^2 of paper that the solvent cannot penetrate the loading. Spot loadings should be spaced at least 2.5 cm apart, centre-to-centre. The spot-loading method

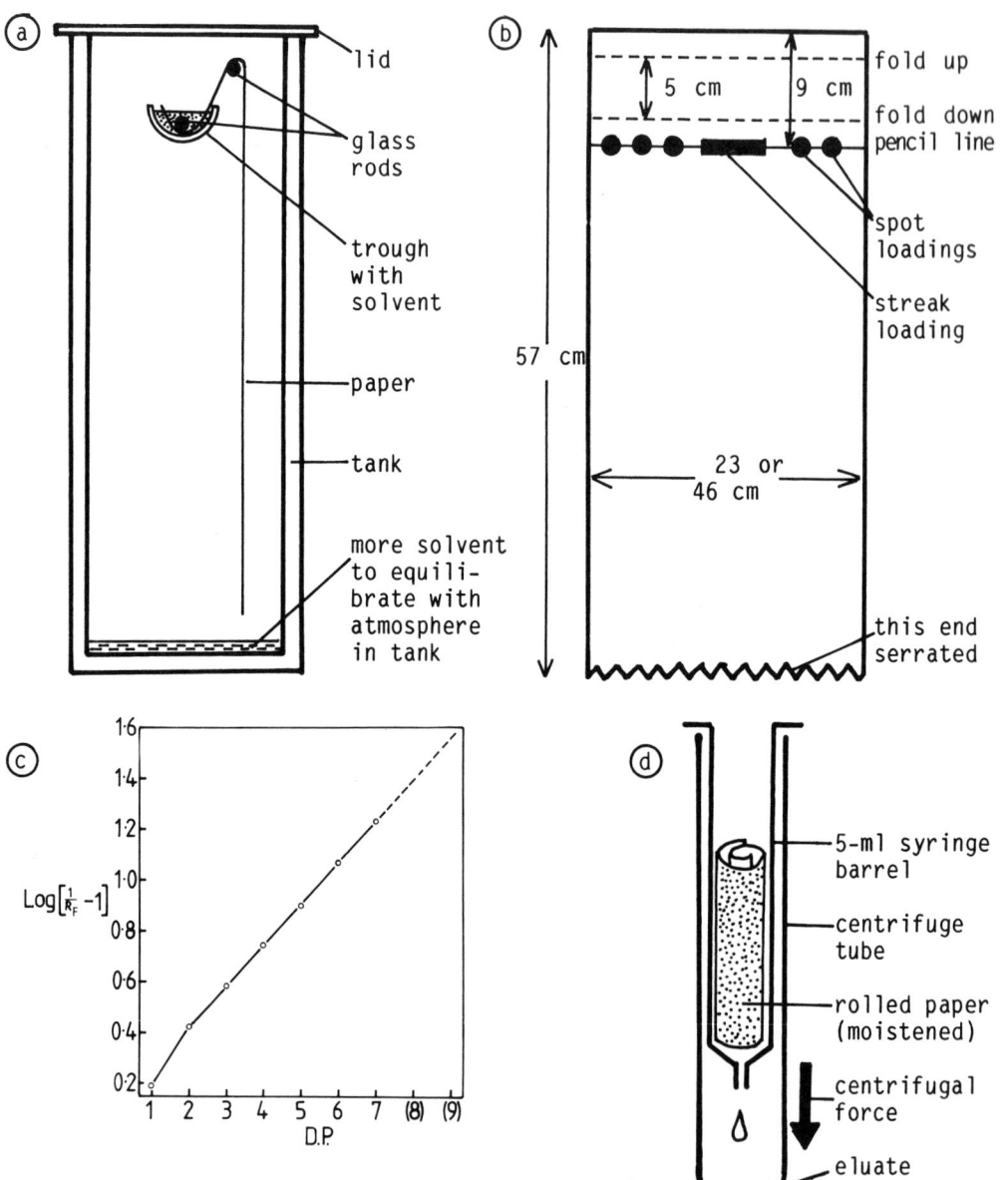

Figs. 4.5.2 a-d: Paper chromatography.
a: The arrangement of the paper in the tank for descending chromatography
(cross-sectional view).
b: Layout of samples on the paper prior to chromatography.
c: The relationship between the chromatographic mobility (expressed as
$\log_{10} [(1/R_F) - 1]$) and size of an oligosaccharide (degree of polymerisation, DP). The compounds tested were glucose (DP 1), maltose (DP 2), maltotriose (DP 3),..... maltoheptaose (DP 7), all of which are available commercially (Sigma Chem. Co.).   Solvent: EtOAc/HOAc/H$_2$O (10:5:6).
d: Elution of compounds from chromatography paper in a minimum volume of
water by centrifugation (see Panel 4.5.2 h).

will accommodate up to 0.4 mg (on Whatman no. 1) or 1 mg (on Whatman 3MM) of sugar. Streaks are used for larger samples, but are only used (a) if the separation is on a preparative scale, (b) if the samples are radioactive [streaks up to about 6 cm wide can be fitted into a scintillation vial for counting after chromatography], and/or (c) if the samples will be eluted from the paper prior to detection of the compounds of interest. If the samples are to be stained in situ on the paper there will be no increase in sensitivity as a result of streak application since the colour is simply spread over a greater width of paper.

The range of chromatography solvents is endless [219,456]. A selection recommended for cell wall work is presented in Panels 4.5.2 a-e. BuOH/HOAc/H$_2$O (12:3:5) [Panel 4.5.2 a] is particularly recommended for exploratory work since most monomers released from walls (Sections 4.2-4.4) will migrate somewhere useful in it (away from the origin but not too close to the solvent front). It also has an excellent tolerance of impurities such as salts or enzymes left over from the degradation method.

Solvents recommended for optimal resolution within particular classes of compounds include the folowing [a-e refer to the solvents described in Panels 4.5.2 a-e respectively]:

For most **neutral monosaccharides: a, b, c.**

For **apiose:** EtOAc/HOAc/H$_2$O (9:2:2) containing 0.25 % phenylboronic acid [chromatography should be limited to 10 h as the apiose moves considerably faster than all other common sugars].

For **uronic acids:** EtOAc/pyridine/HOAc/H$_2$O (5:5:1:3), in a tank equilibrated with the vapour of EtOAc/pyridine/H$_2$O (40:11:6) [219].

For **alditols:** Butanone/HOAc/H$_3$BO$_3$-saturated water (9:1:1) [321]. As an alternative, use EtOAc/pyridine/5 mM H$_3$BO$_3$ (3:2:1) on DEAE-cellulose paper (pre-sprayed to uniform wetness with 0.5 mM borax and dried) [107].

For **short oligosaccharides** (DP 2-4): EtOAc/pyridine/H$_2$O (10:4:3). For a detailed review of solvents, see Bailey and Pridham [34].

For **longer oligosaccharides** (DP 3-15): **d, e.** The R$_F$s of different oligosaccharides (within a particular homologous series and in a given solvent) bear a regular relationship to each other which can best be seen if $\log_{10}$ [(1/R$_F$) - 1] is plotted against degree of polymerisation (DP, i.e. chain-length) (Fig. 4.5.2 c). This can make paper chromatography useful in the estimation of the DP of an unknown oligosaccharide.

For **feruloyl-oligosaccharides: a.**

For **free phenols:** [TLC preferred]. Solvents used on PC include **a,** HOAc/conc HCl/H$_2$O (10:3:30), and propan-2-ol/25 % (w/v) NH$_3$/H$_2$O (8:1:1).

For **amino acids: a, e** (preferably both, run 2-dimensionally).

After chromatography, the paper is hung up to dry in a fume cupboard. Residual traces of pyridine, acetic acid or formic acid can be efficiently removed by spraying with methanol and re-drying. This is particularly important if the chromatographed material will be fed to living cells. Removal of pyridine is also necessary for maximum colour yield with the aniline hydrogen-phthalate stain. If phenol has been used in the solvent, the last traces of it can be removed if the dried paper is washed in diethyl ether (in which sugars and amino acids are insoluble) and re-dried.

Separated compounds are detected in a number of ways:

(a) Monitoring radioactivity. Of the various methods for detection of ^{3}H and ^{14}C (Section 2.8.2), liquid scintillation-counting (LSC) is the most sensitive, and is usually done on strips cut from the chromatogram and placed directly into scintillant. Strips are routinely 2 x 4 cm (the 2 cm edge cut parallel with the solvent flow, and the first strip extending 1 cm behind the origin). After LSC in non-Triton scintillant (Section 2.8.2), strips from any ill-resolved zones of the chromatogram are removed from the scintillant, dried and cut into four 0.5 x 4 cm strips which are individually re-counted. All sugars and amino acids remain on the paper strip in toluene scintillant, and will be counted (without elution) with similar efficiency. Any phenolics and other relatively non-polar compounds that dissolve into the toluene will be counted with higher efficiency since the paper interferes less. Also counted with higher efficiency will be soluble polysaccharides and glycoproteins (R_F 0.00), which during loading on to the paper bind to the surface of the fibres rather than penetrating them: the polymers therefore come into closer contact with the scintillant, and the β-particles from them are detected more efficiently than β-particles from low-M_r compounds that do penetrate the paper fibres.

If possible, radioactive samples are mixed with a stainable quantity of the non-radioactive form of the compound of interest as an **internal marker**: after LSC, scintillant is washed off the paper with toluene (or the faster-drying methylcyclohexane) and the strip is dried and stained to reveal the internal marker. Internal markers are better than **external markers**, i.e. non-radioactive compounds run alongside the radioactive sample, since neighbouring tracks never quite coincide. When internal markers are used, the stained strips can be assayed colorimetrically, and the plots of radioactivity and colour-intensity compared quantitatively.

Then if the two peak centres fail to coincide by as little as 0.5 cm, the radioactive compound was not what it had been suspected to be.

A quick alternative to the above method is to stain the internal markers lightly before LSC. [Heavier staining quenches LSC.] This has the advantage that the exact area of interest is evident before strip-cutting. For example, if the paper is dipped in $1/20$ strength aniline H-phthalate (Panel 4.5.2 f), 50 μg of internal marker glucose can just be detected and the counting efficiency of LSC for [^{3}H]glucose in the same sample is still at least 95 % that of the unstained control.

(b) Staining. A vast array of stains is available for the detection of colourless compounds on chromatography paper [99,123]. A selection particularly recommended for cell wall studies is given in Panel 4.5.2 f. Quantification of stained spots is possible by a number of techniques. Some coloured products can be eluted and assayed by spectrophotometry (Panel 4.5.2 f). Alternatively, chromatograms can be scanned in situ by use of a densitometer (or by eye, giving estimates on a scale 1,2,4,8....). In any of these methods, external markers of known quantity are run on the same sheet and are stained simultaneously to provide a standard curve: this must be done for each compound to be quantified since even closely related compounds may differ in colour yield (e.g. Gal and Glc stained with aniline H-phthalate). Blanks should also be run and 'stained' to indicate any background colour in the relevant zone of the chromatogram.

(c) Autofluorescence and UV-absorbance. Some phenolics fluoresce under UV light, providing a sensitive means of detection (Panel 4.5.2 g). Lamps are usually used that emit UV at 366 nm [for some compounds, e.g. dityrosine, 254 nm is more effective]. Eye protection should be worn. The colour of the fluorescence is an aid to identification, as is any transient change seen upon exposure to NH$_3$ fumes. Compounds that do not visibly fluoresce may nevertheless absorb UV light. They appear as dark spots against the pale background fluorescence of the paper. This effect is enhanced if the back of the paper is sprayed with fluorescein (0.005 % in 0.5 M NH$_3$) and the front is examined under 254-nm UV light.

Elution of samples from the paper. Unstained chromatograms can act as one step in a multi-step purification scheme. Samples are washed off the paper, concentrated if necessary, and analysed further. A simple, efficient and very quick method of elution [149] is /[contd on p. 168]

PANELS **4.5.2** a-e: CHROMATOGRAPHIC MOBILITIES OF AUTHENTIC STANDARDS ON WHATMAN No 1 PAPER IN FIVE RECOMMENDED SOLVENT SYSTEMS.

The left hand vertical column gives **either** R_F (= mobility of compound relative to that of solvent front) **or** R_{Rha} (= mobility of compound relative to that of rhamnose). The footnote to each Panel indicates the speed at which the solvent front and/or rhamnose move under conditions of descending chromatography at room temperature. The abbreviations used are:

Monosaccharides:

All, Allose;   **Api**, Apiose;   **Ara**, Arabinose;   **dGal**, 2-Deoxy-galactose; **dGlc**, 2-Deoxy-glucose;   **dRib**, 2-Deoxy-ribose;   **Ery**, Erythrose;   **Fru**, Fructose;   **Fuc**, Fucose;   **Gal**, Galactose;   **GalA**, Galacturonic acid;   **GalN**, Galactosamine [= 2-Amino-2-deoxy-galactose];   **GalNAc**, N-Acetyl-galactosamine;   **Glc**, Glucose;   **GlcA**, Glucuronic acid;   **GlcAL**, Glucuronic acid lactone;   **GlcN**, Glucosamine [= 2-Amino-2-deoxy-glucose];   **GlcNAc**, N-Acetyl-glucosamine;   **Glycerald**, Glyceraldehyde;   **GulA**, guluronic acid;   **KDO**, Keto-deoxy-octulosonic acid;   **Lyx**, Lyxose;   **Man**, Mannose;   **ManA**, Mannuronic acid;   **ManAL**, Mannuronic acid lactone;   **ManN**, Mannosamine [= 2-Amino-2-deoxy-mannose;   **ManNAc**, N-Acetyl-mannosamine;   **MeGlcA**, 4-O-methylglucuronic acid;   **Rha**, Rhamnose;   **Rib**, Ribose;   **Tag**, Tagatose;   **Xyl**, Xylose.

Alditols:

AraH, Arabinitol;   **EryH**, Erythritol;   **GalH**, Galactitol [= Dulcitol];   **GlcH**, Glucitol [= Sorbitol];   **ManH**, Mannitol;   **RibH**, Ribitol;   **XylH**, Xylitol.

Oligosaccharides:

Cel2, Cellobiose [= D-Glcp-β-(1 → 4)-D-Glc];   **Fer-Ara2**, Feruloyl-arabino-biose [= 3-O-Feruloyl-L-Arap-α-(1 → 3)-L-Ara];   **Fer-Gal2**, Feruloyl-galacto-biose [= 6-O-Feruloyl-D-Galp-β-(1 → 4)-D-Gal];   **GalA2**, Galacturonobiose [= D-GalpA-α-(1 → 4)-D-GalA];   **GalA3**, Galacturonotriose [= D-GalpA-α-(1 → 4)-D-GalpA-α-(1 → 4)-D-GalA];   **Gen**, Gentiobiose [= D-Glcp-β-(1 → 6)-D-Glc;   **Gal→Ara**, D-Galp-β-(1 → 3)-D-Ara;   **Gal→Man**, D-Galp-β-(1 → 4)-D-Man;   **HA4**, Hydroxyproline-tetra-arabinoside [= L-Araf-α-(1 → 3)-L-Araf-β-(1 → 2)-L-Araf-β-(1 → 2)-L-Araf-β-(1 → 4)-L-Hyp];   **Lac**, Lactose [= D-Galp-β-(1 → 4)-D-Glc];   **Lam2**, Laminaribiose [= D-Glcp-β-(1 → 3)-D-Glc];   **Lam3**, Laminaritriose [= D-Glcp-β-(1 → 3)-D-Glcp-β-(1 → 3)-D-Glc];   **Mel**, Melibiose [= D-Galp-α-(1 → 6)-D-Glc];   **Mlt2**, Maltose [= D-Glcp-α-(1 → 4)-D-Glc];   **Mlt3**, Maltotriose [= D-Glcp-α-(1 → 4)-D-Glcp-α-(1 → 4)-D-Glc];   **Mlt4**, Maltotetraose;   **Mlt7**, Malto-heptaose;   **Raff**, Raffinose [= D-Galp-α-(1 → 6)-D-Glcp-α-(1 ↔ 2)-β-D-Fruf];   **Stachy**, Stachyose [= D-Galp-α-(1 → 6)-D-Galp-α-(1 → 6)-D-Glcp-α-(1↔2)-β-D-Fruf];   **Sucr**, Sucrose [= D-Glcp-α-(1 ↔ 2)-β-D-Fruf];   **Treh**, Trehalose [= D-Glcp-α-(1↔1)-α-D-Glcp];   **XG2**, Isoprimeverose [= D-Xylp-α-(1 → 6)-D-Glc].

Amino acids:

Ala, Alanine;   **Arg**, Arginine;   **Asn**, Asparagine;   **Asp**, Aspartic acid;   **Cys**, Cysteine;   **DiT**, Dityrosine;   **Gln**, Glutamine;   **Glu**, Glutamic acid;   **Gly**, Glycine;   **HA4**, Hydroxyproline-tetra-arabinoside [see oligosaccharides]; **His**, Histidine;   **Hyp**, 4-Hydroxyproline;   **Idt**, Isodityrosine;   **Ile**, Isoleucine;   **Leu**, Leucine;   **Lys**, Lysine;   **Met**, Methionine;   **Phe**, Phenylalanine; **Pro**, Proline;   **Ser**, Serine;   **Thr**, Threonine;   **Trp**, Tryptophan;   **Tyr**, Tyrosine;   **Val**, Valine.

Phenolics:

p-Benz, p-Hydroxybenzoic acid;   **Caff**, trans-Caffeic acid;   **Cinn**, trans-Cinnamic acid;   **o-Cou**, trans-o-Coumaric acid;   **p-Cou**, trans-p-Coumaric acid; **Fer**, trans-Ferulic acid;   **Gall**, Gallic acid;   **Sin**, trans-Sinapic acid; **Syringald**, Syringaldehyde [= 4-Hydroxy-3,5-dimethoxybenzaldehyde].

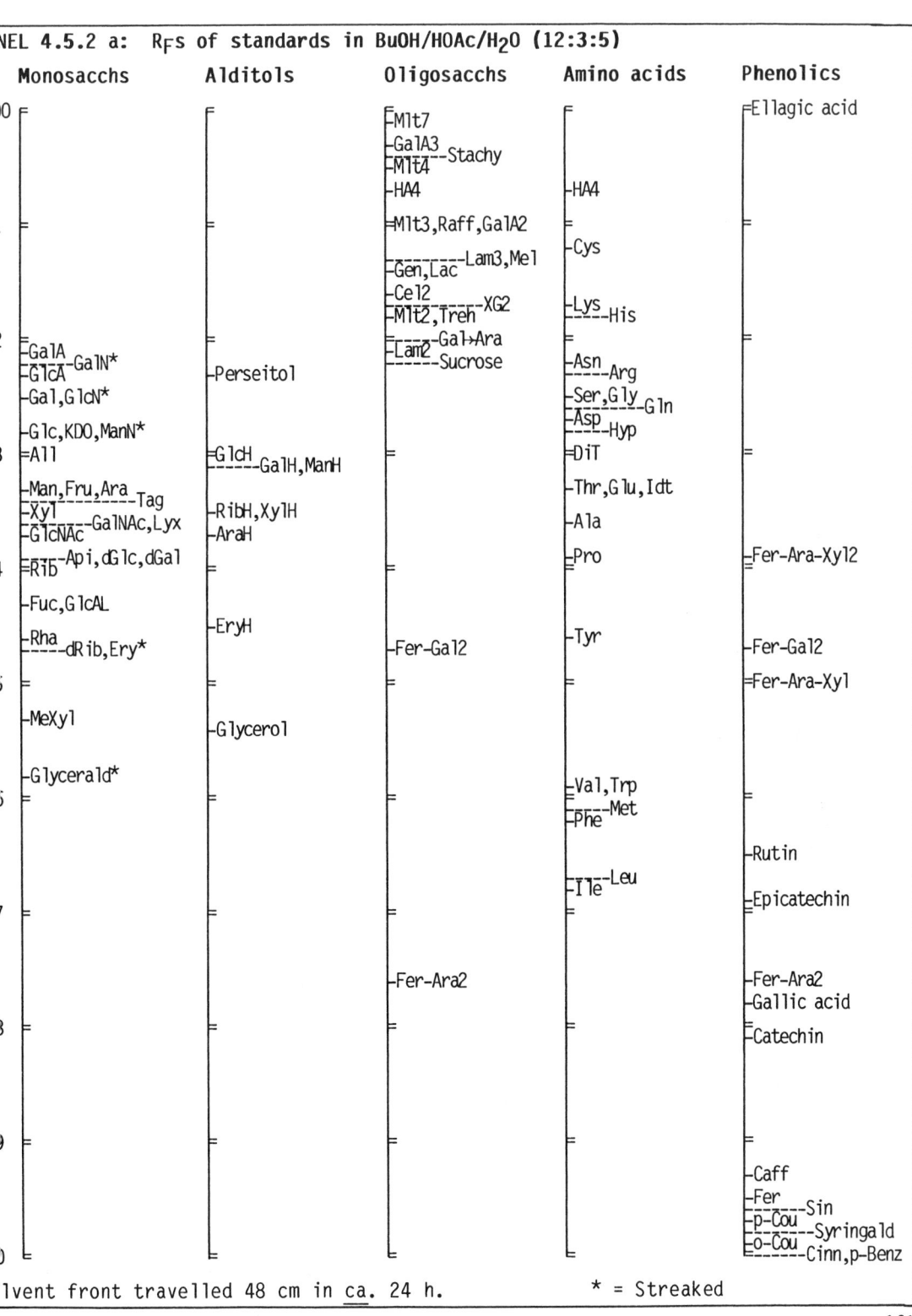

Monosacchs	Alditols	Oligosacchs	Amino acids	Phenolics

Monosacchs:
- GalA, GalN*
- GlcA
- Gal, GlcN*
- Glc, KDO, ManN*
- All
- Man, Fru, Ara, Tag
- Xyl
- GlcNAc, GalNAc, Lyx
- Rib, Api, dGlc, dGal
- Fuc, GlcAL
- Rha, dRib, Ery*
- MeXyl
- Glycerald*

Alditols:
- Perseitol
- GlcH, GalH, ManH
- RibH, XylH
- AraH
- EryH
- Glycerol

Oligosacchs:
- Mlt7
- GalA3, Stachy
- Mlt4
- HA4
- Mlt3, Raff, GalA2
- Lam3, Mel
- Gen, Lac
- Ce12
- Mlt2, Treh, XG2
- Lam2, Gal→Ara, Sucrose
- Fer-Gal2
- Fer-Ara2

Amino acids:
- HA4
- Cys
- Lys, His
- Asn, Arg
- Ser, Gly, Gln
- Asp, Hyp
- DiT
- Thr, Glu, Idt
- Ala
- Pro
- Tyr
- Val, Trp, Phe-Met
- Ile-Leu

Phenolics:
- Ellagic acid
- Fer-Ara-Xyl2
- Fer-Gal2
- Fer-Ara-Xyl
- Rutin
- Epicatechin
- Fer-Ara2
- Gallic acid
- Catechin
- Caff
- Fer, Sin
- p-Cou, Syringald
- o-Cou, Cinn, p-Benz

PANEL 4.5.2 b: R$_{Rha}$s of standards in EtOAc/pyridine/water (8:2:1).

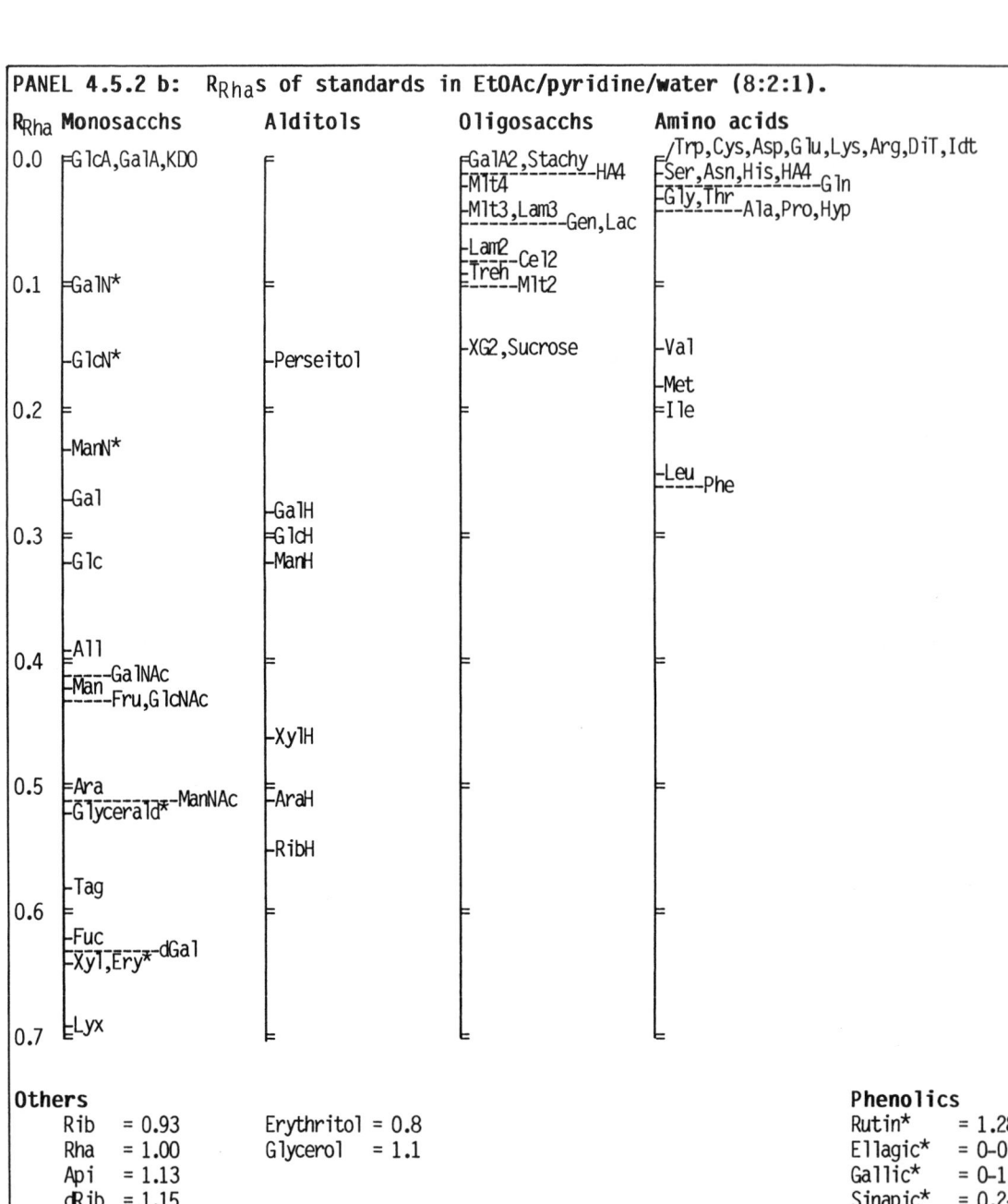

R$_{Rha}$	Monosacchs	Alditols	Oligosacchs	Amino acids
0.0	GlcA,GalA,KDO		GalA2,Stachy ─────── HA4	Trp,Cys,Asp,Glu,Lys,Arg,DiT,Idt
			Mlt4	Ser,Asn,His,HA4
			Mlt3,Lam3 ─────── Gen,Lac	Gly,Thr ───── Ala,Pro,Hyp / Gln
			Lam2 ── Cel2	
			Treh ── Mlt2	
0.1	GalN*			
	GlcN*	Perseitol	XG2,Sucrose	Val
				Met
0.2				Ile
	ManN*			
	Gal			Leu ── Phe
0.3		GalH		
	Glc	GlcH		
		ManH		
0.4	All			
	Man ── GalNAc			
	── Fru,GlcNAc			
		XylH		
0.5	Ara	AraH ── ManNAc		
	Glycerald*			
		RibH		
	Tag			
0.6	Fuc			
	Xyl,Ery* ── dGal			
0.7	Lyx			

Others

Rib	= 0.93	Erythritol	= 0.8
Rha	= 1.00	Glycerol	= 1.1
Api	= 1.13		
dRib	= 1.15		
ManAL	= 1.37		
GlcAL	= 2.68		

Phenolics

Rutin*	= 1.28
Ellagic*	= 0-0.
Gallic*	= 0-1.
Sinapic*	= 0.2-
Ferulic*	= 0.2-
Caffeic*	= 0.3-
Cinnamic*	= 0.3-
p-Coumaric*	= 0.9-
Syringald	= 2.8
Catechin	= 2.8
Epicatec'n	= 2.8
Coumarin	= 3.0

*Streaked

Solvent front travelled 48 cm in ca. 5 h.   R$_F$ of Rha = 0.28.

162

PANEL 4.5.2 c: $R_{Rha}s$ of standards in BuOH/HOAc/water (12:3:5) followed by EtOAc/pyridine/water (8:2:1), 16 h in each.

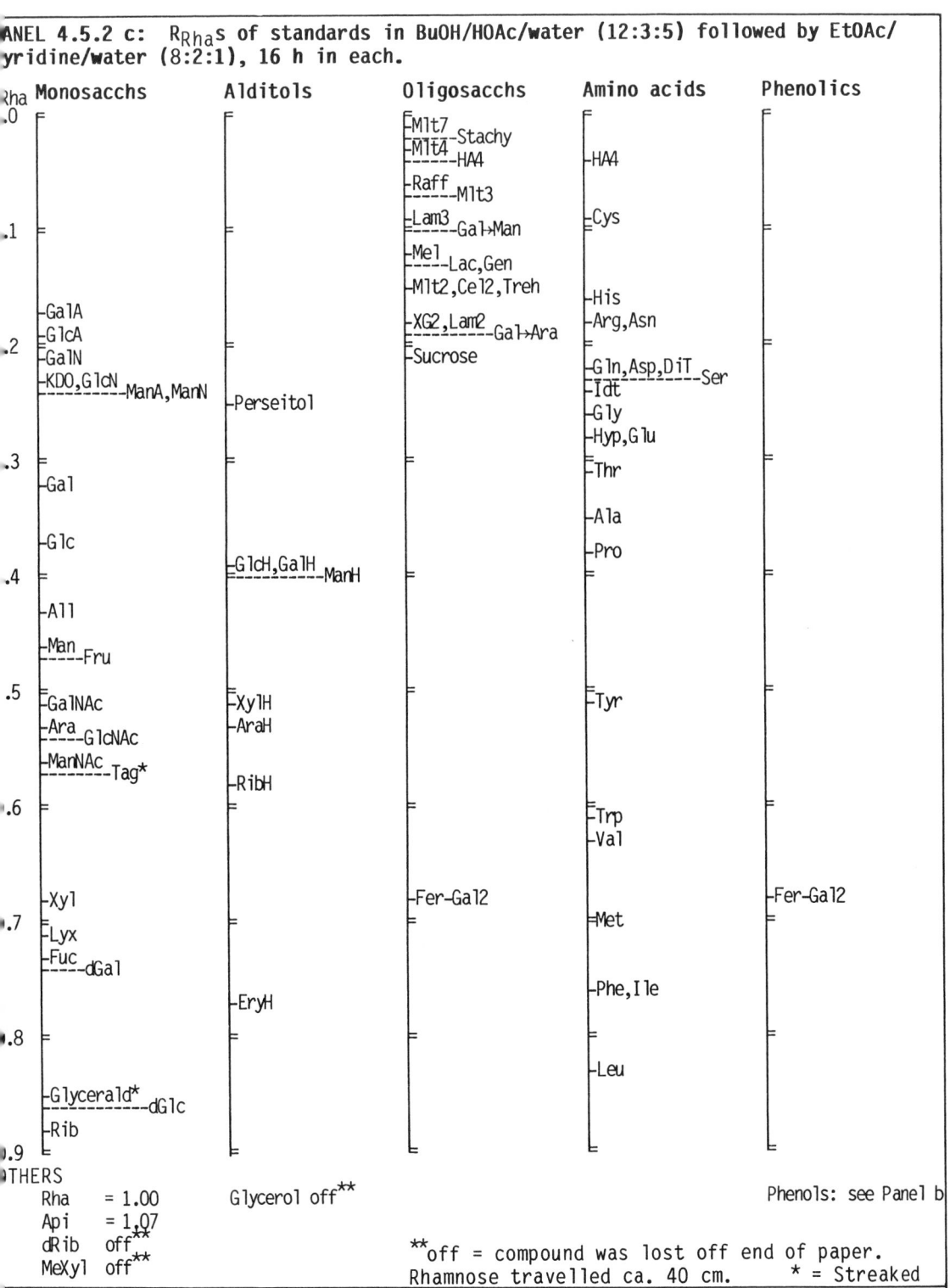

Rha	Monosacchs	Alditols	Oligosacchs	Amino acids	Phenolics
.0			Mlt7, Mlt4—Stachy, -----HA4, Raff—Mlt3	HA4	
.1			Lam3—Gal→Man, Mel-----Lac,Gen, Mlt2,Cel2,Treh	Cys	
	GalA, GlcA, GalN, KDO,GlcN—ManA,ManN		XG2,Lam2-----Gal→Ara, Sucrose	His, Arg,Asn	
.2		Perseitol		Gln,Asp,DiT—Ser, Idt, Gly, Hyp,Glu	
.3	Gal			Thr, Ala	
	Glc	GlcH,GalH—ManH		Pro	
.4	All, Man—Fru				
.5	GalNAc, Ara—GlcNAc, ManNAc—Tag*	XylH, AraH, RibH		Tyr	
.6				Trp, Val	
.7	Xyl, Lyx, Fuc—dGal		Fer-Gal2	Met, Phe,Ile	Fer-Gal2
.8		EryH		Leu	
	Glycerald*-----dGlc, Rib				
.9					

OTHERS

Rha	= 1.00
Api	= 1.07
dRib	off**
MeXyl	off**

Glycerol off**

Phenols: see Panel b

**off = compound was lost off end of paper.
Rhamnose travelled ca. 40 cm.   * = Streaked

163

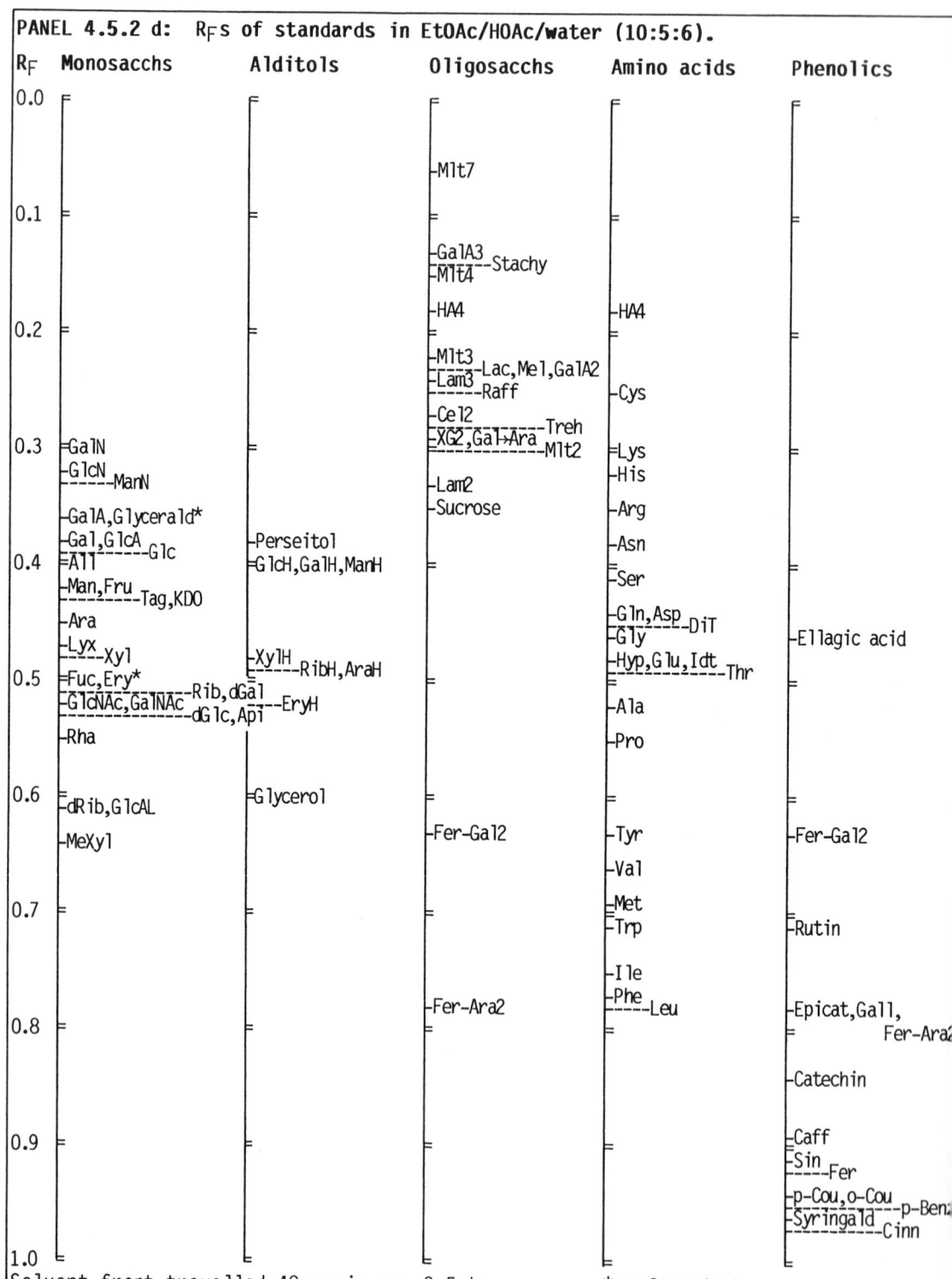

R$_F$	Monosacchs	Alditols	Oligosacchs	Amino acids	Phenolics
0.0					
			⊢Mlt7		
0.1					
			⊢GalA3		
			⊢Mlt4––Stachy		
			⊢HA4	⊢HA4	
0.2					
			⊢Mlt3		
			⊢Lam3––Lac,Mel,GalA2		
			⊢–––––Raff	⊢Cys	
			⊢Cel2		
			⊢XG2,Gal→Ara––Treh		
0.3	⊢GalN		⊢––––––––Mlt2	⊢Lys	
	⊢GlcN			⊢His	
	⊢–––––ManN		⊢Lam2		
			⊢Sucrose	⊢Arg	
	⊢GalA,Glyceraldᐟ			⊢Asn	
	⊢Gal,GlcA	⊢Perseitol			
0.4	⊢All–––––Glc	⊢GlcH,GalH,ManH		⊢Ser	
	⊢Man,Fru				
	⊢–––––––Tag,KDO			⊢Gln,Asp	
	⊢Ara			⊢Gly––DiT	⊢Ellagic acid
	⊢Lyx			⊢Hyp,Glu,Idt	
	⊢–––Xyl	⊢XylH		⊢–––––––––Thr	
0.5	⊢Fuc,Eryᐟ	⊢–––RibH,AraH			
	⊢GlcNAc,GalNAc			⊢Ala	
	⊢–––––––––Rib,dGal				
	⊢dGlc,Apiᐟ––EryH			⊢Pro	
	⊢Rha				
0.6	⊢dRib,GlcAL	⊢Glycerol			
	⊢MeXyl		⊢Fer-Gal2	⊢Tyr	⊢Fer-Gal2
				⊢Val	
0.7				⊢Met	
				⊢Trp	⊢Rutin
				⊢Ile	
			⊢Fer-Ara2	⊢Phe	⊢Epicat,Gall,
0.8				⊢–––––Leu	Fer-Ara2
					⊢Catechin
0.9					⊢Caff
					⊢Sin
					⊢–––––Fer
					⊢p-Cou,o-Cou
					⊢Syringald––p-Benz
					⊢–––––––Cinn
1.0					

Solvent front travelled 48 cm in <u>ca</u>. 8.5 h.   * = Streaked

PANEL 4.5.2 e: R_Fs of standards in BuOH/pyridine/water (4:3:4).

Monosacchs	Alditols	Oligosacchs	Amino acids	Phenolics

```
.0  ⊏              ⊏              ⊏                ⊏              ⊏

.1  ⊏              ⊏              ⊏                ⊏              ⊏
                                  ⊢GalA2*
                                                   ⊢Lys
.2  ⊢GalA          ⊏              ⊨Mlt7             ⊨Cys
    ⊢GlcA                         ⊢Stachy           ⊢Asp,Dil---Arg      ⊏
    ⊢ManA                         ⊢HA4              ⊢HA4,Glu
                                                    ⊢Asn
                                                    ⊢Ser           ⊢Ellagic**
.3  ⊨KDO           ⊏              ⊨Mlt4             ⊢Gln,Gly---Thr,His ⊨Gallic**
                                                    ⊢Ala
                                  ⊢Mel              ⊢Pro,Idt---Hyp
                                  ⊨Mlt3--Lac,Raff
                                   ------Gal→Man
.4  ⊏              ⊢Perseitol     ⊨Treh             ⊏              ⊏
                                  ⊢Mlt2,Gal→Ara
    ⊢Gal,GalN                     ⊢Cel2,XG2,Sucr
    ⊢Glc,GlcN      ⊢XylH
.5  ⊢Fru           ⊢AraH-GlcH,GalH,ManH=            ⊏              ⊏
    ⊨All-Ara,ManN*  ------RibH                      ⊢Val
    ⊢ ----Tag                                       ⊢Met
    ⊢Man,Xyl
                   ⊢EryH,ThrH
    ⊢GalNAc                                                        ⊨Sin*
.6  ⊨Rib,GlcNAc---Fuc,dGal                          ⊨Tyr
    ⊢Lyx          ⊢Glycerol                         ⊢Phe,Ile---Trp
    ⊨ManAL--dGlc,Api                                 ------Leu      ⊢Caff*
    ⊢Rha--dRib
    ⊢GlcAL                                                          ⊢Fer
.7  ⊢ ----MeXyl    ⊏              ⊏                ⊏              ⊨p-Cou
    ⊢Ery
    ⊢ --Glycerald                                                  ⊢Cinn
                                                                   ⊢o-Cou

                                                                   ⊢p-Benz
.8  ⊏              ⊏              ⊏                ⊏              ⊨Rutin

.9  ⊏              ⊏              ⊏                ⊏              ⊨Coumarin

                                                                   ⊢Syringald
                                                                    -------Epicat'n
                                                                   ⊢Catechin
.0  ⊏              ⊏              ⊏                ⊏              ⊏
```

Solvent front travelled 48 cm in ca. 17 h.   * = Streaked.

PANEL 4.5.2 f: Stains for paper chromatograms

ANILINE HYDROGEN-PHTHALATE (for monosaccharides and reducing disaccharides)
  Solution: Stock = 16 g Phthalic acid in 490 ml acetone[3], 490 ml diethyl ether and 20 ml H_2O [stable for years in a tightly capped bottle]. Immediately before use, mix 100 ml stock with 0.5 ml aniline. [Note: to stain a sample lightly, before scintillation-counting, use 20-fold less aniline and phthalic acid.]
  Method: [Can be used on chromatograms run in solvents containing borate. For maximum colour yield, make sure traces of pyridine are removed.] Dip paper rapidly through solution. Dry 3-5 min in fume cupboard. Heat 5 min in an oven at 105°C.[1]
  Results: Uronic acids — orange; hexoses and 6-deoxyhexoses — brown; pentoses — dull red; reducing disaccharides — various. Very faint spots are more readily seen by UV fluorescence (366 nm lamp), especially hexoses and 6-deoxyhexoses. Stained spots stable in dark, but colour **differences** vanish over 2-3 days.
  Detection limit:[2] ca. 0.4 µg arabinose; somewhat poorer with most reducing disaccharides; very poor for maltose; no reaction with sucrose.
  Quantification: Stained spots can be eluted with 2 % $SnCl_2$ in methanol, and assayed by A_{370}. Sugars vary in colour yield.

SILVER NITRATE / SODIUM HYDROXIDE (for monosaccharides, oligosaccharides, alditols and saccharinic acids; also stains phenols)
  Solutions: **A:** Dissolve 0.2 g $AgNO_3$ in 0.4 ml H_2O; add to 26 ml acetone with rapid stirring. If a precipitate appears, re-dissolve it by addition of a minimum amount of H_2O. **B:** Mix 1.25 ml 10 M NaOH with 100 ml absolute ethanol. [Prepare both solutions fresh on day of use.]
  Method: [Cannot be used on chromatograms run in borate-containing solvents, or on aluminium-backed TLC plates] Dip through solution **A**. Dry 3-5 min. Dip rapidly through solution **B**. Re-dry 10-20 min. Repeat dipping through **B** and drying 1-4 more times depending on sensitivity required.
  Results: Brown spots. Rapid with monosaccharides (except mannose); slower with oligosaccharides and alditols; weak with sucrose. Store chromatogram in dark; record results same day as paper eventually goes black.
  Detection limit:[2] ca. 0.1 µg arabinose

POTASSIUM PERMANGANATE (for alditols, monosaccharides, oligosaccharides)
  Solution: 1 g $KMnO_4$ and 2 g Na_2CO_3 in 100 ml H_2O.
  Method: [Can be used on chromatograms run in presence of borate.] Spray the solution evenly. Dry.
  Results: Yellow spots on a purple background. Spots rapidly fade.
  Detection limit:[2] ca. 10 µg glucitol (sorbitol).

MOLYBDATE (for sugar-phosphates and nucleotides)
  Solution: Dissolve 2 g ammonium molybdate in 17 ml H_2O. Add 3 ml concentrated HCl (dropwise, with rapid stirring). Add 6 ml 60 % perchloric acid (dropwise, with rapid stirring). Continue stirring until precipitate has completely re-dissolved. Add the mixture to 180 ml acetone. [Solution stable for weeks in tightly capped bottle.]
  Method: Dip paper rapidly through solution. Dry 5-10 min. Mark any spots. Expose to direct sunlight or UV (366 nm lamp) for 10 min. Mark new spots. Heat at 100°C for 2 min.[1] Mark any further spots.
  Results: Yellow and blue spots.
  Detection limit:[2] ca. 2 µg glucose 6-phosphate.

NINHYDRIN (for amino acids and de-acetylated amino sugars)

      Solution: 0.5 % Ninhydrin in acetone. [Store dark; fresh weekly.]

      Method: Dip paper quickly. Dry 5 min. Heat at 105°C for 5 min.[1]

      Results: Violet spots for most amino compounds. Yellowish spots for proline and hydroxyproline.

      Detection limit:[2] ca. 0.05 µg serine.

ISATIN / NINHYDRIN (for hydroxyproline, hydroxyproline-arabinosides, proline, and amino acids)

      Solution: Dissolve 270 mg ninhydrin, 130 mg isatin and 2 ml triethylamine in 100 ml acetone.[3]

      Method: Dip. Dry 5 min. Heat at 80°C;[1] examine at 1-min intervals.

      Results: Hydroxyproline — pink spot rapidly, hydroxyproline-arabinosides and proline — orange spots fairly rapidly, amino acids — blues, violets and greens slowly.

      Detection limit:[2] ca. 0.02 µg hydroxyproline.

      Quantification: Wash paper in water to de-colorise background. Dry. Elute spots in acetone / ethanol (1:2). Measure $\underline{A}_{515}$.

FOLIN & CIOCALTEU'S PHENOL REAGENT (for phenolics)

      Solution: Available commercially. Store refrigerated, not frozen.

      Method: Spray paper evenly. Wait 1-2 min. Note any spots. Hang in a closed tank containing a beaker of concentrated NH_3 solution until yellow background is de-colorised. Note any further spots.

      Results: Blue-black spots. Those that form immediately possess o- or p-dihydroxybenzene groups; those that require NH_3 do not. Spots stable in dark. Reagent also stains guanine (sea-green), and various other non-phenolics may produce blue spots upon storage.

      Detection limit:[2] ca. 0.05 µg ferulic acid.

      Quantification: [Requires standardisation of spraying: spray evenly and almost to saturation, i.e. so that the yellow colour just seeps through to the unsprayed side of the paper but no puddles form.] Expose to NH_3 vapour. Dry paper. Elute spots with 2.5 M NaOH. Centrifuge down any precipitate. Read A_{740} (or at the highest wavelength offered by the spectrophotometer if lower).

BROMOTHYMOL BLUE (non-specific stain for acids)

      Solution: Dissolve 40 mg bromothymol blue in 100 ml 10 mM NaOH.

      Method: [Remove any acetic or formic acid from paper by repeated spraying with methanol.] Spray the stain evenly. Ring any spots in pencil as soon as they appear.

      Results: Transient yellow spots on blue background. [If background is yellow, acids from chromatography solvent had not been completely removed.]

      Detection limit:[2] ca. 4 µg galacturonic acid.

[1]Temperatures should be adhered to in quantitative work, but good qualitative results are obtained by heating the paper evenly with a hair-drier.

[2]Detection limit was estimated on Whatman no. 1 paper after descending chromatography for 16 h in $BuOH/HOAc/H_2O$ (12:3:5).

[3]Acetone is much quicker-drying than the usually-recommended butanol.

PANEL 4.5.2 g: Fluorescence of phenolic compounds under 366 nm UV light

Compound	Fluorescence at neutral or acidic pH	Fluorescence during fuming with NH_3 vapour
Dityrosine	blue	brighter blue
Isodityrosine, Tyrosine	invisible	invisible
Ferulic acid	blue	brighter blue
Feruloyl esters	blue	bright blue-green
p-Coumaric acid	invisible	violet
p-Coumaroyl esters	invisible	blue
Caffeic acid	blue	light blue
Sinapic acid	blue	blue-green
p-Hydroxybenzoic acid	invisible	invisible

PANEL 4.5.2 h: Elution of compounds from paper chromatograms

Step 1:   Cut out the zone containing the compound of interest.

Step 2:   Roll or fold the paper and press it down into the barrel of a 5-ml disposable syringe. Hang the barrel in a suitably sized centrifuge tube such that the nozzle of the syringe has about 1-2 cm clearance from the bottom of the tube (Fig. 4.5.2 d — page 156).

Step 3:   Add just enough water (or other suitable eluent) to wet the paper; no surplus water should emerge from the nozzle. Centrifuge the assembly (5 min at 2,500 **g**), thereby wringing out the paper. Eluate collects in the centrifuge tube.

Step 4:   Repeat step 3 until no further material is eluting (normally 3-5 times).

described in Panel 4.5.2 h. This method yields the chromatographed compound dissolved in the minimum of solvent, eliminating the need to concentrate large volumes. The eluent chosen depends on the solubility of the compound, but it should preferably be volatile: often water is used; 50 % methanol may be preferred for free phenolics; 1 M NH_3 is recommended for tyrosine and isodityrosine as they have low solubility in water.

4.5.3: Thin-layer chromatography (TLC)

TLC [471,506] is a smaller-scale method than PC. The stationary phase is a thin layer (often 0.2 mm thick) of silica gel or cellulose powder, bound on to an inert plate (glass, aluminium or plastic, routinely

20 x 20 cm). The sample is dried on to the thin layer, 1-1.5 cm from one edge, and the plate is placed in a closed tank containing about 50 ml of solvent. The solvent rises up the plate by capillarity. Different compounds migrate with the solvent flow at different rates (R_F) relative to the solvent front. The R_F values are less consistent from run to run than in paper chromatography, but the order in which a set of markers migrate is usually reproducible. If all the compounds of interest are known to have low R_Fs, resolution can be improved by clipping a wad of filter paper to the top edge of the plate, so that the ascent of solvent will continue after the front has migrated 20 cm.

TLC on cellulose powder can separate polar compounds in a similar way to PC. TLC is quicker, typical run times being 1-8 h, and is therefore chosen when results are needed the same day. However, TLC is more demanding — both of manual dexterity in loading the sample without disrupting the delicate layer of cellulose powder, and of sample purity. The maximum spot-loading of a pure sugar on cellulose TLC is about 25-50 µg, and the sample must be relatively free of salts, proteins and other contaminants. Resolution is similar to that on PC, although on TLC the separation occupies 20 cm (cf. 50 cm on paper); this may make TLC preferable if samples are to be analysed by autoradiography or fluorography, or if the total amount of sample available is very small (the limit of detection of arabinose on TLC is about 0.08 µg by staining with aniline H-phthalate).

Solvents for cellulose TLC include many devised for PC. An excellent solvent system for TLC of the major **monosaccharides** of the plant cell wall is **(a)** freshly prepared butan-1-ol/acetic acid/H_2O (3:1:1), which is run until the solvent has risen almost 20 cm (which takes 6-7 h), followed, after drying, by **(b)** re-chromatography in the same direction in ethyl acetate/pyridine/H_2O (10:4:3) until the solvent has again run nearly 20 cm (2-3 h). Merck plates have given the best results (Fig. 4.5.3). The sugars separated (R_{Rha} values in parentheses) are:

GalA (0.32), Gal (0.51), Glc (0.58), Man (0.66), Ara (0.69), Xyl (0.78), Fuc (0.84), Rib (0.90), Rha [1.00], MeXyl (1.15), MeFuc (1.18).
Api co-chromatographs with Rha and GlcA co-chromatographs with GalA.

Silica-gel TLC is better than cellulose TLC for non-polar compounds. Thus, whereas cellulose is recommended for TLC of monosaccharides, oligosaccharides and phenol—sugar complexes (Sections 4.4.1-2), silica gel is better for free phenolics, methyl-sugars (Section 4.2.8) and lipids. Amino acids have intermediate polarity and can be run on either cellulose or silica-gel.

A suitable solvent for separation of the common **phenolic acids** on silica gel is benzene/acetic acid (9:1), in which the individual acids run in the order (approx. R_F values in parentheses):

caffeic (0.08), diferulic (0.14), p-coumaric [and p-hydroxybenzoic, not resolved] (0.23), sinapic (0.28), ferulic (0.41), and cinnamic (0.58). These compounds (except cinnamic acid) often occur as a mixture of the cis- and trans-isomers. TLCs run in the dark yield a double spot for each compound owing to resolution of the cis- and trans-isomers. Since the ratio of cis-:trans-isomers is not usually of interest, it is helpful to run the chromatogram under constant exposure to UV light by shining a 366-nm UV lamp through the glass sides of the chromatography tank [glass transmits light of this wavelength.] UV light causes trans ↔ cis isomerisation and thus ensures that each acid migrates as a rapidly interconverting mixture of the two isomers and therefore as a single spot.

(i)   (ii)

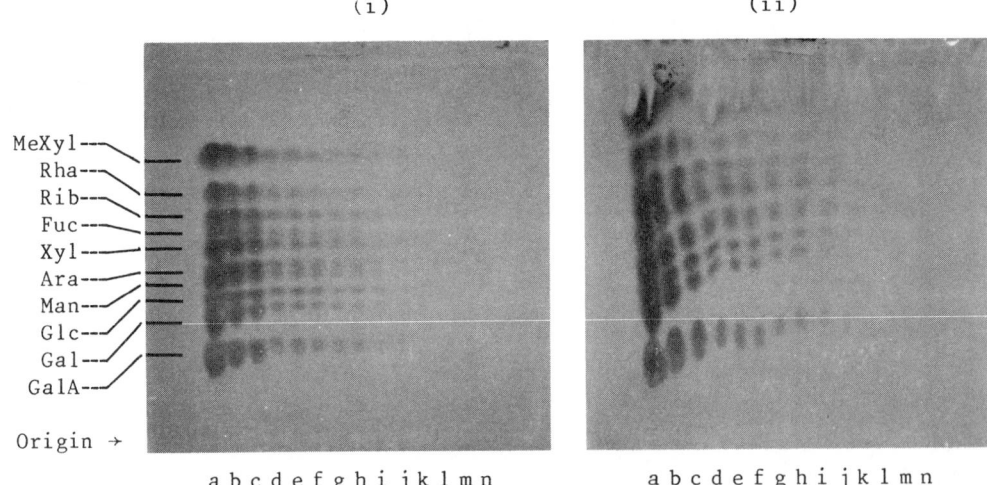

MeXyl
Rha
Rib
Fuc
Xyl
Ara
Man
Glc
Gal
GalA

Origin →

a b c d e f g h i j k l m n   a b c d e f g h i j k l m n

Fig. 4.5.3: Thin-layer chromatography of ten common monosaccharides at various loadings.

Two cellulose plates [layer 0.2 mm] were used, as supplied by Merck **(i)** or Machery-Nagel **(ii)**, and run simultaneously in the same tank.

The loadings contained **(a)** 80 µg, **(b)** 40 µg, **(c)** 20 µg, **(d)** 10 µg, **(e)** 5 µg, **(f)** 2.5 µg, **(g)** 1.25 µg, **(h)** 0.63 µg, **(i)** 0.31 µg, **(j)** 0.16 µg, **(k)** 0.08 µg, **(l)** 0.04 µg, **(m)** 0.02 µg, and **(n)** 0.01 µg of each sugar. On the original plate, the spots were visible at loadings down to 0.08 µg.

Solvents: BuOH/HOAc/H_2O (3:1:1) followed in the same direction by EtOAc/pyridine/H_2O (10:4:3). Stain: aniline hydrogen phthalate.

A suitable solvent for **amino sugars** [413] is acetonitrile/acetic acid /ethanol/H_2O (13:1:2:4);   the solvent is allowed to ascend 3 times, with thorough removal of the acetic acid between each run.   'R_F' values (after the 3 runs) are:

  GalN = 0.33, GlcN = 0.40, ManN = 0.48.

Suitable solvents for **methyl-sugars** are:

(i)   acetone/4.5 M NH_3 (500:9) [49],   when   the   order   of   migration   of trimethyl-galactoses is the following (with $R_{2,3,4,6-tetramethylgalactose}$ in parentheses):

   3,4,6-trimethylgalactose (0.54), 2,3,4-trimethylgalactose (0.62), 2,4,6-trimethylgalactose (0.75), 2,3,6-trimethylgalactose (0.85), 2,3,4,6-tetra-methylgalactose [1.00];

(ii) butanone/H_2O (85:7) [553],   when   the   order   of   migration   of   the methylarabinoses is (with $R_{2,3,4-trimethylarabinose}$ values in parentheses):

   3-methylarabinose (0.31), 3,4-dimethylarabinose (0.44), 2-methylarabinose (0.56), 5-methylarabinose (0.94), 2,3,4-trimethylarabinose [1.00], 2,3-dimethylarabinose (1.03), 3,5-dimethyl arabinose (1.47), 2,3,5-trimethyl-arabinose (1.67);

(iii) benzene/EtOH/H_2O/conc NH_3 (200:47:15:1) [553], when selected R_F values are:

   2,4-dimethylglucose (0.05), 2,6-dimethylglucose (0.05), 2,4,6-trimethyl-glucose (0.13), 2,3-dimethylxylose (0.15), 2,4,6-trimethylgalactose (0.17), 2,3,6-trimethylglucose (0.18), 2,3,4,6-tetramethylglucose (0.38).

Suitable solvents for **amino acids** are

(i) $CHCl_3$/MeOH/17 % (w/v) NH_3 (2:2:1),

(ii)   Phenol/H_2O (3:1, w/v),

(iii) Butanone/pyridine/HOAc/H_2O (70:15:2:15)

(iv) BuOH/HOAc/H_2O (4:1:1).

Useful 2-dimensional systems (which are essential for protein hydrolysates) are: **(iv)** followed by **(ii)** [394], and **(i)** followed by **(ii)** [376].

A suitable solvent for the aliphatic lipid components of **cutin and suberin** is $CHCl_3$/EtOAc (7:3) — see Section 4.4.4.

Although silica-gel is not ideal for **underivatised neutral sugars**, a fair separation of common monosaccharides can be obtained very rapidly on silica-gel in EtOAc/pyridine/HOAc/H_2O (6:3:1:1) [353];   see also [99].

TLCs can be stained with most of the reagents quoted for PC in Panel 4.5.2 f, but aluminium-backed plates are unsuitable for corrosive stains e.g. $AgNO_3$/NaOH.   Glass-backed silica-gel plates are extremely resistant, and a useful general spray for any organic compound is 10 % H_2SO_4 in EtOH followed by heating in an oven at 110°C. Organic compounds char, giving spots which often have characteristic colours. Another wide-range stain, especially suitable for organic compounds with double bonds, is iodine vapour: the plate is placed in a dry tank with a few crystals of iodine, and the vapour evolved interacts with the compounds on the plate.

Compounds can be eluted from thin layer with suitable solvents, prior to further analysis or scintillation-counting. In the case of glass plates, the appropriate area of the thin layer is scraped off (dry) and suspended in eluent or scintillant. On plastic- or aluminium-backed plates, the appropriate area can be cut out with scissors.

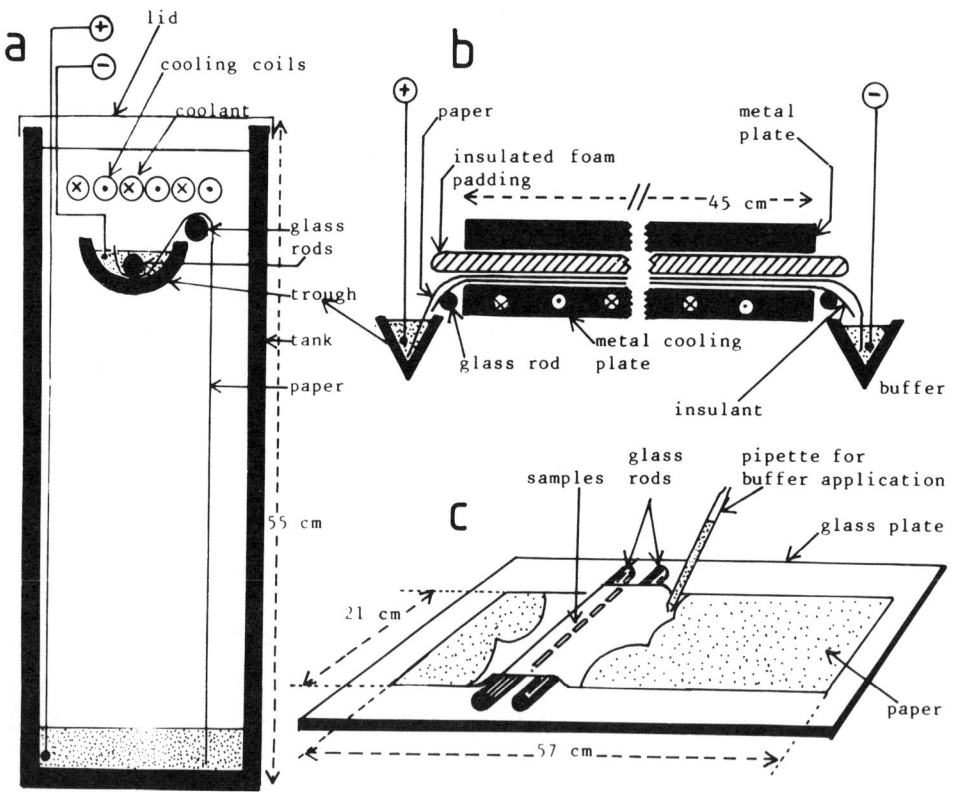

Fig 4.5.4: Paper electrophoresis.
(a) Diagrammatic representation of an apparatus for high-voltage PE using a liquid coolant e.g. white spirit.
(b) Flat plate apparatus for high-voltage PE.
(c) Method of wetting the paper with electrophoresis buffer.

  + = anode, **–** = cathode, **x · x · x · x ·** = cooling coils with cold tap water flowing through. The electrophoresis buffer is shown stippled.

4.5.4: PAPER ELECTROPHORESIS (PE) [457]

PE was introduced in Section 3.6.2 with reference to polysaccharides and glycoproteins. Here the emphasis is on low-M_r compounds. PE separates charged molecules, or those that can be given a charge by specific binding to an ion, e.g. borate or molybdate. Although in some ways less convenient than ion-exchange chromatography (which separates on similar parameters), it has important advantages. Thus, several samples can be run simultaneously under identical conditions and PE has most of the other advantages of paper methods (Section 4.5.2). Also, PE avoids the artefact that a compound might bind to an ion-exchange resin non-ionically and yet be released by increasing ionic strength or changing pH. In contrast, it is only possible for a compound to migrate in PE (relative to a neutral marker) if it really does possess a charge.

The paper used for PE of low-M_r compounds is often Whatman no. 3 or 3MM, or no. 1 (which is thinner). The samples are dried on to the paper as spots (1-1.5 cm diameter) or streaks (e.g. 3 x 1 cm) — see Fig. 4.5.4 c. If both anions and cations are of interest, the samples are loaded across the centre of the paper; if only neutral compounds and anions are of interest, the samples are loaded nearer to the cathode end — but not too near because neutral compounds move towards the cathode by electro-endo-osmosis (EEO: Section 3.6.2). A neutral marker (preferably internal), e.g. glucose, should be run so that the extent of any EEO can be monitored. The amount of sample spot-loaded on to Whatman no. 1 should not exceed about 50 μg of pure uronic acid or about 0.25 μmol of total non-volatile ions; samples should therefore be de-salted if possible. The loaded paper is laid on a sheet of glass and wetted with buffer (see below) delivered from a pipette. This is done carefully to prevent the samples running into one another. It is useful to have the area of paper bearing the samples raised between two glass rods and to leave the wetting of this part of the paper until last (Fig. 4.5.4 c). Excess buffer is lightly blotted from the electrophoresis paper with dry tissue paper (take care not to crease the PE when doing this).

Several types of equipment are available for PE. **(a)** The best for most purposes is high-voltage PE (up to 5 kV, at which voltage a 23 x 57 cm sheet of Whatman no. 3 paper typically draws a current of about 100-200 mA, giving a considerable ½-1 kW heating effect) with the paper (57 cm edges vertical) suspended in a large tank filled with immiscible coolant e.g. white spirit (Fig. 4.5.4 a). The top of the paper is held in a trough containing about 250 ml of the buffer and a platinum cathode; the bottom dips into another layer of buffer (about 500 ml) containing a platinum anode. Coils with running tap water keep the white spirit below 30°C. Disadvantages of this method are that non-polar compounds e.g. phenols may partition into the coolant and be lost, and that it is inconvenient to change the bottom buffer. There is also a fire risk from the white spirit, and sophisticated extinguishers and fume cupboards are required. **(b)** High-voltage PE can also be performed on a flat-bed system (Fig. 4.5.4 b): the paper (ca. 100 x 25 cm) is laid along an insulated metal plate (containing cooling coils) with the ends of the paper dipping into troughs containing buffer (about 150 ml each) and platinum electrodes. An insulated and padded lid is pressed down on top of the paper to maximise uniform contact with the cooling plate. This system has the advantages over (a) that compounds cannot be lost from the paper, and that it is easy to change over to a different buffer, but it does not provide such uniform cooling and therefore some irregularities in migration rate may occur across the width of the paper. **(c)** A simple low-voltage version (up to 250 V) of (b) can be made by clamping the paper (30 x 19 cm) between two air-cooled 20 x 20 cm glass plates [172]. This method gives poorer but still useful resolution.

High-voltage PE is a very fast separation method, typically taking 20-60 min. It is useful to include coloured markers (e.g. picric acid, bromophenol blue, xylene cyanol or orange G as anions; ethidium bromide, methyl green or crystal violet as cations) so that the course of PE can be monitored visually. It is difficult to find coloured neutral markers: alternatives include the feruloyl-oligosaccharides described in Section 4.4.2, which are neutral and can be seen under UV, and glucose, which can be stained with aniline H-phthalate (Panel 4.5.2 f).

Volatile buffers are chosen if possible so as to facilitate their removal after electrophoresis. Some examples are:

pH 0.6	formic acid/acetic acid/water (67:50:134)
pH 2.0	formic acid/acetic acid/water (1:4:45)
pH 3.5	acetic acid/pyridine/water (10:1:189)
pH 5.2	acetic acid/pyridine/water (1:2:97)
pH 6.5	acetic acid/pyridine/water (1:33:300)
pH 8.9	1 % (w/v) ammonium carbonate
pH 9.5	ethanolamine/acetic acid/water (5:2:118).

The charges borne by compounds at various pHs depend on the pK_as of the ionisable groups. These are roughly:

phosphate	pK_a = 2 (first ionisation)
——COOH	pK_a = 3-5
phenolic ——OH	pK_a = 9-10
——NH$_2$	pK_a = 10-13
imidazole ring	pK_a = 6.

The mobility (m) of a substance in PE is often quoted relative to a marker, e.g. $m_{picrate}$ or $m_{ethidium}$. Methods of detection and elution are the same as for PC (Section 4.5.2), but note that borate precludes the use of AgNO$_3$/NaOH staining.

At pH 2 (Panel 4.5.4 a) all amino acids (except cysteic acid) and oligopeptides have a net positive charge; they migrate towards the cathode with an m value which bears a predictable relationship to the number of full +ve charges and the M_r [136,387]. Hydroxyproline stands out as the slowest-migrating common amino acid, and hydroxyproline mono-, di-, tri- and tetra-arabinosides migrate progressively slower still; these compounds also stain in a characteristic way with isatin/ninhydrin (Panel 4.5.2 f). PE at pH 2 followed by staining with Folin & Ciocalteu's phenol reagent (Panel 4.5.2 f) is useful for the detection of tyrosine and isodityrosine, since these are among the very few cationic phenols. Polyamines (spermine, spermidine and putrescine), which may be found in cell wall hydrolysates [511], migrate more rapidly towards the cathode than any common amino acids (R_{lysine} ca. 1.6), and stain with ninhydrin. GlcA shows slight movement towards the anode (relative to glucose), whereas GalA is immobile. Most sugar-phosphates and nucleotides migrate towards the anode since the first pK_a of the phosphate group is low; but pH 2 is not recommended for phosphorylated compounds as some are acid-labile.

At pH 3.5 (Panel 4.5.4 b) most amino acids still migrate towards the cathode, except glutamic, aspartic and cysteic acids. Carboxylic acids e.g. uronic acids move towards the anode; PE at pH 3.5 is one of the best ways of separating GlcA from the slower-migrating GalA [566]. Phosphates and nucleotides, including Co-A derivatives, migrate rapidly towards the anode, and are more stable than at pH 2: only those with a glycofuranosyl ——phosphate bond (e.g. UDP-apiose) are appreciably hydrolysed at pH 3.5.

At pH 6.5 aspartic, glutamic and cysteic acids move towards the anode; lysine, arginine and histidine towards the cathode, and other amino acids are immobile. Carboxylic acids and phosphorylated compounds both migrate rapidly towards the anode.

Higher pH values are used more rarely. At pH 9.5 feruloyl-oligosaccharides acquire a negative charge by ionisation of the phenolic —OH group, and are thus separated from non-feruloylated neutral oligosaccharides, which are immobile at pH 9.5. Cooling during PE is especially important here to avoid hydrolysis of the ester linkages.

Electrophoresis in the presence of binding anions. Another use of PE is to test the ability of neutral carbohydrates to bind anions e.g. borate and molybdate; this ability can give a great deal of structural information. The buffer is supplemented with the anion, which binds to the carbohydrate and gives it a negative charge. Two useful buffers are:

1.9 % $Na_2B_4O_7.10H_2O$ (borax), adjusted to pH 9.4 with NaOH;

2.0 % $Na_2MoO_4.2H_2O$, adjusted to pH 5.0 with H_2SO_4.

Neutral sugars and alditols do not migrate at these pHs in the absence of binding anions. **Borate** binds to suitable pairs of —OH groups (Sections 3.1.3 and 3.5.5): such pairs occur in most sugars and alditols, giving electrophoretic mobility; the number of borate ions bound per unit mass of carbohydrate determines mobility. The method is useful for distinguishing oligosaccharides that differ in position of the bond [(1→2), (1→3), (1→4) etc.] [369]. There is often also separation of pairs of oligosaccharides that differ only in anomeric configuration (α versus β), e.g. maltose and cellobiose. Mobilities of many oligosaccharides, monosaccharides and alditols are known (Panel 4.5.4 c) [542]. Where the mobility of an oligosaccharide is unknown, some idea of the effect of changing α to β or p to f can be gained from the mobilities of the α/β,p/f methylglycosides (see Panel 4.5.4 c, right-hand column; abbreviations used include Me-α-Galf = methyl-α-D-galactofuranoside etc.). A good immobile (non-borate-binding) marker for correction for EEO in borate PE is 2,3,4,6-tetra-O-methylglucose (TMG). TMG is stained by spraying with aniline H-phthalate (cf. Panel 4.5.2 f). A coloured mobile marker is picric acid, which approximately co-migrates with arabinose.

Molybdate PE (Panel 4.5.4 d) gives valuable information on sugar—sugar linkages [542]. Molybdate-binding is possible in either of two main arrangements, requiring 3 or 4 correctly orientated —OH groups and giving rise to moderate and high PE mobility respectively. Arrangements permitting

PANELS 4.5.4 a-c: Mobilities ($\underline{m}$) of standards in electrophoresis.

$\underline{m}$	[a] pH 2.0 ($\underline{m}_{lysine}$)	[b] pH 3.5 ($\underline{m}_{ppi}$)	[c] Borate, pH 9.4 ($\underline{m}_{glucose}$)	
0.0	⊨Glc	⊨Glc	⊨TMG	⊨Me-α/β-Xylp
				⊢Me-α/β-Araf
0.1	⊢	⊢	⊨Sucrose	⊢Me-α-Glcp
		⊢AMP		
	⊢Hyp-Ara$_4$		⊢Propan-1,2-diol	
	⊢Hyp-Ara$_3$			
	⊢Hyp-Ara$_2$		⊨Trehalose	⊨Me-β-Glcp
0.2	⊨Hyp-Ara$_1$	⊨CDP-Glc		
			⊦Glc-β1→4-Glc	
			⊦Glc-β1→2-Glc	
	-Xylene Cyanol	⊢GulA, ADP-Glc		
		⊢GalA		
0.3	⊨Hyp	⊨MeGlcA	⊨	⊨ Me-β-Galf, Me-β-Manp
	⊢**ε-DNP-lysine**		⊦Glc-α1→2-Glc,	⊢Me-β-Xylf
		⊢ManA	Glc-α1→4-Glc	
	⊢Asp---Tyr	⊢GlcA		⊦Me-α/β-Arap,
	⊢Idt	⊢GDP-Glc		⊨ Me-α/β-Galp
0.4	⊨Phe, Pro	⊨UMP	⊨	⊢Me-α-Galf
	⊢Glu			⌐Me-α-Manp
	⊢Met----Thr	⊢Glc-1-P		
		⊢Ara-1-P, Xyl-1-P		
		-Picric acid	⊦Glycerol,	
0.5	⊨Leu -Ile	⊨UDP-Glc, UDP-Gal	⊨ $\underline{myo}$-Inositol*	⊨
	⊨Ser --Val	⊢ATP	⊢Rha	
		⊢UDP-Ara, UDP-Xyl		
	-Crystal Violet			⊢Me-α-Xylf
0.6	⊨	⊨UDP-GalA	⊨	⊨
	⊢Ala	⊢UDP, GalA-1-P		
0.7	⊨	⊨UDP-GlcA	⊦Glc-α/β1→3-Glc,	⊨
			⊨ Glc-α1→6-Glc	
			⊢Man	
	⊢Gly, Tyramine	⊢GlcA-1-P	⊦Glc-β1→6-Glc,	⊢Me-α-Glcf
			⊢Rib EryH, ThrH	
0.8	⊨	⊨UTP	⊨Glycerald, XylH	⊨
			⊢GlcH	
			⊢RibH	
			⊢AraH	
0.9	⊨His	⊨	⊨Fuc	⊨
			⊨ManH -Fru	
	⊢Arg		⊢Gal	
			⊢Ara----GalH	**-Picric acid**
1.0	⊨Lys	⊨pyrophosphate (PP$_i$)	⊨Glc, Xyl	⌐

FOR ABBREVIATIONS SEE PAGE 160

* = Streaked; Other $\underline{m}_{lysine}$ values at pH 2.0: cysteic acid = -0.29, polyamines $\underline{ca}$. +1.6.

177

binding are found in few monosaccharides (examples are mannose and rhamnose), but an arrangement permitting the 'high-mobility' type of binding occurs in all alditols except threitol and glycerol. The blocking of an —OH group in the alditol (e.g. with a non-binding sugar residue or with a methyl group or by complete removal of a particular —OH group as in a deoxyalditol) may, <u>depending on which —OH group is involved</u>, have one of three principal effects:

(a) no effect on molybdate binding (if the —OH group blocked was not one of the four required for high mobility) [e.g. glucitol substituted at position 1, 2, 5 or 6],

(b) restrict binding so that only the moderate-mobility arrangement is possible [e.g. glucitol substituted at position 4], or

(c) prevent binding altogether [e.g. glucitol substituted at position 3].

This means that PE mobility in molybdate buffer can indicate which —OH group of an alditol is substituted. This can be exploited in structural studies of an oligosaccharide to work out the <u>linkage between the penultimate sugar residue and the reducing terminus</u>. First it is established that the oligosaccharide itself is immobile in molybdate. Then the reducing terminal monosaccharide is converted to the corresponding alditol and all traces of borate produced from the NaBH$_4$ are removed (Panel 2.4.6). Finally the glycosyl-alditol so obtained is electrophoresed in molybdate buffer and its mobility compared with that of the free alditol (commercially available). Glucosyl-(1→6)-glucitol has a high mobility similar to that of 6-O-methyl-glucitol, whereas glucosyl-(1→4)-glucitol* has about half this mobility, similar to 4-O-methyl-glucitol, and glucosyl-(1→3)-glucitol is immobile, like 3-O-methyl-glucitol.

---

*Since alditols, unlike monosaccharides, do not have a C=O group to distinguish the two ends of the molecule, there are fewer alditols than aldoses. Thus, glucitol is the same as gulitol, and so 'galactosyl-(1→4)-glucitol' (trivial name lactitol; the product of treating lactose with NaBH$_4$) should strictly speaking be called galactosyl-(1→3)-**gulitol** since names are preferred that give the lowest numerals. In Panel 4.5.4 d, this nomenclatural rule is ignored to make it easier to relate the alditol units to the reducing termini of plausible oligosaccharides (gulose being unknown in plant cell walls). Weigel's [542] more complete lists of electrophoretic data do follow the rules, and it is important to realise that:

1-O-substituted galactitol	=	6-O-substituted galactitol;
1-O̅-substituted mannitol	=	6-O̅-substituted mannitol;
1-O̅-substituted gulitol	=	6-O̅-substituted glucitol;
1-O̅-substituted lyxitol	=	5-O̅-substituted arabinitol;
1-O̅-substituted xylitol	=	5-O̅-substituted xylitol.

PANEL 4.5.4 d: Mobilities of standards in molybdate electrophoresis

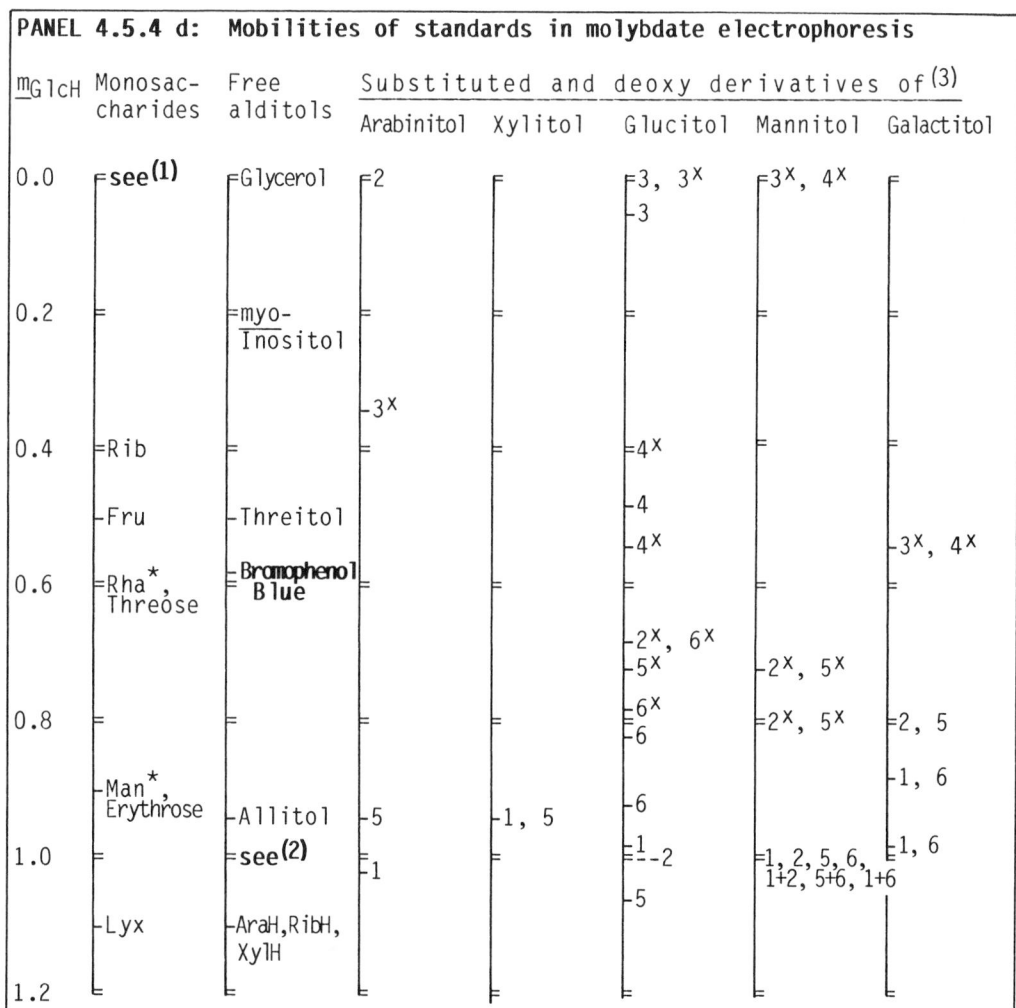

m_{GlcH} | Monosaccharides | Free alditols | Substituted and deoxy derivatives of [3] — Arabinitol, Xylitol, Glucitol, Mannitol, Galactitol

m_{GlcH}	Monosaccharides	Free alditols	Arabinitol	Xylitol	Glucitol	Mannitol	Galactitol
0.0	see[1]	Glycerol	2		3, 3^x / 3	3^x, 4^x	
0.2		myo-Inositol					
			3^x				
0.4	Rib				4^x		
	Fru	Threitol			4 / 4^x		3^x, 4^x
0.6	Rha*, Threose	Bromophenol Blue					
					2^x, 6^x / 5^x	2^x, 5^x	
0.8					6^x / 6	2^x, 5^x	2, 5
	Man*, Erythrose	Allitol	5	1, 5	6		1, 6
1.0		see[2] / 1			1 --2 / 5	1, 2, 5, 6, 1+2, 5+6, 1+6	1, 6
	Lyx	AraH, RibH, XylH					
1.2							

[1]Compounds at $m_{glucitol}$ = 0.0 include glyceraldehyde, Ara, Xyl, All, Glc, Gal, Fuc; oligosaccharides of any of these; and sucrose.

[2]Alditols at $m_{glucitol}$ = 1.00 are EryH, (GlcH), ManH, GalH, FucH and RhaH.

[3]The Panel shows $m_{glucitol}$ values of alditols, and the effect of blocking certain —OH groups of the alditol. Mobilities of glycosyl-alditols (i.e. NaBH$_4$-reduced disaccharides) are indicated by **x**; all others are O-methyl-alditols and deoxy-alditols. Thus, an entry '6' in the GlcH column could be 6-O-methyl-glucitol or 6-deoxy-glucitol, and the entry '6^x' could be galactosyl-α-(1→6)-glucitol (= NaBH$_4$-reduced melibiose); '1+6' indicates a di-substituted alditol. Glycosyl-alditols have slightly lower mobilities than the corresponding methyl- or deoxy-alditols, but certain trends still emerge, regardless of the substituent.

*Streaked.

179

Suitable immobile markers for molybdate PE are TMG (stained with aniline H-phthalate), glycerol (stained with AgNO$_3$/NaOH) or feruloyl-galactobiose (see Panel 4.4.2 a; located by UV fluorescence). A useful mobile marker is bromophenol blue, which has $\underline{m}$galactitol = 0.56.

4.5.5: Charcoal (carbon) chromatography

Activated charcoal adsorbs sugars from aqueous solution, and will subsequently desorb them into dilute ethanol. The concentration of ethanol required for related oligosaccharides rises with increasing DP. This forms the basis of a column chromatographic method for isolating oligosaccharides. The method has unusual advantages: **(a)** very large volumes of dilute aqueous samples can be applied to the charcoal column since the initial binding in water is strong; **(b)** highly impure samples can be handled since concentrated salts have little effect on binding and they can be selectively washed off the column with pure water; **(c)** a small amount of charcoal will bind a large amount of oligosaccharide.

Different batches of charcoal differ in binding properties: some will bind monosaccharides in water, others not. The chemical make-up of an oligosaccharide, as well as its DP, influences the ethanol concentration required for desorption. Clogging of the charcoal column is prevented by use of a 1:1 (w/w) mixture of charcoal and diatomaceous earth (e.g. Celite) rather than pure charcoal. The ethanol is applied to the column as a series of shallow steps, e.g. 0, 2.5, 5.0, 7.5, 10, 15, 20, 30, 40, 50 % ethanol. Flow rate should be slow, and gradients shallow since desorption from the charcoal is rather slow. A 10-ml bed volume column would easily accommodate a 100-mg sample of oligosaccharides and would typically be eluted with 50 ml of each ethanol concentration [554]. An example of the application of charcoal chromatography to the separation of xyloglucan oligosaccharides is given by Kato & Matsuda [279].

4.5.6 Gel-permeation chromatography (GPC)

GPC has already been described as a method for separating polysaccharides and glycoproteins (Section 3.5.2, q.v.). It is also very useful for separating oligosaccharides since it resolves compounds largely on the basis of their M_rs. The best medium for oligosaccharides in the range DP 2-12 is Bio-Gel P-2, which can give near-baseline resolution of a homologous series over this range [294,572]. For good resolution, sample size should be kept below 5 % of the bed volume; for the best possible resolution, the column should be water-jacketed at about 65°C. Uronic acid-containing oligosaccharides do not always behave in the same way as neutral oligosaccharides; but by use of an eluent buffered at about pH 4.5 this problem is minimised. A suitable eluent is pyridine/acetic acid/water (1:1:23) since it is volatile and can therefore be easily removed after chromatography. Even so, there is still slight variation in elution volume between different sugars: for instance monosaccharide deoxyhexoses elute slightly before pentoses, and these slightly before hexoses. For oligosac- charides of DP greater than 12, other grades of Bio-Gel P are used, e.g. P-4 and P-6. Useful markers are the malto-oligosaccharides, which are available commercially (Sigma Chem. Co.).

4.5.7: High-pressure liquid chromatography (HPLC)

An advantage of liquid chromatography over paper and thin-layer methods is that relatively volatile compounds (e.g. methanol or acetic acid, which may be present in wall hydrolysates) can be assayed. In addition, accurate quantification of separated compounds is easier. Liquid chromatography can give excellent resolution of many compounds of biochemical interest, but the sugars have yet to catch up with the phenolics and amino acids.

The smaller the particle size of a column packing material, the more efficiently does it interact with solutes in the sample, and this is the basis of the high resolution. Beds of very small particles require a high pressure to maintain solvent flow, and HPLC provides this. HPLC is a rapid form of chromatography, typical run times being 15-60 min, but only one sample can be analysed at a time. In some HPLC methods, the composition of the solvent is gradually changed during the run (gradient chromatography), but others use continuous delivery of a single solvent (isocratic chromatography). All HPLC systems require the following basics:

(a) Pump(s). The pumping system must be able to deliver the solvent at a pressure of up to about 5,000 lb/in^2. Isocratic chromatography requires one pump; gradient chromatography requires two — one for the initial eluent and one for the final, the ratio of pumping rates changing gradually during the run, preferably under the control of a computer.

(b) Pressure monitor. This is useful to confirm that the plumbing system is neither clogging nor leaking.

(c) An injection valve (e.g. Rheodyne) is an important piece of engineering since the sample (typically 20 µl) must be introduced into a solvent stream which is under a very high pressure.

(d) Detector(s). Compounds eluting from the column are either collected on a fraction collector and assayed later, or, more conveniently, are detected continually, as they elute, by use of a flow-through detector. Carbohydrates are usually detected by **refractive index** (RI) measurements by use of a flow-through RI monitor. All compounds can be detected in this way, and the limit of detection of glucose after chromatography is about 0.5-5.0 µg, depending on the quality of the detector and the chromatographic separation. RI monitoring is not possible with gradient elution as the background RI varies too widely. Other compounds are detected with an **ultra-violet (UV) absorbance** monitor, e.g. phenolics (reliable assay of 1 ng ferulic acid by A_{320}) and amino acid derivatives (e.g. dansyl- or PTH-amino acids). Most carbohydrates absorb UV only at very short wavelengths, below 200 nm, where many solvents and impurities also absorb. However, very clean samples can be analysed for carbohydrate by A_{190}. Other more specialised flow-through monitors are also available, e.g. based on UV-fluorescence, redox reaction, or solid-state scintillation-counting.

(e) If flow-through monitors are used, a means of continually recording their output is required: monitors can be wired to a moving-pen chart-recorder, or to an electronic integrator. The latter has the advantage that the results can be subsequently re-drawn on a different scale, useful if the 'pen' went off scale or showed only low peaks; and it can measure the area under a peak, facilitating quantification.

In addition to the above basics, 'optional extras' are available. A dynamic mixing unit is useful in gradient chromatography to ensure efficient mixing of the two solvents before they pass through the column. A column oven is useful for certain types of carbohydrate chromatography where resolution is improved by heating.

The chemistry of the column packing is the subject of a wide choice. Some of the more useful types are outlined here:

(i) Normal phase silica. This works on the same principle as silica-gel TLC, and is useful for chromatography of hydrophobic compounds e.g.

182

lipids. Non-polar compounds elute from the column first, and the more polar, which adsorb to the silica gel, elute last. Alkaline solvents will greatly shorten the useful life of the column.

(ii) Reversed phase silica. The silica is chemically substituted with alkyl side chains, usually octyl (C_8) or octadecyl (C_{18}), which enable it to adsorb <u>non-polar</u> compounds. The order of elution of polar and non-polar compounds is the reverse of (i). The starting solvent is relatively polar (e.g. water-rich) and is gradually changed during the run to a less polar one (e.g. methanol-rich). This mode of HPLC has been widely applied, especially to phenolics and to derivatised amino acids, but also with great success to oligosaccharides of DP 2-30 after <u>in vitro</u> acetylation [545].

(iii) Ion-exchange. Both cation- and anion-exchange columns are available: the principles of operation are the same as described in Section 3.5.3 but the resolution is improved and run-time curtailed. Anion-exchange HPLC on a silica support substituted with quaternary ammonium groups is suitable for analysis of uronic acids [531], the various acids derived from periodate oxidation (Section 4.2.7), and sugar-phosphates and -nucleotides. The compounds are eluted by increasing the ionic strength of the solvent. [With RI monitoring, a single ionic strength must be chosen at which all the compounds of interest will eventually elute; a purge with high-ionic strength eluent is possible between runs to cleanse the column.]

(iv) Amino-substituted silica adsorbs sugars and can be used for the chromatography of monosaccharides and oligosaccharides [524].

(v) Ion-moderated partition HPLC is performed on a bed of sulphonated polystyrene bearing suitable counter-ions e.g. Ag^+, Pb^{2+}, Ca^{2+} or H^+ (for last, see [396]. Monosaccharides, oligosaccharides, alditols (and even uronic acids [247]) bind reversibly to these columns and emerge after various times on isocratic elution with hot water (or with 5 mM H_2SO_4 for columns in the H^+ form). Bio Rad's HPX-87P column of this type, which has Pb^{2+} as counter-ion, is claimed to resolve Glc, Xyl, Gal, Ara, and Man; however, it is difficult to achieve base-line resolution of the last two, and Fru, Fuc and Rha each co-chromatographs with one of the previous five.

<u>Recommended HPLC methods</u> for specific classes of compounds likely to be met in cell walls are:

Phenolic acids [221]: Underivatised. Reverse-phase on C_8-silica. Initial eluent: H_2O/MeOH/HOAc (94:5:1), final eluent: H_2O/MeOH/HOAc (49:50:1), linear gradient, 2 ml/min. Separation time ca. 30 min. Monitor: UV at 270 nm. This is a compromise wavelength that detects all phenolics and is suitable for general surveys; use the absorption maximum of a particular compound of interest for higher selectivity and sensitivity.

Amino acids [106]: Derivatised by conjugation with PITC (Panel 4.3.3 b). Reverse-phase HPLC on C_8-silica. Dissolve sample in soln **S** [710 mg Na_2HPO_4, 900 ml H_2O, pH adjusted to 7.4 with 10 % H_3PO_4, volume made up to 950 ml with H_2O; all added to 50 ml acetonitrile]. Initial eluent: soln **D** [11.5 g NaOAc.3H_2O, 940 ml H_2O, 0.47 ml triethylamine, pH adjusted to 6.1 with HOAc; all added to 60 ml acetonitrile]. Final eluent: 54 % soln **D** : 46 % soln **E** [60 % acetonitrile]. Convex gradient, 1 ml/min. Separation time ca. 15 min. Monitor: UV at 269 nm.

Monosaccharides: It has not so far been possible to separate all the common monosaccharides of the plant cell wall in a single HPLC system. For a detailed account of progress reported up to 1983, see [255]. Samples can be run first on an amino-substituted silica column with acetonitrile/H_2O (3:1) as solvent, and the partially-resolved peaks pooled, dried and re-run on HPX-87P in H_2O to give complete resolution [57].

Oligosaccharides: Underivatised oligosaccharides can be resolved on Ag^+-moderated partition columns (e.g. Bio Rad's Aminex HPX-42A), with water as isocratic eluent, column temperature 85°C, and detected by RI. Underivatised oligosaccharides can also be resolved very effectively on amino-substituted silica (with gradient elution, e.g. 80 → 40 % CH_3CN in 15 mM NaH_2PO_4, pH 5.2) or by reversed-phase HPLC on C_{18}-substituted silica (gradient elution, 0 → 10 % acetonitrile in H_2O [359]. With any system of gradient elution, RI-monitoring is impossible, so the oligosaccharides must be detected by measurements of A_{190} or radioactivity. Oligosaccharides can be more easily detected after reversed-phase HPLC in a gradient system if they are first benzoylated; the per-O-benzoyl derivatives can be detected by UV measurements at higher wavelengths, where fewer impurities interfere [117]. However, benzoylation produces **two** derivatives (α and β) from each oligosaccharide, complicating the chromatography.

Acids, alcohols (including methanol) and the products of periodate oxidation [396]: H^+-moderated partition chromatography with 5 mM H_2SO_4 as isocratic eluent, and RI-monitoring.

4.5.8: Gas chromatography (GC, GLC)

Unlike the forms of chromatography so far discussed, where the mobile phase was a liquid, GC uses a flow of gas (e.g. helium) to keep the compounds moving past the stationary phase. The stationary phase is packed into a long, coiled column, which is heated in an oven. The standard (**packed**) columns are about 3 m x 3 mm, whereas **capillary** columns, which give better resolution and very high sensitivity, are about 15-25 m long and less than 0.5 mm in internal diameter. The sample is injected on to the column via a heated port: different volatile compounds pass along the column at different rates, eluting after characteristic retention times. The eluate is usually monitored with a flame-ionisation detector (FID). (For GC/MS see [443,533].) Separation is rapid, resolution is often excel-

```
┌─────────────────────────────────────────────────────────────────────────┐
│ PANEL 4.5.8:  Analysis of neutral sugars by GC of alditol-acetates(1)     │
│                                                                           │
│ Sample:         Mixture of monosaccharides(2) (0.1 mg each of Glc, Gal,   │
│                 Man, Ara, Xyl, Fuc, and Rha).  Alternatively, a TFA-hydro-│
│                 lysate of cell walls (see Panel 4.2.1 a).                 │
│                                                                           │
│ Step 1:         Add a known amount of myo-inositol as internal standard.  │
│ Reduce the monosaccharides to alditols [Panel 4.2.6].(2) Dry the sample.  │
│                                                                           │
│ Step 2:         To acetylate alditols: add 100 µl acetic anhydride (Ac2O)/│
│ pyridine (1:1) and heat at 100°C for 1½ h.                                │
│                                                                           │
│ Step 3:         Cool.  Gradually add saturated aqueous Na2CO3 to neutality.│
│                                                                           │
│ Step 4:         Add 1 ml CH2Cl2 and shake well to extract alditol-acetates│
│ into the organic (lower) layer.                                           │
│                                                                           │
│ Step 5:         Centrifuge (5 min, 2,500 g).  Collect & dry lower phase.(2)│
│                                                                           │
│ Step 6:         Re-dissolve the product in 50 µl acetone.                 │
│                                                                           │
│ Step 7:         Inject ca. 1 µl on to the gas chromatograph.  Column:     │
│ methylsilicone—fused silica [7,443,533].  Carrier gas: helium.  Port      │
│ temperature:  250°C.  Oven temperature:  140°C for 3 min, rising to 240°C │
│ over 17 min, followed by 5 min at 240°C [7,533].                          │
│                                                                           │
│ (1)See reference [55a] for a convenient alternative method, in which H2SO4│
│ hydrolysis of cell walls, alkalification with NH3, reduction of the sugars│
│ with NaBH4 in DMSO, re-acidification with HOAc, acetylation of the alditols│
│ with Ac2O + 1-methylimidazole, and extraction of the acetylated alditols  │
│ into CH2Cl2 are all performed in a single tube, minimising losses.        │
│                                                                           │
│ (2)The same method can also be used for analysis of methylated monosaccha-│
│ rides produced as described in Panel 4.2.8.  In this case, the reduction to│
│ alditols (step 1) should be performed with NaBH4 in 1 M NH3 in 95 % ethanol│
│ instead of water, and the methyl-alditol-acetates should be dried only    │
│ briefly (step 5) as these compounds are volatile.  Again, see also [55a]. │
└─────────────────────────────────────────────────────────────────────────┘
```

lent, and detection is sensitive and quantitative. Disadvantages are that most samples of cell wall origin require derivatisation to make them volatile, it is difficult to recover a sample for further analysis after GC, and quantitation of radioactive compounds is troublesome. Nevertheless, GC has become a standard method for quantitative analysis of non-radioactive, neutral sugars (Panel 4.5.8). The method described is not suitable for uronic acids. These are usually reduced to the corresponding sugar (e.g. GalA → Gal), which can be analysed as its alditol-acetate. Reduction can be achieved, after methyl-esterification, with LiBH4 [443].

4.6 SEQUENTIAL ANALYTICAL METHODS

Much information about the structure of a wall polysaccharide can be obtained by sequential analyses on the same sample, substantially cutting down the amount of sample needed. The product of one analytical step acts as the starting material for the next. Non-destructive methods of detection are required, e.g. scintillation-counting (Section 4.5.2), refractive index, or UV-absorbance or -fluorescence (Section 4.5.7). The use of radioactive samples is particularly powerful because (a) it greatly decreases the sample size required, thereby eliminating the need to remove reagents after each analysis (Section 4.5.1), (b) it is easy to ensure that all reaction products have been quantitatively accounted for, including any unsuspected ones, and (c) it is impossible to confuse the reaction-products, which are radioactive, with reagents or their contaminants [e.g. sugars released by autolysis of enzymes (Section 4.2.5) if the latter are glycoproteins or are contaminated with polysaccharides].

Much information can be deduced from changes in chromatographic behaviour in response to a given treatment, without any requirement for an authentic standard identical with the compound under investigation. For instance, if an oligosaccharide decreases in R_F upon treatment with pure β-galactosidase, then it had a non-reducing terminal β-galactose residue.

An example of the use of sequential methods can be taken from work on the structure of feruloyl-oligosaccharides (Section 4.4.2) [168]. Spinach cells were grown for several cell cycles on [U-^{14}C]glucose so that all organic components were labelled to a given specific radioactivity, measured as Ci or Bq of ^{14}C per mole (gramme-atom) of carbon. An unknown feruloyl-oligosaccharide (compound 29) was isolated and purified by paper chromatography. The sequence of treatments then used was:

1. Cmpd 29 [R_F = 0.29 in BuOH/HOAc/H_2O (15:3:5) (= BAW)] was hydrolysed by 0.5 M NaOH to compounds with R_F 0.04 and 0.88, the latter co-chromatographing with authentic ferulic acid and the former presumably being the de-feruloylated oligosaccharide. No ^{14}C was lost, showing that no acetyl groups had been present (they would have been released as [^{14}C]acetic acid and lost on drying). The ratio of ^{14}C recovered in the two products was about 6:5, i.e. for every ferulic acid (C_{10}) unit, the oligosaccharide must have had 12 carbon atoms. Compound 29 was thus probably a feruloylated C_{12} oligosaccharide [or possibly a bis-feruloylated C_{24} oligosaccharide, etc.].

2. The R_F-0.88 cmpd from step (1), upon re-chromatography on TLC in benzene/HOAc (9:1), ran with authentic ferulic acid, confirming its identity.

3. A small portion of the de-feruloylated oligosaccharide from step (1)

was heated with 0.05 M TFA, and the products were re-chromatographed in BAW: the R_F was unchanged, indicating that no furanose (highly acid-labile) residues had been present. The oligosaccharide was eluted from the paper, and re-hydrolysed in 2 M TFA: this time a single product was obtained, which ran with authentic galactose (R_F 0.10). When eluted and rerun in EtOAc/pyridine/H_2O (8:2:1), it again co-migrated with authentic galactose. Thus the oligosaccharide was composed of two (or possibly a multiple thereof) galactose units, agreeing with its possession of a multiple of 12 carbon atoms. The non-reducing residue(s) must have been in the relatively acid-resistant pyranose form.

4. The [^{14}C]galactose from step (3) was eluted again and treated with ATP + an extract of mung beans (known to contain D-galactokinase but no L-galactokinase — see Section 5.4.1). Electrophoresis at pH 3.5 showed the [^{14}C]galactose to have acquired a negative charge (i.e. been phosphory-lated), and thus to have been the **D**-isomer.

5. The rest of the oligosaccharide from step (1) was electrophoresed in borate buffer, and found to have approximately the same electrophoretic mobility as reported for D-Gal<u>p</u>-β-(1→4)-D-Gal [70]. This disaccharide is not available commercially, so its identity could not easily be confirmed by co-electrophoresis with an authentic internal marker.

6. The oligosaccharide was eluted from the paper in step (5) and treated with $NaBH_4$ to convert the reducing terminus into galactitol. Electro-phoresis in molybdate showed that the reduced oligosaccharide had a mobility about equal to that of bromophenol blue and thus the galactitol unit had another sugar linked to it at C-3 or C-4 (Section 4.5.4).

7. The reduced oligosaccharide from step (6) was eluted, treated with pure β-galactosidase, and re-chromatographed in butanone/HOAc/H_3BO_3-satu-rated H_2O (9:1:1) (Section 4.5.2). Two ^{14}C-spots were found, corresponding to galactose and galactitol. This shows that the linkage in the oligosac-charide was β-, not α-. The two products were obtained in <u>equal</u> amounts (^{14}C basis); thus the oligosaccharide had had one reducing terminus and one non-reducing unit, i.e. it was a <u>disaccharide</u>, D-Gal<u>p</u>-β-(1→4)-D-Gal.

This example illustrates the use of a battery of simple analytical techniques, many of which are applied in sequence, minimising the amount of sample required, to deduce the complete structure of a novel cell wall component. The entire analysis was done on less than 10 ng of sample.

5 Wall biosynthesis

5.1: BACKGROUND TO WALL BIOSYNTHESIS
5.1.1: Subcellular site

Cell wall polysaccharides are not synthesised in the wall itself. Matrix polysaccharides (except callose) are synthesised mainly or entirely in the endomembrane system, especially in the Golgi bodies, from whence vesicles appear to carry the newly-synthesised polymers to the plasma membrane. The vesicles fuse with the plasma membrane, expelling their contents into the wall. This is known mainly as a result of autoradiography at the EM level. Tissue is briefly fed [3H]sugars and then, after a defined time, free [3H]sugar and low-M_r [3H]metabolites are extracted with a fixative, which simultaneously immobilises the polysaccharides [384]. [3H]Polysaccharide first appears in the Golgi bodies, next in the vesicles, and finally in the cell wall. The [3H]polysaccharide rapidly disappears from the endomembrane system during a chase (Section 2.1.5) with non-radioactive sugar. From the kinetics of such pulse—chase experiments, it can be estimated that the transit time for the movement of polysaccharide molecules from Golgi to plasma membrane is usually ca. 10-30 min [177,426].

Back-up evidence for the idea that wall polysaccharides are synthesised in the Golgi bodies comes from the isolation of these subcellular organelles from samples of tissue taken during a pulse-labelling experiment: [3H]polysaccharide appears in the Golgi body, Golgi-derived vesicle and cell wall fractions in that order. Although this approach has less cytological precision because the isolated organelles are impure, it has greater chemical precision since it provides enough of the [3H]polysaccharide for detailed radiochemical characterisation and thus for demonstration that the [3H]polysaccharides detected are identical with wall components [72,88]. Further support comes from the demonstration that isolated Golgi bodies contain enzymes capable of synthesising wall-like polysaccharides (Section 5.5).

Cellulose and callose, on the other hand, are synthesised in the plasma membrane, directly at their final wall destination [125,563].

The polypeptide backbones of wall glycoproteins are thought to be synthesised by the ribosomes on the rough endoplasmic reticulum (RER) and extruded into the lumen of the RER. Extensin may be synthesised with a 'leader sequence' of hydrophobic amino acids, which is post-translationally removed from the polypeptide [97]. Proline residues are probably hydroxylated in the RER, yielding a hydroxyproline-rich protein [425]. Oligoarabinose side-chains also apparently begin to be attached to the Hyp residues within the RER [425]; this process may continue in the Golgi bodies [391]. There is a suggestion that certain extracellular polysaccharides are assembled as side-chains on polypeptide 'primers': if so, the process may also begin in the RER [382].

The precursors of lignin (coniferyl alcohol etc.) are synthesised intracellularly and exported into the wall, where they are polymerised [164]. The ferulic acid residues of spinach pectin are attached to the nascent polysaccharide's sugar residues almost simultaneously with sugar incorporation [177], indicating that feruloylation also occurs in the Golgi bodies.

Cutin monomers are synthesised intracellularly and probably exported into the wall as CoA-thioesters, but there is some evidence that further monomer synthesis and/or CoA-thioesterification occurs in the wall. The cutin polymer is formed in the wall and/or cuticle by transesterification of the CoA-thioesters [253].

5.1.2: Precursors and intermediates

Sugars are activated, prior to polysaccharide and glycoprotein synthesis, by conversion to nucleoside diphosphate sugars (NDP-sugars), mainly UDP- and GDP-sugars. CDP- and ADP-sugars have been reported in plants, but no rôle for them in the biosynthesis of wall polysaccharides or glycoproteins has been established [155]. The linkage between the sugar and the NDP is an α-glycosidic bond in the case of D-sugars and a β-bond in the case of L-sugars. [This means that the NDP group always points 'down' from carbon 1 of the sugar: see Section 3.1.2.] In most NDP-sugars the sugar residue is in the pyranose ring form [179], the only known exception in higher plants being UDP-Apif (apiose cannot form a pyranose ring because its main-chain has only 4 carbon atoms); some fungi contain UDP-Galf [187], and UDP-Araf has been synthesised chemically [27].

189

The starting material for biosynthesis of NDP-sugars can be regarded as the pool of hexose monophosphates, which are in rapid equilibrium by the action of the enzymes (1) phosphoglucomutase, (2) glucose 6-phosphate isomerase, (3) mannose 6-phosphate isomerase and (4) phosphomannomutase. The five hexose monophosphates act as a reservoir which can be tapped at three main points (a,b,c):

Glc-1-P $\underset{(1)}{\leftrightarrows}$ Glc-6-P $\underset{(2)}{\leftrightarrows}$ Fru-6-P $\underset{(3)}{\leftrightarrows}$ Man-6-P $\underset{(4)}{\leftrightarrows}$ Man-1-P.
 (a)   (b)   (c)

  ↕   ↓   ↕

NDP-Glc   Ino-1-P   GDP-Man

(a)   **Glc-1-P**   **+**   **UTP**   ↔   **UDP-Glc**   **+**   **PP$_i$**   (UDP-Glc pyrophosphorylase)
[and smaller amounts of:
(a') Glc-1-P + GTP ↔ GDP-Glc + PP$_i$ (GDP-Glc pyrophosphorylase) and
(a") Glc-1-P + ATP ↔ ADP-Glc + PP$_i$ (ADP-Glc pyrophosphorylase)]
(b)   **Glc-6-P**   →   **Ino-1-P**   (inositol 1-phosphate synthase)
(c)   **Man-1-P**   **+**   **GTP**   ↔   **GDP-Man**   **+**   **PP$_i$**   (GDP-Man pyrophosphorylase)

The two pyrophosphorylases (**a** and **c**) form NDP-sugars directly. The inositol 1-phosphate can be de-phosphorylated and oxidised to free glucuronic acid (GlcA), which is then re-phosphorylated to GlcA-1-P and acted upon by a third pyrophosphorylase to yield the NDP-sugar [325]:

       Ino-1-P + H_2O →   Ino + P$_i$ (Inositol 1-phosphatase)
             Ino + O_2 → GlcA + H_2O (Inositol oxygenase)
    GlcA   +   ATP   →   GlcA-1-P   +   ADP   (Glucuronokinase)
GlcA-1-P   +   UTP   ↔   UDP-GlcA   +   PP$_i$ (UDP-GlcA pyrophosphorylase)

UDP-GlcA can be formed by oxidation of UDP-Glc [155] as well as via inositol: it is unclear which route predominates in vivo [173].

The three NDP-sugars formed de novo (UDP-Glc, UDP-GlcA and GDP-Man) can act as precursors for the other NDP-sugars required in wall biosynthesis (Fig. 2.5). Major reactions involved in NDP-sugar interconversions (Fig. 2.5) include the following, which are practically irreversible:

UDP-Glc + 2 NAD$^+$ → UDP-GlcA + 2 NADH + H$^+$ (UDP-Glc dehydrogenase)
            UDP-GlcA → UDP-Xyl + CO_2 (UDP-GlcA decarboxylase)
            UDP-GlcA → UDP-Api + CO_2 (UDP-Api synthase)
                UDP-Glc → UDP-Rha   (multi-enzyme reaction)
               GDP-Man → GDP-Fuc   (multi-enzyme reaction)

and the following, which are freely reversible:

              UDP-Glc ↔ UDP-Gal   (UDP-Glc 4-epimerase)
            UDP-GlcA ↔ UDP-GalA   (UDP-GlcA 4-epimerase)
              UDP-Xyl ↔ UDP-L-Ara   (UDP-Xyl 4-epimerase)
            GDP-Man ↔ GDP-L-Gal   (GDP-Man 3,5-epimerase)

and possibly:

             GDP-Glc ↔ GDP-Man   (GDP-Glc 2-epimerase).

A number of sugar kinases occur in plants. These can phosphorylate free monosaccharides, at the expense of ATP, to form:

D-Glucose 6-phosphate	(Glucokinase; hexokinase)
D-Mannose 6-phosphate	(Mannokinase)
D-Fructose 6-phosphate	(Fructokinase)
D-Glucosamine 6-phosphate	(Glucosamine kinase)
N-Acetyl-D-glucosamine 6-phosphate	(N-Acetylglucosamine kinase)
D-Galactose 1-phosphate	(Galactokinase)
D-Glucuronic acid 1-phosphate	(Glucuronokinase)
D-Galacturonic acid 1-phosphate	(Galacturonokinase)
L-Arabinose 1-phosphate	(Arabinokinase)
L-Fucose 1-phosphate	(Fucokinase)

Kinases are not generally present that act on L-rhamnose or D-apiose. Xylose may be isomerised to xylulose, which is then phosphorylated to xylulose 5-phosphate, but this pathway is usually of low activity. The sugar 6-phosphates are readily converted to 1-phosphates (see above), and these, as well as the five other 1-phosphates listed, can be converted to NDP-sugars by pyrophosphorylases. These enzymes, which are also known as nucleotidyltransferases, catalyse the general reaction:

Sugar-1-P + NTP $\leftrightarrow$ NDP-Sugar + PP$_i$,

both the P atoms of the PP$_i$ arising from the NTP. Known pyrophosphorylase activities include those necessary to make the following NDP-sugars [155]:

UDP-α-D-Glc	GDP-α-D-Glc	ADP-α-D-Glc
UDP-α-D-Gal	GDP-α-D-Man	dTDP-α-D-Glc
UDP-α-D-GlcA	GDP-β-L-Fuc	CDP-α-D-Glc
UDP-α-D-GalA		
UDP-α-D-Xyl		
UDP-β-L-Ara		
UDP-α-D-GlcNAc		

By the action of kinases and pyrophosphorylases, cells can thus make NDP-sugars out of a wide range of free monosaccharides. These pathways are often of high capacity (Section 2.1.3), but most of them do not play a primary rôle in the de novo formation of wall material since free monosaccharides are rare in plant tissues. Rather they act as scavenger pathways, re-cycling monosaccharides produced by hydrolysis of polysaccharides, NDP-sugars and translocated oligosaccharides e.g. raffinose. Such pathways are nevertheless extremely valuable experimentally as they allow in vivo labelling of polysaccharides (Ch 2).

Some sugars occur in several different NDP-sugars, e.g. UDP-glucose, ADP-glucose and GDP-glucose. These probably serve different biosynthetic rôles. UDP-Glc is the major sugar donor for wall-bound glucose residues.

ADP-Glc is formed in the plastids, and used there for starch synthesis. GDP-Glc has long been suspected to serve a special rôle in cellulose synthesis [127], although this has never been confirmed and there is now good evidence, at least in secondary walls, that UDP-glucose plays the major rôle [93]. GDP-Glc is probably mainly involved in the synthesis of mannans [116]. In contrast to the NDP-glucoses, the NDP-pentoses contributing to wall polysaccharide synthesis are virtually confined to UDP-Ara and UDP-Xyl [179].

NDP-Sugars are probably the direct sugar-donors for wall polysaccharide biosynthesis. Supporting this, when [^{3}H]sugars were fed in vivo, polysaccharides started to accumulate radioactivity as soon as the corresponding NDP-sugar had become labelled [179]. Suggestions that sugar residues are transferred on to a membrane-bound lipid before being transferred to the nascent polysaccharide have not been substantiated [330].

There have been suggestions that wall polysaccharides are assembled as side-chains on polypeptide backbones, i.e. that glycoproteins are intermediates in polysaccharide synthesis [77]. There is no definitive evidence of this for any of the structural polysaccharides of the wall, though there is for non-structural polysaccharides such as the slime that root caps secrete to lubricate their passage through the soil. In maize, the slime is a fucose-rich polysaccharide. Since it is extracellular, it must pass through the cell wall and therefore is at least transiently a wall component. The polysaccharide is synthesised as a glycoprotein [the sugar —amino acid bond was suggested to be xylose—threonine] from which the protein moiety is subsequently cleaved [201]. The evidence for this was based on in vivo feeding with radioactive fucose in pulse—chase experiments; polysaccharides and glycoproteins were distinguished by CsCl centrifugation (Section 3.5.4).

Enzymic glycoproteins in the cell wall frequently possess mannose-rich side-chains: these are probably assembled on a lipid carrier [e.g. dolichol (Dol)] in the ER or Golgi bodies, in the way established for many animal glycoproteins. Key intermediates include Dol-P-Man, Dol-P-Glc and Dol-P-P-oligosaccharides. A hydroxyproline-rich glycoprotein, synthesised in response to fungal elicitors (Ch 9), may be arabinosylated via lipid intermediates [66]. Some polymers tentatively identified as arabinogalactan proteins (AGPs) may also be synthesised via lipid intermediates [239,330].

5.1.3: Glycosyltransferases and polysaccharide synthases

An enzyme that transfers sugar residues from one molecule to another is called a glycosyltransferase, e.g.

NDP-Sugar + X → NDP + X-Sugar.

In the special case where the product is a polysaccharide, the enzyme is called a polysaccharide synthase:

NDP-Sugar + (Sugar)$_n$ → NDP + (Sugar)$_{n+1}$

The glycosyltransferases involved in the biosynthesis of wall polysaccharides and glycoproteins are bound to ER, Golgi or plasma membranes (Section 5.1.1) [88,202,244,391]. Little is known of the enzyme molecules: since they are membrane-bound their purification in active form is difficult. This makes it difficult to characterise them, e.g. with regard to specificity of sugar-donor and -receptor. It is thus impossible to estimate, even to an order of magnitude, how many different glycosyltransferases are required to build the cell wall. Glycosyl-transferases that have been solubilised in active form include a β-glucan synthase [244,498], which in one case retained enzymic activity when electrophoresed in a polyacrylamide gel [498], and a glucuronosyl-transferase capable of adding GlcA residues to nascent xylan [535]. A monoclonal antibody raised against Golgi membranes inhibits membrane-bound arabinosyltransferase activity, but the same antibody binds to five different Golgi proteins, precluding definitive identification of one particular protein as the arabinosyltransferase [68]. It is intriguing that even cellulose synthase, the enzyme that produces the most abundant organic chemical on Earth, has never been isolated, nor even merely identified as a 'band on a gel'. In fact, cellulose synthase has proved to be one of the most difficult polysaccharide synthase activities even to demostrate in crude cell homogenates [125].

The difficulty of demonstrating cellulose synthase activity in plant extracts _in_ _vitro_ contrasts with the ease of demonstrating callose synthase. This has led to the suggestion that the same protein molecule can catalyse both reactions in plants, but that its cellulose synthase activity is dependent on the cell's structural integrity and is therefore lost during the preparation of a cell-free extract [125]. Compatible with this idea is the fact that callose synthesis is a widely observed wound response, activated extremely rapidly upon physical injury of living plant cells [74].

Little is known of the chemistry of the polysaccharide synthase reaction. For instance, it is unclear whether the polysaccharides grow towards their reducing or non-reducing termini. It is unknown whether the enzymes can build a polysaccharide de novo or whether a primer (e.g. oligosaccharide or protein) is required as a foundation. There is ambiguity as to whether the synthesis of polysaccharide backbones and side-chains is obligatorily coupled [535]; in the case of xyloglucan some evidence suggests that the α-xylose residues are attached simultaneously with the elongation of the β-glucan backbone [242], but other evidence suggests that the xylose residues can be added to a long, pre-formed glucan backbone [411]. The β-galactosyl and α-fucosyl side-chains can be added to a pre-formed xyloglucan backbone [89].

Although all known NDP-D-sugars and NDP-L-sugars have α- and β-glycosidic linkages respectively, they act as precursors for polysaccharides with both α- and β-glycosidic linkages. Thus some glycosyltransferases apparently retain the anomeric linkage of the NDP-sugar during the transfer reaction and others reverse it. On simple chemical grounds one would expect a reversal; those transferases that do not appear to effect a reversal have therefore sometimes been suspected of proceeding via two transfers (e.g. NDP-α-sugar → lipid-β-sugar → polysaccharide-bound α-sugar), thus restoring the original anomeric form [382]. Similarly, although all known NDP-sugars in plants (except UDP-apiose) have pyranose rings, some glycosyltransferases contract the ring to the furanose form, apparently during the transfer reaction [179].

The study of polysaccharide synthases would benefit from the availability of specific inhibitors, but few have been found. One notable exception is cellulose synthase, which is specifically inhibited in vivo and probably also in vitro by the herbicide 2,6-dichlorobenzonitrile. The affinity of cellulose synthase for 2,6-dichlorobenzonitrile suggests novel ways of tagging and thus identifying this protein [128]. A second inhibitor is the fluorochrome of aniline-blue [available commercially from Biosupplies Australia Pty], which inhibits callose synthase. These two inhibitors, and any others that are discovered, will also be useful in testing the biological rôles of wall polysaccharides. A potential future approach to this problem would be the study of mutants defective in

194

specific polysaccharide synthases.

Membrane-bound polysaccharide synthase activities have been detected by _in vitro_ assay (Section 5.5) that use NDP-sugars to incorporate the following sugar residues into nascent polysaccharides:

Sugar residue	Donor(s)	Possible polysaccharide(s) involved
β-(1→4)-D-Glcp	UDP-Glc	Cellulose [59], Xyloglucan [242], β-(1→3),(1→4)-Glucan [244]
β-(1→4)-D-Glcp	GDP-Glc	Glucomannan [116]
β-(1→3)-D-Glcp	UDP-Glc	β-(1→3),(1→4)-Glucan [244], callose [74]
β-(1→4)-D-Galp	UDP-Gal	Galactan [393], Pectins
β-D-Galp (terminal)	UDP-Gal	Xyloglucan [89]
β-(1→4)-D-Manp	GDP-Man	Mannan [460], Glucomannan [116]
β-(1→4)-D-Xylp	UDP-Xyl	Xylan [67], Arabinoxylan, Glucurono-xylan [534]
α-D-Xylp (terminal)	UDP-Xyl	Xyloglucan [242]
α-L-Araf	UDP-Ara	Arabinan [386], Pectins [67], Extensin side-chains [391]
α-L-Fucp	GDP-Fuc	Xyloglucan [88,89], Maize slime [201]
α-(1→4)-D-GalpA	UDP-GalA	Pectins [528]
α-D-GlcpA (terminal)	UDP-GlcA	Glucuronoxylans [534]
D-Apif (terminal)	UDP-Api	Apiogalacturonan [392]

5.1.4 Post-synthetic modification of polysaccharides

Sugar residues of polysaccharides are believed to be methyl-esterified and -etherified, acetylated, feruloylated and p-coumaroylated **intra-cellularly,** before the polysaccharide has been secreted into the wall [173,177]. Non-sugar groups are probably added directly to the nascent polysaccharide rather than to NDP-sugars. Methyl-ester (and ether?) groups are added from S-adenosylmethionine (SAM) [286], and acetyl groups may arise from acetyl-CoA [156]. The feruloyl donor is unknown: naturally occurring candidates include feruloyl-CoA [329,578] and feruloyl-β-D-glucoside [503]. We have little information about the methyl-, acetyl- and feruloyl-transferases, their subcellular location, substrate specificities or control. There are also no data on how or where polyamines [511], silanoates [438] or gibberellins [479] become linked to wall polymers.

5.1.5: Secretion

The polysaccharides and glycoproteins that are synthesised in the ER and Golgi bodies are thought to be carried to the wall in Golgi-derived vesicles. It is unclear how these vesicles find their way to the correct position, adjacent to the appropriate section of the cell surface. Once they arrive there, they appear to fuse with the plasma membrane, discharging their contents by exocytosis into the wall. Isolated Golgi vesicles will fuse with isolated fragments of plasma membrane in vitro, as can be shown by changes in buoyant density. This fusion depends on Ca^{2+}, and on the presence in the plasma membrane of a specific protein, possibly a Ca^{2+}-dependent ATPase [43].

The functioning of the Golgi-transport system can be inhibited by the ionophore monensin [493]. This provides a method for blocking the deposition of matrix polysaccharides in the wall [83], although the specificity of monensin should be checked for each tissue on which it is used.

Cellulose and callose, apparently synthesised by enzymes in the plasma membrane itself, are delivered to the wall directly. The enzymes take intracellular UDP-glucose, and deliver extracellular polysaccharides, somehow manoeuvring the hydrophilic sugar residues through the hydrophobic membrane [125]. The individual cellulose molecules are initially produced free, but very quickly aggregate to form semi-crystalline microfibrils; this aggregation can be blocked by application of substances that intercalate with cellulose, e.g. Congo Red [424].

5.1.6: Regulation of wall polysaccharide synthesis

Regulation is an important consideration because changes in wall composition occur during the cell cycle [11], growth [83] and development [399,501]. Regulation is conceivable at a variety of levels, including:

(a) Supply of NDP-sugars. The general carbon and energy balance of the cell will govern its ability to make NDP-sugars. Thus, it has been shown that cell suspension cultures grown in a chemostat under conditions of carbon-limitation produce far less wall material than those under nitrogen-limitation [291].

(b) Interconversion of NDP-sugars. UDP-GalA was found in larch cambial cells actively synthesising GalA-rich pectins, but not in the young xylem cells where synthesis of xylan greatly predominated over that of pectin [112]. This suggests control of polysaccharide synthesis at the level of UDP-sugar supply. Another example is in the synthesis of apiogalacturonan, a major polysaccharide of duckweed fronds that is virtually absent from the resting buds (turions). Fronds but not turions can convert [^{3}H]GlcA into UDP-[^{3}H]apiose, whereas both can convert it into UDP-xylose

196

[327]. This indicates control over the pathway by which UDP-GlcA is converted into UDP-apiose, a pathway only required in fronds.

(c) Subcellular site of NDP-sugars. Virtually nothing is known of where in the cell the NDP-sugars are synthesised and interconverted. Also, it is not known whether the Golgi membrane-bound polysaccharide synthases require their NDP-sugars to be present in the Golgi lumen or in the cytosol. The latter is possible by analogy with the plasma membrane-bound polysaccharide synthases, which require cytosolic NDP-sugars [74] even though their polysaccharide products are churned out on the other side of the membrane. Sub-cellular distribution of NDP-sugars could potentially control polysaccharide synthesis; methods for the investigation of this are being developed [87].

(d) Polysaccharide synthase activity. This is widely regarded as the most general control mechanism. Activities of specific polysaccharide synthases, assayed *in vitro*, change in parallel with the rate of polysaccharide synthesis during the differentiation of cambium into xylem [115, 383] or phloem [216]. Control could be at the level of *de novo* formation of enzyme molecules, or of activation of existing ones (including provision of an appropriate primer [556]). There is evidence for the former [68].

(e) Secretion. It has been suggested that polysaccharide secretion is a further control point. Since Ca^{2+} is required for fusion of Golgi-derived vesicles with the plasma membrane [43], cytosolic Ca^{2+} levels, which are themselves tightly controlled, could regulate polymer secretion [477]. An accumulation of polysaccharide in the endomembrane system might secondarily block further polysaccharide synthesis [362].

5.1.7: Biosynthesis of glycoproteins

The carbohydrate side-chains of glycoproteins are discussed in Sections 5.1.1-6. The polypeptide backbones are synthesised on ER-bound 80-S ribosomes, using aminoacyl-tRNAs as donors, by mRNA translation. Most plant cells can synthesise all 20 genetically-encoded amino acids *de novo* but can also take them up from the medium if supplied artificially.

Amino acids (R—COOH) are activated by binding to the 3'-end of a tRNA to form an ester bond, in some cases *via* an intermediary adenylyl anhydride:

R—COOH + ATP ↔ R—CO—O—**P**—O—adenosine + PP$_i$,
R—CO—O—**P**—O—adenosine + H̄Ō—tRNA ↔ R—CO—O—tRNA + AMP.

The ester bond has a high enough free energy of hydrolysis to 'drive' the formation of a peptide bond, a well known reaction occurring on the ribosomes and relying on an mRNA template.

Cytosolic protein synthesis can be inhibited in certain plant tissues e.g. spinach cell suspensions by cycloheximide, e.g. at 10 μM. A more generally-applicable inhibitor, which is more specific for protein synthesis, is MDMP [= D-2-(4-methyl-2,6-dinitroanilino)-N-methylpropionamide], e.g. at 15 μM [270].

5.1.8: Biosynthesis of diferulate bridges, lignin and related phenolic coupling products

Phenolic monomers for wall synthesis are formed intracellularly. The ultimate raw materials are sugars, which are converted via the shikimate pathway to phenylalanine and tyrosine. All cells active in protein synthesis need these two amino acids, which thus represent the branch points at which material may be diverted from primary to secondary metabolism. A rate-limiting step for total phenolic synthesis is that catalysed by phenylalanine ammonia-lyase (PAL) [269],

$$\text{Phenylalanine} \rightarrow \text{trans-Cinnamate} + NH_3,$$

and the levels of this enzyme in the cell respond dramatically to stimuli e.g. hormones, oligosaccharins (Ch 9) and even the simple dilution of a cell suspension culture [218]. However, the supply of substrate (phenylalanine) can also limit phenolic synthesis [334].

Some Monocotyledons possess a second enzyme leading into phenolic synthesis, tyrosine ammonia-lyase (TAL):

$$\text{Tyrosine} \rightarrow \text{trans-p-Coumarate} + NH_3.$$

PAL is specifically and strongly inhibited both in vivo and in vitro by micromolar concentrations of amino-oxyphenylpropionic acid (AOPP) [269]. Only at much higher doses does AOPP affect other enzymes (transaminases). Low doses of AOPP block lignification in developing wood [12] showing that AOPP is a useful tool for probing the biological rôles of wall phenolics.

The trans-cinnamate formed by PAL has two fates:
(a) it is hydroxylated (by O_2) and methoxylated (by S-adenosylmethionine) to form various hydroxycinnamates

$$\text{Cinnamate} \xrightarrow{O_2} \text{p-Coumarate} \xrightarrow{O_2} \text{Caffeate} \xrightarrow{SAM} \text{Ferulate} \xrightarrow{O_2, SAM} \text{Sinapate};$$

(b) cinnamate and hydroxycinnamates are conjugated with glucose and CoA (and, in some tissues, quinate, malate, shikimate, malonate, sucrose etc.):

p-Coumarate + UDP-Glucose → p-Coumaroyl-β-glucoside + UDP   [329,503]
p-Coumarate + CoA-SH + ATP → p-Coumaroyl-S-CoA + AMP + PP$_i$ [578].

It is likely that many of the reactions listed under (a) actually involve (hydroxy)-cinnamate groups that have been conjugated as in (b).

These conjugates probably include the donors for the synthesis of hydroxycinnamoyl-polysaccharides (Section 5.1.4), although it is unclear which conjugate plays the major rôle. CoA-thioesters are also the precursors of the hydroxycinnamyl alcohols used in lignin synthesis, e.g. [212]:

Feruloyl-S-CoA + 2 NADH + 2 H$^+$ → Coniferylalcohol + 2 NAD$^+$.

In some lignifying cells, e.g. in gymnosperm wood, the hydroxycinnamyl alcohols react with UDP-glucose to form 'coniferin' (trans-coniferyl-4-0-β-D-glucopyranoside). Coniferyl alcohol and/or coniferin are secreted into the cell wall. The glucose residue of coniferin is removed by a β-glucosidase (Section 6.1.1), and the free coniferyl and related alcohols are then polymerised to form lignin. In this process, they are first oxidised by peroxidase + H$_2$O$_2$ (see p. 225) to form free radicals (Fig. 5.1.8 a). The free radicals are non-enzymically interconverted to yield a mixture of isomers (mesomers) (Fig. 5.1.8 b), which couple non-enzymically to form dimers (Fig. 5.1.8 c). The dimers are then subject to further oxidative coupling, eventually forming a complex, random, insoluble polymer: lignin [212].

The initially formed dimers are often unstable, especially those with a quinone-methide structure:

Quinone-methides are reactive, being attacked by water to form benzyl alcohols, by alcohols to form benzyl ethers, and by acids to form benzyl esters (Fig. 5.8.1 e) [50,316,492]. The reaction can be intra-molecular (if sterically favourable), forming a ring (Fig. 5.1.8 d), or inter-molecular, e.g. with part of a different phenolic or polysaccharide molecule, forming a cross-link (Ch 7). It seems plausible that, via quinone methide intermediates, some initially-formed dimers of feruloyl-polysaccharides can graft on to one (or two) non-feruloylated polysaccharide molecules, forming stable ether cross-links (Fig. 5.1.8 f) [50].

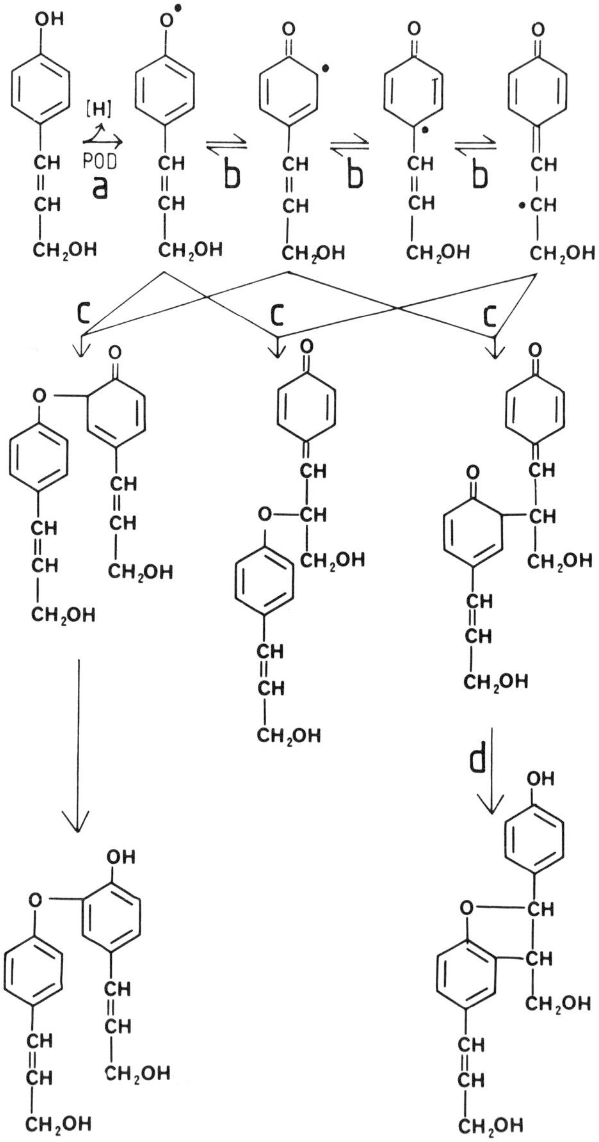

Fig. 5.1.8 a-d: Lignin synthesis, illustrated by one of its three main precursors, p-coumaryl alcohol. **(a)** Formation of free radicals by peroxidase (POD) + H₂O₂, **(b)** interconversion of mesomeric free radicals, **(c)** random coupling, **(d)** intramolecular reaction of a quinone methide type dimer.

200

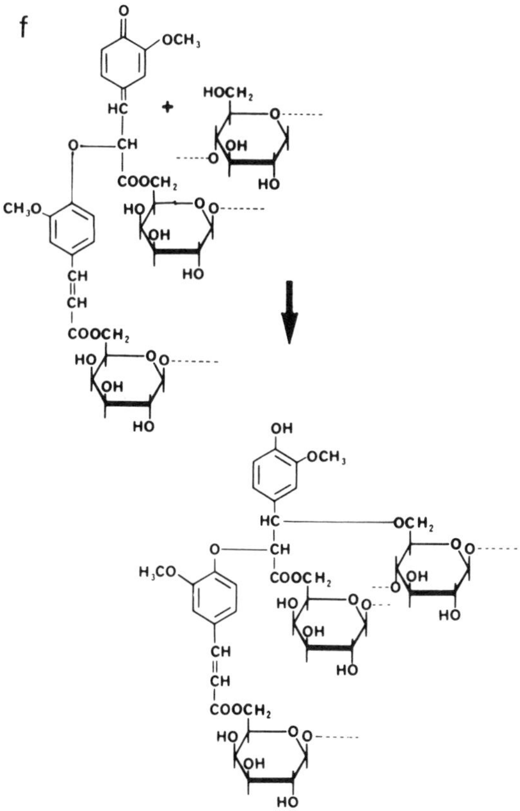

Fig. 5.1.8 e-f: Reactions of quinone methides.   (e) Generalised reaction of a quinone methide with water, an alcohol or an acid;   (f) Possible formation of ether cross-links between a quinone methide type dimer of feruloyl-polysaccharides and a third polysaccharide molecule.

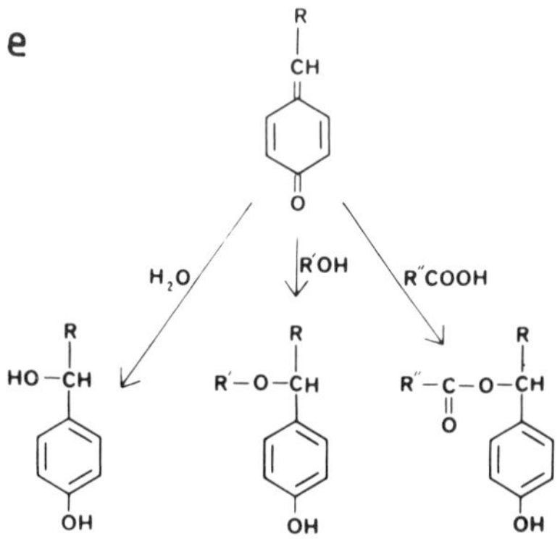

5.2: IN VIVO ASSAY OF WALL POLYSACCHARIDE AND GLYCOPROTEIN SYNTHESIS

5.2.1: Rate of synthesis

Methods for studying the rate of polymer synthesis include:

(a) Simple assay of the difference in mass of wall components in two equivalent samples taken at different times. The time points need to be widely spaced, and only net wall synthesis can be recorded, since the possibility that polymers undergo turnover [simultaneous synthesis and breakdown (Ch 8)] is ignored.

(b) In vivo feeding of radioactive precursors. This allows study of synthesis over short time intervals. By use of pulse—chase experiments, turnover can be allowed for. Pitfalls, and strategies for their avoidance, are mentioned in Section 2.6.2. In particular it is important to take into account the fact that endogenous non-radioactive wall precursors will be present in addition to the radioactive molecules; cells use both sources.

(c) In vitro assay of polysaccharide synthases (Section 5.5). This approach may indicate the mechanism for any change in rate of synthesis, but should not be relied on to indicate the existence of the change since synthase activity may not always be limiting for synthesis (Section 5.1.6).

5.2.2: Route of synthesis

Pulse-labelling (Section 2.1.4) is a powerful tool to investigate the pathway of polysaccharide synthesis, i.e. to identify intermediates. A radioactive precursor which will be incorporated into the polymer of interest is fed to the cells. Labelled compounds detected are potential intermediates. However, it is not sufficient to show that these compounds are labelled during polymer synthesis. It is possible that such compounds are intermediates in the synthesis of a different polysaccharide, or that they are not intermediates at all but metabolic end-products in their own right. Evidence for an intermediary rôle [179,201,350,351] can include:

(a) Comparison of the kinetics of labelling of the putative intermediate and of the polymer of interest. When tracer levels (e.g. 1 μM) of exogenous [^{3}H]arabinose are fed to cultured spinach cells, incorporation of ^{3}H into polysaccharides is detectable within 1 min, and reaches a linear rate within about 5 min. This means that all obligatory intermediates in the pathway Ara $\rightarrow$ $\rightarrow$ $\rightarrow$ $\rightarrow$ polysaccharide will be labelled to a plateau value of specific activity within 5 min. Such kinetics are observed for UDP-Ara and UDP-Xyl, suggesting that they are polysaccharide precursors [179]. [Since the tracer levels of [^{3}H]arabinose used are insufficient to affect the pool size of cellular UDP-Ara, and therefore the total amount of UDP-Ara (in μmol/g fresh weight) remains nearly constant during the labelling, the amount of UDP-[^{3}H]Ara (in μCi/g fresh weight) is essentially equivalent to a specific radioactivity.] If there are several suspected intermediates, the kinetics with which they reach their respective plateaux can indicate the sequence of the pathway.

(b) Pulse-chase experiments. These kinetics, while consistent with an intermediary rôle for UDP-Ara, would also be consistent with UDP-[^{3}H]Ara being formed, in small amounts, as an unstable by-product unconnected with polysaccharide synthesis. The agreement of kinetics could be coincidence. Evidence that the proposed intermediate is not an irrelevant by-product can be obtained from a pulse—chase experiment (Section 2.1.5) by demonstration that after the chase is started the amount of ^{3}H lost from UDP-[^{3}H]Ara is gained by continued synthesis of a small amount of [^{3}H]polysaccharide. For the latter to be detectable, the chase must be started quickly, before much [^{3}H]polysaccharide has accumulated.

(c) Use of inhibitors. To eliminate further the risk of coincidence, it would be valuable to apply a specific inhibitor of the synthesis of the polysaccharide or glycoprotein of interest. Such an inhibitor would be expected either to block the labelling of the proposed intermediate, or to suppress the loss of radioactivity from the intermediate during the chase, depending on the inhibitor's particular target reaction:

Precursor -----x------→ Intermediate -----x------→ End product.
             | |
      inhibitor 1   inhibitor 2

Unfortunately, few specific inhibitors are available (see Section 5.1.3).

Some workers have attempted to provide evidence for particular pathways by addition of a high concentration of the non-radioactive form of a suspected intermediate (B) in the presence of an undisputed, earlier, ^{3}H-labelled intermediate (A). The aim is to distinguish between the pathways

A → B → polysaccharide   and   A → C → polysaccharide.

The idea is that the [^{3}H]B molecules formed by the cells from the exogenous [^{3}H]A are swamped by the large excess of added non-radioactive B, thereby blocking incorporation of the ^{3}H into the polysaccharide. However, the concentration of added B will probably be non-physiologically high. The evidence obtained may indicate that **B** <u>can</u> participate in polymer synthesis, if it is supplied, but not that it <u>does</u> so under normal conditions. For instance, the incorporation of ^{3}H from 1 μM [^{3}H]glucose into arabinan <u>in vivo</u> is blocked by addition of 1 mM arabinose (without affecting the rate of incorporation of radioactivity into [^{3}H]cellulose). One might therefore propose the pathway:

Glucose → → → → → Arabinose → → → → → **Arabinan.**

However, it is highly unlikely that free arabinose is a normal intermediate in the **primary** synthesis of arabinan [173]. The competing effect of cold arabinose is probably due to a separate **scavenger** pathway which comes into play only when Ara-containing polymers are being broken down to any appreciable extent:

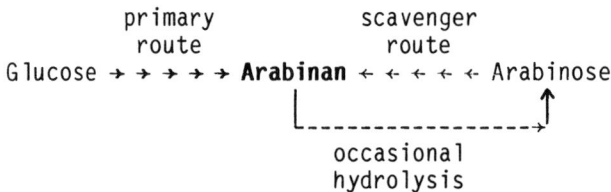

The same argument can be applied to the fact that excess inositol blocks the incorporation of radioactive glucose into UDP-GlcA (and its products — uronans, xylans, arabinans, etc.) [19,58,423]. Application of non-radioactive inositol therefore does not provide evidence for or against the inositol pathway as being the primary route of UDP-GlcA biosynthesis (Section 5.1.2) — i.e. it does not distinguish which of the following two pathways predominates in the living cell under normal circumstances:

Glucose → → → Inositol → → → UDP-GlcA

versus:

```
            primary                      scavenger
            route                        route
Glucose →   →   →  UDP-Glc  →  UDP-GlcA  ←   ←   ←  Inositol
    |                                                   ↑
    L→ → → → Inositol polyphosphates ------------------→|
             and phospholipids              occasional
                                            hydrolysis
```

5.3: IN VIVO ANALYSIS OF INTERMEDIATES IN WALL SYNTHESIS

The analysis of intermediates requires their extraction from plant tissues, and a recommended general extractant for low-M_r intermediates is $CHCl_3$/MeOH/H_2O (10:10:3). Originally developed for the extraction of glycolipids [47], it also gives good recoveries of inositol, phosphates, nucleotides [179] and phenolics. It does not extract polysaccharides or glycoproteins. Thus, a standard extraction protocol (Panel 5.3.1, steps 1-3) is common to many of the following Panels; this is helpful as it makes it possible to perform many different analyses on a single tissue extract. The material not extracted can subsequently be analysed for polymers.

5.3.1: Inositol

Since myo-inositol is a key intermediate in one of the proposed pathways to UDP-GlcA, its assay is of interest in studies of wall synthesis [325]. Unfortunately, there are no simple colour tests to distinguish inositol from other cell constituents. However, it has useful diagnostic properties: it is neutral, relatively stable to boiling acid and alkali, and does not bind to activated charcoal. After such treatments, its R_F on paper chromatography identifies it reliably (Panel 5.3.1).

5.3.2: Sugar phosphates

The sugar phosphates in the $CHCl_3$/MeOH/H_2O-extract are purified by electrophoresis or ion-exchange chromatography (high or low pressure), techniques that exploit the strong negative charge of the phosphate group. They are then identified by hydrolysis and chromatography of the resulting free sugars. Sugar 1-phosphates, which are of particular interest as wall polysaccharide precursors, can be distinguished from sugars phosphorylated on other positions (e.g. glucose 6-phosphate) by the high susceptibility of the former to acid hydrolysis and by the much higher susceptibility of the latter to hot alkali [21]:

Compound	Reaction products obtained upon treatment with	
	50 mM KOH, 100°C, 30 min	100 mM HCl, 100°C, 20 min
Glucose	Saccharinic acids, lactate	No reaction
Glucose-1-P	No reaction	Glucose, P_i
Glucose-6-P	Saccharinic acids, lactate, P_i	No reaction
Glucose-1,6-P_2	No reaction	Glucose-6-P, P_i

PANEL 5.3.1:   Extraction and assay of inositol

Sample: Cultured rose cells, actively growing. Detection of intermediates is aided if cells are grown with glycerol as sole
carbon source, and then fed a trace of $[^{14}C]$glucose (e.g.
10 μCi at 300 mCi/mmol) for <u>ca</u>. 1 h prior to harvesting.

Step 1:   Plunge the freshly-harvested tissue (exactly 1 g)[1] or
whole cell suspension (exactly 1 ml) into 7 ml $CHCl_3$/MeOH (1:1) at 4°C
followed (in the case of ^{14}C-labelled cells) by 50 μl of 2 % non-radioactive marker (inositol in this case) and stir at 4°C for 15-30 min.

Step 2:   Centrifuge (2,500 **g**, 5 min) or filter, collecting the clear
solution.

Step 3:   Add 1 ml H_2O, shake well, and centrifuge (2,500 **g**, 5 min)
to separate the two phases.

Step 4:   Collect the upper (aqueous) phase, and add to it H_2SO_4 to
give a final concentration of 0.1 M. Heat in a boiling water bath for 20
min to hydrolyse sucrose. Cool.

Step 5:   Add an equal volume of 0.18 M [= saturated] $Ba(OH)_2$ to
precipitate H_2SO_4 as $BaSO_4$ and leave an excess of alkali.   Heat in a
boiling water bath for 30 min to oxidise monosaccharides to acids. Cool.

Step 6:   Neutralise with CO_2 (see Panel 4.3.2 step 4).

Step 7:   Centrifuge down the $BaCO_3$ and $BaSO_4$ (2,500 **g**, 5 min).

Step 8:   Pass **(i)** the supernatant from step 7, and then **(ii)** 2 ml
H_2O through a 1-ml bed volume column of thoroughly water-washed Dowex-50
(H^+ form) packed in a Pasteur pipette. Collect the whole eluate, which
contains inositol but no cations e.g. amino acids.

Step 9:   Pass the eluate through a 1-ml bed volume column of
thoroughly water-washed Dowex-1 [OH^- form, **freshly** prepared by treating
Dowex-1 (Cl^- form) with 2 M NaOH for 1 h and then with water until the
washings are neutral] in a Pasteur pipette, followed by 2 ml H_2O. Collect
whole eluate, which will contain inositol but no anions e.g. sugar acids.

Step 10:   Shake the solution with 0.5 g of a mixture of activated
charcoal + diatomaceous earth (1:1, w/w) for 5 min.

Step 11:   Centrifuge (2,500 **g**, 5 min). Dry the supernatant <u>in vacuo</u>.

Step 12:   Analyse by chromatography. This can be by GC (after
acetylation — see Panel 4.5.8 steps 2-7), HPLC (RI detection), TLC or PC.
A suitable solvent for PC (ends of paper serrated) is EtOAc/HOAc/H_2O (10:5:
6); run time 24 h. External marker inositol can be stained with $AgNO_3$/
NaOH or $KMnO_4$ (Panel 4.5.2 f). If the peak of radioactivity appears to be
contaminated with partially-resolved shoulders, the sample can be re-
chromatographed in acetone/water (17:3) for 48 h.

[1]It is important to keep the ratio of solvent components constant. The
fresh weight of the tissue is taken as H_2O, and thus 1 g fresh weight plus
50 μl aqueous marker solution per 7 ml of $CHCl_3$/MeOH gives the overall
ratio of $CHCl_3$/MeOH/H_2O (10:10:3); on mixing, a single phase is formed.
If more than the stated amount of tissue (H_2O) is added, two phases will be
present and the extraction will be less efficient.

PANEL 5.3.2: Extraction and assay of sugar phosphates [179,21]

Sample: As Panel 5.3.1.

Step 1:   As Panel 5.3.1, steps 1-3, but replacing the 50 µl marker solution with 50 µl 2 M NaF (N.B. poisonous) as a phosphatase inhibitor.

Step 2:   Shake aqueous phase with 25 mg of activated charcoal for 30 min to bind NDP-sugars. Centrifuge (2,500 **g**, 10 min) to remove charcoal.

Step 3:   Load a sample (e.g. 0.1 ml, concentrated in vacuo if necessary) of the supernatant on to Whatman 3MM paper and electrophorese[1] at pH 3.5 (Section 4.5.4) until a picric acid marker has moved $3/4$ of the way to the anode (e.g. 30 min at 5 kV on a 40-cm effective paper length). Cut the dried paper into 1-cm strips, and assay ^{14}C in non-Triton scintillant. [Phosphates of neutral sugars have about the same mobility as picrate; those of uronic acids move somewhat faster: see Panel 4.5.4 b.]

Step 4:   Wash radioactive strips with toluene, dry, and elute with a minimal volume of water (Panel 4.5.2 h). Dry each eluate down, re-dissolve in 0.1 ml H_2O, and divide into two 50-µl portions (**a** & **b**).

Step 5:   To portion **a**, add 50 µl of pure acid-phosphatase [e.g. Boehringer-Mannheim grade I from potato;   25 µg/ml in buffer **A** (0.1 M pyridine, pH adjusted to 5.0 with acetic acid)], and incubate at 25°C 16 h.

Step 6: To portion **b**, add 5 µl 0.5 M NaOH, heat in a capped tube at 100°C for 30 min to destroy reducing sugar-phosphates (e.g. Glc-6-P). Cool. Add 5 µl 2 M TFA, and heat at 100°C for 20 min.

Step 7:   Add internal marker monosaccharides (e.g. 50 µg each of GalA, Gal, Glc, Man, Ara, Xyl, and Rha) to samples **a** and **b**, load on to Whatman 3MM paper and chromatograph as in Panel 4.5.2 c.

Step 8:   Assay ^{14}C in the separated sugar zones. For preliminary work, the internal marker sugars can be stained with dilute aniline H-phthalate prior to measurment of ^{14}C (see Section 4.5.2).

Interpretation: In effect, this procedure gives a 2-dimensional separation: (i) electrophoresis[1] to separate sugar phosphates from other sugar-derivatives, and (ii) chromatography to identify the released sugar moieties. Sugars found in **a** are derived from total sugar phosphates; those in **b** are from sugar 1-phosphates [if radioactive labelling is used, care must be taken with **b** to distinguish sugars from saccharinic acids and lactate]. Reducing sugar phosphates (e.g. Glc-6-P) are estimated by difference, **a** minus **b**; this is often satisfactory because the non-reducing phosphates (e.g. Glc-1-P) occur at lower concentrations in vivo.   Any contaminating NDP-sugars would be left undigested in sample **a**, and would be detected as a peak of ^{14}C near R_F 0.00, which thus provides a check for the efficiency of step 2.

[1]Ion-exchange chromatography is an alternative to electrophoresis. Adjust pH to 3.5 with formic acid and pass through a column (1 x 10 cm) of anion-exchanger (e.g. S Sepharose Fast Flow or DEAE-Bio-Gel A), in the formate form [=previously soaked for 1 h in 1 M sodium formate and rinsed copiously with solution **C** (10 mM formic acid, pH to 3.5 with pyridine)]. Elute with a linear gradient, **C → D** [**D** = 4 % pyridine, pH adjusted to 3.5 with formic acid].

Sugar phosphates can be distinguished from NDP-sugars by the fact that the phosphates are hydrolysed by phosphatase and not by phosphodiesterase,

$$\text{Sugar-O-}\underline{\text{P}} + H_2O \xrightarrow{\text{phosphatase}} \text{Sugar} + P_i,$$

whereas NDP-sugars are hydrolysed by phosphodiesterase and not phosphatase,

$$\text{Sugar-O-}\underline{\text{P}}\text{-O-}\underline{\text{P}}\text{-O-nucleoside} + H_2O \xrightarrow{\text{phosphodiesterase}} \text{Sugar-O-}\underline{\text{P}} + \text{NMP}$$

[A mixture of phosphodiesterase + phosphatase hydrolyses NDP-sugars **and** sugar-phosphates to free sugars.] Also, NDP-sugars will bind to activated charcoal (via the nucleoside residue), whereas sugar phosphates will not [126,488]. Schedules for the extraction and identification of sugar phosphates and NDP-sugars are given in Panels 5.3.2 and 5.3.3 respectively.

5.3.3: NDP-sugars

The extraction and work-up of NDP-sugars can follow essentially the same strategy as for sugar phosphates, the material of interest being elec-trophoresed after elution from the charcoal (Panel 5.3.3). An alternative to charcoal is to treat the total extract with pure phosphatase so as to destroy sugar phosphates (Panel 5.3.2, step 5); remaining phosphorylated derivatives are then predominantly NDP-sugars and other phosphodiesters. If this approach is adopted, the NaF must be omitted from step 1.

Alternatives to electrophoresis for separation of NDP-sugars include:
(i) Paper chromatography in EtOH/**AA** (5:2) [**AA** = 1 M ammonium acetate, pH adjusted to 3.8 with HOAc] separates largely on the basis of the nucleoside moiety of the NDP-sugar [393]. Representative $R_{UDP-Gal}$ values are:

GDP-galactose	0.56
ADP-galactose	0.73
CDP-galactose	0.73
UDP-galactose	1.00
TDP-galactose	1.42
Galactose-1-P	2.15
Galactose	4.50

(ii) Paper chromatography in EtOH/butanone/**MB** (7:2:3) [where **MB** = 12.3 g H_3BO_3 + 8.7 g morpholine, dissolved and made up to 100 ml in 10 mM EDTA] separates on the basis of both nucleoside and sugar moiety [90].
(iii) Chromatography on polyethyleneimine-treated paper with 0.4 M LiCl as solvent separates largely on the basis of nucleoside [523].
(iv) HPLC on an anion-exchange system [488] and UV detection at 260 nm.

PANEL 5.3.3: Extraction and assay of NDP-sugars

Sample: As Panel 5.3.1.

Step 1: As Panel 5.3.1, steps 1-3, but replacing the 50 µl marker solution with 50 µl 2 M NaF (N.B. poisonous) to inhibit phosphodiesterase.

Step 2: Shake aqueous phase at 0°C with 25 mg of activated charcoal for 30 min to bind NDP-sugars. Centrifuge (2,500 **g**, 10 min, 0°C). Wash the charcoal in 10 ml H_2O at 0°C three times. Resuspend charcoal in 2 ml of H_2O/EtOH/conc NH_3 (150:50:1) [90] at 0°C. Recentrifuge. Collect supernatant. Quickly[1] add 0.25 g (wet weight) of CM-Sephadex (H^+ form) to remove NH_4^+ and restore neutrality. Dry below 25°C. Redissolve in H_2O (e.g. 1 ml).

Step 3: As soon as possible[1], preferably within 2 h of killing the cells, load a sample (e.g. 0.1 ml) on to Whatman 3MM paper and electrophorese[2] at pH 3.5 (Section 4.5.4) until a picric acid marker has moved about three-quarters of the way to the anode (e.g. 30 min at 5 kV on a paper of path length 40 cm). Cut the dried paper into 1-cm strips, and assay ^{14}C in non-Triton scintillant. [Approximate $m_{picrate}$ values of NDP-sugars are: UDP-sugar = 1.1, GDP-sugar = 0.8, ADP-sugar = 0.55, CDP-sugar = 0.45. NDP-uronic acids and NDP-hexoses move somewhat faster and slightly slower respectively than the corresponding NDP-pentoses — Panel 4.5.4 b.]

Step 4: Wash radioactive strips with toluene, dry, and elute with about 1 ml 0.1 M trifuloroacetic acid[3] (Panel 4.5.2 h). Add internal markers (e.g. 50 µg each of GalA, Gal, Glc, Man, Ara, Xyl, and Rha).

Step 5: Heat on a boiling water bath for 20 min, dry _in vacuo_ or under a stream of cold air, and re-dissolve in 30 µl H_2O.

Step 6: Load on Whatman 3MM and chromatograph as in Panel 4.5.2 c.

Step 7: Assay ^{14}C in the separated sugar zones. For preliminary work, the internal marker sugars can be stained lightly with **dilute** aniline H-phthalate prior to measurment of ^{14}C (see Section 4.5.2).

Interpretation: In effect, a two-dimensional separation is obtained in which the first dimension (electrophoresis or ion-exchange) separates on the basis of nucleoside moiety and the second (chromatography) on the basis of sugar moiety. That the compounds separated on the electrophoretogram are not sugar phosphates can be demonstrated by lack of effect of pure phosphatase (Panel 5.3.2, step 5).

[1]Speed is important, especially with certain NDP-sugars e.g. UDP-Gal and UDP-Xyl which form sugar 1,2-cyclic phosphates on storage, especially at high pH [466]. One NDP-sugar, UDP-apiose, is so alkali-labile that it is largely converted to apiose 1,2-cyclic phosphate by the NH_3 treatment in step 2; UDP-Api is also highly acid-labile, even at pH 5-6 [208,290,392].

[2]As an alternative to electrophoresis, see Panel 5.3.2, footnote (1).

[3]If the possible occurrence of NDP-**oligo**saccharides is of interest, use H_2O instead of TFA, and incubate the eluate in a mixture of phosphodiesterase + phosphatase [use pure enzymes so as to minimise hydrolysis of glycosidic bonds].

5.3.4: Glycolipids involved in glycosyltransferase activities

Polyprenyl-phosphate- and -pyrophosphate-sugars are soluble in CHCl$_3$/MeOH/H$_2$O (10:10:3) [47,61]. On addition of more H$_2$O, to form 2 phases, glycolipids partition into the CHCl$_3$-rich phase and/or bind to the glassware. They can be purified by ion-exchange chromatography, hydrolysed with acid or enzymes, and the free sugar moiety can be analysed by chromatography (Panel 5.3.4; for more details and alternatives see [75,330,546]). The glycolipids discussed here are the phosphorylated polyisoprenols, e.g. of the dolichol (Dol) type, which play a rôle in the biosynthesis of

---

PANEL 5.3.4: Extraction and analysis of glycolipids of the dolichol type

Step 1:   As Panel 5.3.1, steps 1-3, but collect the lower (CHCl$_3$-rich) phase in step 3. Add 50 µl 2 M NaF to the CHCl$_3$/MeOH/H$_2$O mixture during step 1 to inhibit enzymes that cleave phosphate esters and diesters. Add non-radioactive Dol-P-Glc and Dol-P-P-Glc internal markers, dissolved in soln **A** [CHCl$_3$/MeOH/H$_2$O (10:10:3)].

Step 2:   Load lower phase on to Whatman 3MM paper as a 2 x 8 cm streak. Rinse the tube thoroughly with soln **A** to re-dissolve any adsorbed lipids, and load washings. Run in 10 % EtOH by descending chromatography overnight. [Lipids stay at origin; any contaminating sugars, phosphates, NDP-sugars and inorganic salts are swept down the paper.]

Step 3:   Dry paper at 25°C. Elute lipids from origin with soln **A**. Apply to a column (e.g. 1.5 x 10 cm) of DEAE-Sepharose, previously washed in CHCl$_3$/MeOH/H$_2$O/pyridine/HOAc (10:10:3:3:3) and rinsed thoroughly in soln **A**. When the whole sample has been applied, apply 25 ml soln **A**. [Neutral lipids eluted.]

Step 4:   Apply a linear gradient, 100 % **A** →→→ 100 % **B** [= 10 mM pyridine + 10 mM HOAc in soln **A**]. [Dol-P-sugars eluted.] Then apply a second linear gradient going from 100 % **B** to 100 % **C** [400 mM pyridine + 400 mM HOAc in soln **A**]. [Dol-P-P-sugars eluted.] Collect fractions.

Step 5: To locate internal non-radioactive markers, dry a portion of each fraction on to a silica-gel TLC plate, and run in CHCl$_3$/MeOH/H$_2$O (65:25:4) alongside pure markers. Stain by standing the plate in a dry tank containing a few crystals of iodine; warming speeds the staining reaction.

Step 6:   Dry a further 100 µl of each fraction on to a 2.5-cm disk of Whatman GF/A glass fibre paper. Assay ^{14}C in non-Triton scintillant [N.B. CHCl$_3$ quenches scintillation-counting and must be removed by thorough drying.] Select for further analysis those fractions containing ^{14}C and corresponding approximately to the authentic markers. [They can be further purified by preparative TLC in CHCl$_3$/MeOH/H$_2$O (65:25:4).]

Step 7:   Dry fractions of interest on to Whatman 3MM paper (4 x 4 cm), and incubate overnight with pure phosphodiesterase + phosphatase. The paper allows the lipid to dry in a state readily accessible to the enzymes.

Next steps: Analysis of [^{14}C]oligosaccharides (Section 4.2).

---

glycoproteins and possibly some polysaccharides; care should be taken to distinguish them from neutral and sulphated glycolipids.

5.3.5: Aminoacyl-tRNAs

It is valuable [248] in the in vivo labelling of glycoproteins to have a measure of the labelling of the immediate precursors of the polypeptide backbones — the aminoacyl-tRNAs. These can be distinguished from free amino acids by precipitation with ethanol and from proteins by failure to partition into phenol. Mild alkali splits the ester bond of amino acyl-tRNA, releasing the free amino acid for identification, but does not affect peptide bonds (Panel 5.3.5; see also [18]).

PANEL 5.3.5:   Extraction and analysis of aminoacyl-tRNAs

Sample: 1 g Fresh weight of cultured rose cells, previously fed 10 μCi [^{14}C]leucine for 1 h and 24 h[(1)] (see Ch 2).

Step 1:   Plunge freshly harvested tissue (1 g) quickly into an ice-cold mortar containing (all at 0°C):   5 ml of soln **A** [= 0.15 M NaCl + 100 mM HOAc, pH adjusted to 4.7 with 1 M NaOH], 5 ml of soln **B** [= the lower (phenol) phase obtained when solid phenol is shaken with an equal weight of solution **A**] and 0.5 ml of 1 % non-radioactive leucine.

Step 2:   Grind vigorously with a little sand for 1 min.

Step 3:   Centrifuge the homogenate to separate two phases. Proteins pass into the lower (phenol) phase, RNA into the upper phase. To the H_2O phase, add 5 ml of soln **B**. Shake well and re-centrifuge.

Step 4:   Collect the aqueous phase and add 0.1 volume of 1 % non-radioactive RNA [e.g. a crude commercial preparation from yeast] followed by 2.2 volumes of EtOH at 0°C. Incubate at -20°C for at least 1 h.

Step 5:   Filter through a 2.5-cm disk of Whatman GF/C glass fibre paper; wash disk with four 5-ml portions of EtOH/soln **C** (2:1, 0°C). [Soln **C** = 10 mM HOAc, pH adjusted to 4.7 with pyridine.]   [Free leucine is removed; tRNA remains on the disk.] Dry the disk.

Step 6:   Assay the disk for ^{14}C in non-Triton scintillant.

Step 7:   As a double check that the radioactivity originates from leucyl-tRNA, wash the disk in toluene, dry it, cut it in half, incubate one half in 0.5 M NH_3 at 25°C for 2 h and the other half in soln **C**, and load a portion of each half on to Whatman 3MM paper. Chromatograph overnight in BuOH/HOAc/H_2O (12:3:5) alongside marker leucine (Panel 4.5.2 a). Obtain a profile of ^{14}C from R_F 0-1; the major peak should be found at about R_F 0.67, and should only be found in the sample hydrolysed with NH_3.

[(1)]The 1-h sample should give a high yield of [^{14}C]leucyl-tRNA, but the 24-h sample will not because by this time, although ca. 40 % of the [^{14}C]leucine remains, it is almost all compartmentalised in the inactive vacuolar pool.

5.3.6: Hydroxycinnamoyl-CoA and -glucoside derivatives

Although it is not yet known which, if either, of these two groups of compounds plays the major rôle in formation of feruloyl-polysaccharides and related conjugates, it is useful to have methods for their assay. They are readily extracted from tissue by homogenisation in $CHCl_3/MeOH/H_2O$ (10:10:3) and separated into the following groups by electrophoresis at pH 3.5 [171].

(a) neutral esters e.g. feruloyl-β-D-glucoside,
(b) weakly acidic esters ('depsides') e.g. feruloyl-quinate,
(c) strongly acidic thioesters, e.g. feruloyl-S-CoA,
(d) basic amides e.g. feruloyl-putrescine.

Alternatives to electrophoresis include TLC on cellulose in BuOH/HOAc /H_2O (5:2:3) [371] and paper chromatography in the same solvent or in iso-butyric acid/conc NH_3/H_2O (66:1:33) [478].

Feruloyl-, p-coumaroyl-, caffeoyl- and sinapoyl- derivatives are readily detected by their intense UV fluorescence. They are also rendered detectable at extremely low levels, and with high specificity, if the tissue is fed with [14C]cinnamic acid [171].

PANEL 5.3.6: Detection of low-M_r feruloyl compounds in spinach

Sample: 1 g Fresh weight of cultured spinach cells.

Step 1:   Plunge cells (exactly 1 g) into 7 ml $CHCl_3$/MeOH (1:1, 0°C) plus 50 µl 0.8 M aqueous NaF. Stir in a stoppered tube at 0°C for 15 min.

Step 2:   Centrifuge the suspension (2,500 **g**, 5 min), and apply loadings (20, 80 and 320 µl) to Whatman 3MM paper. Loadings should be about $1/3$ of the way along the paper as compounds will migrate both ways.

Step 3:   Electrophorese at pH 3.5 (with the origin nearer the cathode) until marker picric acid has run almost to the anode (Section 4.5.4).

Step 4:   After drying, spray with MeOH and re-dry to remove the last traces of HOAc. View the paper under 366-nm UV light. Record any spots. Expose to NH_3 vapour, and note any colour changes (Panel 4.5.2 g). Approx. $m_{picrate}$ values are: feruloyl-glucoside = 0.0, depsides = 0.2-0.3, feruloyl-CoA = 0.7, feruloyl-polyamines = -0.2.

Step 5:   To confirm feruloyl-glucoside, treat with β-glucosidase and look for free ferulic acid by TLC on silica-gel in benzene/HOAc (9:1).

Step 6:   To confirm feruloyl-CoA, treat the appropriate zone from the electrophoretogram[1] with pure phosphatase (Panel 5.3.2, step 5), which removes one of the **P** groups of CoA, and look for a shift in $m_{picrate}$ upon re-electrophoresis at pH 3.5. Acetyl-CoA is a commercially-available compound suitable for comparison.

[1]The initial electrophoresis is essential to remove the fluoride, which is a potent inhibitor of phosphatases.

5.4: PREPARATION OF RADIOACTIVE PRECURSORS OF WALL BIOSYNTHESIS

5.4.1: Preparation of radioactive NDP-sugars

Some polysaccharide-precursors are commercially available in radio-active form, especially those of mammalian interest. Thus Dol-P-Glc, Dol-P-Man, UDP-Glc, UDP-Gal, UDP-GlcA, UDP-Xyl, UDP-GlcNAc, UDP-GalNAc, GDP-Glc, GDP-Man, GDP-Fuc and ADP-Glc are available from Amersham International and/ or New England Nuclear. Fortunately, UDP-GalA and UDP-Ara, which are mainly of interest to plant scientists, can be prepared in good yield from radioactive monosaccharides by a simple method involving treatment with an excess of ATP + UTP in the presence of kinase and pyrophosphorylase:

$$\text{Sugar} \xrightarrow[\text{ATP}]{\text{kinase}} \text{Sugar-1-P} \xrightarrow[\text{UTP}]{\text{pyrophosphorylase}} \text{UDP-Sugar.}$$

Similar methods can be used for many of the other NDP-sugars listed above, and might be used when larger amounts are required than can be afforded. The kinases [372,373] and pyrophosphorylases [374] are easily extracted from mung bean seeds or seedlings, and can be used with little or no purification. The radioactive NDP-sugars are purified by paper electrophoresis [179] or ion-exchange chromatography [491], followed by paper chromatography. The radioactive monosaccharides needed as starting material can usually be purchased; where larger amounts are required or the labelled sugar is not sold (e.g. GalA), it can be custom-labelled by catalytic exchange with $^3H_2$ gas. This service, offered for example by Amersham (method 'TL7'), costs about £600, and typically yields 0.2-2.0 Ci of the [1-^{3}H]sugar, enough for many years' work. The impure [^{3}H]sugar, as received from Amersham, requires purification, e.g. by paper chromatography as in Panels 4.5.2 a & b. As an alternative means of preparing radioactive monosaccharides, a suspension-culture is grown for several days in [U-^{14}C]glucose as sole carbon source and then the cell walls are hydrolysed with TFA (Panel 4.2.1 a) and the resulting [U-^{14}C]monosaccharides are purified by paper chromatography. If 1 mCi of [^{14}C]glucose is fed to spinach cells, a typical yield is 50 µCi of labelled cell walls, yielding on hydrolysis about 5 µCi [^{14}C]GalA, 10 µCi [^{14}C]Gal, 5 µCi [^{14}C]Glc, 7 µCi [^{14}C]Ara, 4 µCi [^{14}C]Xyl, 0.5 µCi [^{14}C]Fuc and 1 µCi [^{14}C]Rha.

Two NDP-sugars not available commercially, and that cannot be made by the kinase/pyrophosphorylase route, are UDP-[^{14}C]rhamnose and UDP-[^{14}C]apiose. UDP-[^{14}C]Rhamnose can be made from UDP-[^{14}C]glucose by the action of an enzyme preparation from tobacco [35]. Unfortunately, however, the enzyme system is not very active, and the separation of UDP-Rha from unreacted UDP-Glc requires **repeated** chromatography [275]. A possible (small-scale) approach to the latter problem is the use of an α-Glc-specific lectin affinity column; cf. [55]. UDP-[14C]Apiose can be made enzymically from UDP-[14C]glucuronic acid [208,290].

NDP-Sugars can also be synthesised chemically [142,355].

5.4.2: Preparation of radioactive ferulic acid and related compounds

Exogenous hydroxycinnamic acids (ferulic, p-coumaric, caffeic and sinapic) can be used by plant cells in the post-synthetic modification of wall polysaccharides, perhaps via pathways such as:

Ferulic acid + ATP + CoA-SH → Feruloyl-S-CoA + AMP + PP_i
Feruloyl-S-CoA + Polysaccharide → Feruloyl-polysaccharide + CoA-SH.

Radioactive hydroxycinnamic acids are not commercially available, but can be prepared by a simple 'biotransformation'. trans-[14C]Cinnamic acid, which is available commercially, is converted by plant cells into phenolic conjugates (Section 5.3.6). For example, cultured spinach cells accumulate feruloyl- and p-coumaroyl-β-D-glucosides as a high proportion of the supplied [14C]cinnamate [171] and radish cotyledons accumulate sinapoyl esters [503]. Alkaline hydrolysis readily releases the free phenolic acids from these conjugates [171] (Panel 5.4.2).

Hydroxycinnamoyl-CoA thioesters can be made chemically [371,478].

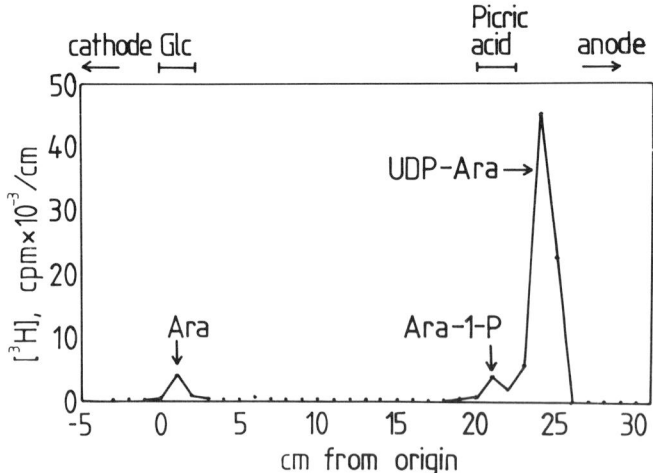

Fig. 5.4.1: Preparation of UDP-[3H]arabinose. [3H]Arabinose was incubated with ATP + UTP in the presence of a crude extract of mung bean seeds, and the products were subjected to electrophoresis at pH 3.5 (see Panel 5.4.1).

PANEL 5.4.1: Enzymic preparation of UDP-[³H]sugars and [³H]sugar-1-Ps[1]

(A) Preparation of kinase and pyrophosphorylase:

Step 1: Soak 30 g mung bean seeds in running tap water. Blot dry and weigh (weight should have increased to <u>ca.</u> 75 g). Cool to 4°C.

Step 2: Homogenise seeds in 100 ml soln **A** [100 mM NaH_2PO_4, 5 mM $MgCl_2$, 1 % Triton X-100, 10 mM 2-mercaptoethanol, final pH to 7.4 with 1 M NaOH] by two vigorous ½-min bursts in a food liquidiser, with re-cooling.

Step 3: Stir the homogenate at 4°C for ½ h to solubilise enzymes.

Step 4: Filter through muslin (4 layers). Centrifuge the filtrate (17 000 **g**, 4°C, ½ h), collect the clear supernatant, and dialyse it against several changes of soln **B** [20 mM Tris-HCl, 5 mM $MgCl_2$, 1 % Triton X-100, 10 mM 2-mercaptoethanol, final pH adjusted to 7.4 with 0.1 M HCl].

Step 5: Centrifuge. Keep the clear supernatant [soln **C**].

(B) Synthesis and purification of UDP-[³H]arabinose:[3]

Step 6: Mix 20 µl soln **C** + 20 µl soln **D** [80 mM KF, 40 mM ATP, 20 mM UTP,[1] final pH to 7.4 with 1 M NaOH] + 5 µl of aqueous solution (less than 2 mM arabinose) of the [³H]arabinose. Incubate at 25°C for 7 h.

Step 7: Add 5 µl picric acid (satd aq soln, as a marker); load the 50 µl sample on to Whatman 3MM paper as a 1 x 5 cm streak. Electrophorese at pH 3.5 (Section 4.5.4) until the yellow picrate has moved <u>ca.</u> ³/₄ of the way to the anode (e.g. 3.5 kV for 40 min on a 40-cm path-length paper).[2]

Step 8: Cut the dried electrophoretogram into strips and assay for ³H (Sections 2.8.2 and 4.5.2). **See Fig. 5.4.1.** Yield 88 % UDP-[³H]Ara.

Step 9: Pool the strips corresponding to UDP-Ara, wash in toluene, dry, and elute the ³H with H_2O (Panel 4.5.2 h).

Step 10: To remove contaminating ATP and UDP, use paper chromatography in EtOH/**AA** (71:29) [AA = 1 M ammonium acetate, containing 0.01 % EDTA and final pH adjusted to 7.0 with a trace of HOAc or NH_3], when the R_{UMP} values are: UDP = 0.42, ATP = 0.20, UDP-Ara = 1.05, UDP-Xyl = 1.16.

Step 11: To test for UDP-[³H]Xyl, which may have been formed by UDP-Xyl 4-epimerase[4] in soln **C**, hydrolyse some of the 'UDP-[³H]Ara' in TFA (Panel 4.2.1 a), separate Ara and Xyl (Panel 4.5.2 b), and assay them for ³H. If necessary, UDP-Ara and UDP-Xyl can be separated by paper chromatography in PrOH/**MB** (13:7) [where **MB** = 0.2 M aqueous H_3BO_3, 0.01 % EDTA, pH adjusted to 8.6 with morpholine], when R_{UMP} values are: UDP-Ara<u>p</u> = 0.71, UDP-Xyl<u>p</u> = 1.07, UDP-Ara<u>f</u> = 1.35, Ara<u>p</u>-1-P = 1.45, Ara = 2.4 [27].

[1]For preparation of [³H]arabinose 1-phosphate, simply omit the UTP from solution **D** (step 6).

[2]Alternatives to PE include PC [393] or ion-exchange [491].

[3]The method is also suitable for synthesis of UDP-Gal, UDP-GlcA and UDP-GalA, and the corresponding 1-phosphates if the appropriate sugar is used in step 6. For the UDP-uronic acids, electrophoresis at pH 2.0 (step 7) gives better resolution than at pH 3.5.

[4]To minimise this, solution (**C**) can be stored at -20°C for a few days: the epimerase is less stable than the kinase and pyrophosphorylase.

p-Hydroxy-[^{14}C]benzoic acid is commercially available (Sigma); it is not known if this is useful for in vivo labelling of the p-hydroxybenzoyl-polysaccharides that are found in many grass cell walls.

PANEL 5.4.2: 'Biotransformation' of [^{14}C]cinnamate to [^{14}C]ferulate and p-[^{14}C]coumarate.

Starting material: trans-[side-chain-3-^{14}C]Cinnamic acid (from Amersham).

Step 1: Thoroughly dry 10 µCi of [^{14}C]cinnamic acid on to the bottom of a 500-ml conical flask. Add 100 ml of standard culture medium (e.g. Panel 2.7.1 a) and autoclave. Inoculate aseptically with spinach cell suspension culture (5 ml PCV) and incubate under standard conditions [168] for 2-3 days. Harvest the cells on muslin, and wash with 500 ml H$_2$O.

Step 2: Add the cells to 20 ml 0.5 M NaOH. Stir at 25°C for 0.5 h, in the dark, preferably under N$_2$.

Step 3: Adjust the pH to 2 by careful addition of HCl. Add 20 ml EtOAc, and shake well. Centrifuge (2,500 **g**, 5 min) to separate the two layers, and recover the (upper) organic layer. Remove residual H$_2$O from this by shaking with anhydrous Na$_2$SO$_4$.

Step 4: Evaporate the EtOAc in vacuo and redissolve the residue in 1 ml 80 % acetone. Apply as a 15-cm streak to a silica-gel TLC plate, and run (ca. 2-3 h) in benzene/HOAc (9:1).

Step 5: Locate [^{14}C]ferulic and p-[^{14}C]coumaric acids by reference to authentic standards (with absolute minimum exposure of the ^{14}C-labelled samples to UV light), and elute with 80 % acetone. Store cold.

[1]Note that ferulate (unlike cinnamate) will readily photoisomerise, producing a mixture of cis- and trans-isomers. To minimise this, and maintain the compounds as trans-isomers, the cells should be grown and the work should be carried out in very subdued light.

5.5: IN VITRO ASSAY OF POLYSACCHARIDE SYNTHASES IN ISOLATED MEMBRANES

Membrane-bound glycosyltransferase activities that use NDP-sugars as the sugar-donor have been detected for most of the sugars of the plant cell wall (Section 5.1.3). Panel 5.5 gives a general indication of the techniques used, based on the work of Dalessandro et al. [116].

Many plant membrane preparations contain epimerases and other enzymes that interconvert NDP-sugars. Thus radioactivity incorporated from UDP-[^{14}C]GalA into a polysaccharide cannot be assumed to be in the form of GalA residues since the substrate may have been converted to UDP-GlcA, UDP-Api,

PANEL 5.5:   Assay of polysaccharide synthase activity in isolated membranes

Sample: 6 g Fresh weight of plant cells[1]

Step 1:   Suspend cells in 10 ml of ice-cold soln **A** [0.1 M KH_2PO_4, 10 mM $MgCl_2$, 1 mM dithiothreitol, 0.4 M sucrose and 1 % bovine serum albumin; all adjusted to pH 7.2 with 1 M KOH].

Step 2:   Collect cells on fine nylon mesh, and wash with a further 50 ml soln **A** at 0°C.

Step 3:   Resuspend the cells in 20 ml of solution **A** at 0°C and homogenise, e.g. with a 'Polytron' in short bursts of progressively increasing power, with re-cooling in an ice bucket between bursts.

Step 4:   Squeeze the homogenate through 6 layers of 'Miracloth', and centrifuge the filtrate briefly at <u>low</u> speed in a cooled centrifuge bucket (500 **g**, 10 min, at 0°C).

Step 5:   Collect the (turbid) supernatant and ultra-centrifuge it (100 000 **g**, 1 h, 2°C).

Step 6:   Resuspend the pellet in 0.5 ml soln **A** (0°C). If it cannot be used immediately, store it in liquid N_2.

Step 7:   Mix 20 µl of membrane suspension + 10 µl of soln **B** [= NDP-[^{14}C]sugar(s) or NDP-[^{3}H]sugar(s), dissolved in soln **A** at 5 µCi/ml[2]], and incubate at 25°C for 1-15 min.

Step 8:   To stop the reaction, dry the whole 30 µl sample on to the origin of a Whatman 3MM paper chromatogram, at a point previously wetted with 5 µl of 50 mM Na_2-EDTA, pH 6.5. Dry in a stream of air. Run the chromatogram overnight in EtOAc/HOAc/H_2O (10:5:6).

Step 9:   Assay the radioactivity at R_F 0.00 [= polysaccharide and/or glycoprotein].   If required, assay the radioactivity at the positions occupied by the NDP-sugar, the corresponding sugar 1-phosphate, and the monosaccharide, all located by reference to markers run in parallel.

Next steps: Radiochemical characterisation of the polymeric products, after washing the R_F 0.00 strip with toluene and drying.   Note that polysaccharides and glycoproteins may be difficult to elute from paper, so re-assay the paper for radioactivity after elution.   As an alternative to elution, TFA-hydrolysis (Panel 4.2.1 a), and many enzymic hydrolyses can be performed on the paper-bound polymer.

[1]The technique was worked out for pine cambial cells, scraped from freshly felled trees [116];   however, it is typical of protocols reported for many other tissues.   If possible, select a tissue without large vacuoles.

[2]The concentration of the NDP-sugar [including any non-radioactive NDP-sugar(s) added — generally 1 µM to 1 mM] can affect not only the rate of incorporation, but also the type of product formed, e.g. (1→3) <u>vs</u> (1→4).

UDP-Xyl and UDP-Ara (Fig. 2.5). It is necessary to hydrolyse the polysaccharide and identify the [^{14}C]sugars chromatographically.

A further question is whether the incorporation of ^{14}C represents true polysaccharide formation or merely the addition of one or a few residues to the terminus of a polysaccharide whose synthesis was already almost complete. Non-reducing termini can often be distinguished from mid-chain residues by treatment with NaIO$_4$ — NaBH$_4$ — TFA (Section 4.2.7): in the case of (1→3)- or (1→4)-pyranose linked polymers of glucose or mannose, the non-reducing terminus yields [^{14}C]glycerol whereas mid-chain residues yield [^{14}C]monosaccharide or [^{14}C]erythritol.

The synthesis of heteropolysaccharides will require the presence of two or more different NDP-sugars. For instance, the incorporation of [^{14}C]xylose residues from UDP-[^{14}C]Xyl into xyloglucan may be promoted by the simultaneous presence of non-radioactive UDP-Glc, and the incorporation of [^{14}C]apiose residues from UDP-[^{14}C]apiose into apiogalacturonan is promoted by the presence of non-radioactive UDP-GalA [392].

The synthesis of some polymers depends on intermediary sugar carriers e.g. glycolipids or glycoproteins. These are often present endogenously in membrane preparations; but some workers add them, either during homogenisation of the tissue or after purification of the membranes. Dolichol is available commercially, and can be applied as an emulsion produced by sonication in solution A of Panel 5.5. Deglycosylated glycoproteins can also be added as sugar carriers or acceptors. Deglycosylation can be achieved by mild acid hydrolysis (Section 4.3.1), enzymic hydrolysis, or treatment with trifluoromethanesulphonic acid [129,462] or anhydrous HF [435].

6 Wall enzymes

6.1: OCCURRENCE AND POSSIBLE FUNCTIONS

The primary cell wall is thought to be a site of metabolic activity, implying the presence of enzymes. A large number of enzymes have indeed been reported from primary walls. They are all hydrolases and oxido-reductases [173,310], i.e. enzymes that operate with simple substrates (H_2O, H_2O_2, O_2, etc.), compatible with the harsh environment of the wall rather than with labile substrates such as ATP, CoA or UDP-Glc. The wall environment is also unkind to the enzymes themselves, and wall enzymes are among the most stable known. They often contain cystine (S-S) bridges, and some are even inactivated by reducing agents e.g. ascorbate and glutathione [84] — compounds that are often added to preparations of intracellular enzymes to stabilise them. Many wall enzymes have pH optima in the region 4-5, whereas intracellular enzymes tend to have optima around pH 8.

6.1.1: Glycosidases

The most numerous enzymes of the primary wall are the glycosidases (Section 4.2.5); the following are well-substantiated examples [173]:

> α- and β-D-Glucopyranosidases
> α- and β-D-Galactopyranosidases
> α- [514] and β- [403] D-Mannopyranosidase
> α-L-Arabinofuranosidase
> α- and β-D-Xylopyranosidases
> α-D-Galacturonopyranosidase
> β-D-Glucuronopyranosidase
> N-Acetyl-β-D-glucosaminidase [403]
> β-D-Fructofuranosidase.

There is also evidence for the effects of α-L-fucopyranosidase [Baydoun & Fry, unpub.] and α-D-glucuronopyranosidase [119]. This battery of enzymes is sufficient to attack most of the known wall polysaccharides.

Glycosidases nibble polysaccharides from the non-reducing ends, re-leasing monosaccharides in the sequence shown (1,2,3...):

Glc——Glc——Glc——Glc——Glc——Glc——Glc——Glc——Glc——Glc——Glc——Glc...
　　(1)　　(2)　　(3)　　(4)　　(5)　　(6)　　(7)　　　etc...

In addition, there is some evidence for a cellobiohydrolase in plant cell

walls [233];   this enzyme would attack cellulose from the non-reducing
terminus and release the disaccharide, cellobiose:

Glc—Glc——Glc—Glc——Glc—Glc——Glc—Glc——Glc—Glc—Glc—Glc...
     (1)   (2)   (3)   (4)   etc...

    Glycosidases are usually very specific for the sugar **residue** hydro-
lysed, but are relatively non-specific for the **aglycone** (Sections 3.1.2,
4.2.5).   Thus β-glucopyranosidases can attack cellulose, callose, β-(1→3),
(1→4)-glucan, methyl-β-D-glucopyranoside, p-nitrophenyl-β-D-glucopyranoside and coniferin, although the wall contains several different isozymes
of β-D-glucopyranosidase, and these differ in preferred aglycone [273].   A
more extreme example of aglycone-specificity is the wall-bound α-D-xylopy-
ranosidase, which will cleave α-xylose residues from xyloglucan and from
its repeating heptasaccharide, but not from D-Xyl-α-(1→6)-D-Glc, nor from
the chromogenic model substrate, p-nitrophenyl-α-D-xylopyranoside [299].
The number of reported glycosidases (or isozymes thereof) will probably
increase in the next few years as more specific substrates are tested.

    Glycosidases are unlikely to cause wall-loosening (Ch 10) by any
simple chopping mechanism (e.g. cutting of links in a chain of structural
components) because they hydrolyse polymers from the termini and are thus
unlikely to have much impact on the mean chain length of the polymers.
There is no consistent correlation between glycosidase levels and growth
rate [402]. Nevertheless, they may have a substantial effect on the physical properties of a polysaccharide or glycoprotein by stripping short side-
chains off the backbone.   For example, α-L-arabinofuranosidase + α-D-glucuronopyranosidase would convert arabinoglucuronoxylan into xylan [119]:

Xyl—Xyl—Xyl—Xyl—Xyl—Xyl—Xyl—Xyl—Xyl—Xyl—Xyl—Xyl...
   |← |← |← →| →| →| |← |← |←
  Ara Ara Ara GlcA   GlcA GlcA   Ara Ara Ara

← = site of action of α-L-arabinofuranosidase.
→ = site of action of α-D-glucuronopyranosidase.

In vitro this would make the polysaccharide precipitate; in vivo it might
profoundly affect the physical properties of the wall, though in this case
the effect would probably be a tightening rather than a loosening.

    Another plausible rôle for wall-bound glycosidases is the hydrolysis
of translocated oligosaccharides prior to uptake.   Walls often contain β-
fructofuranosidase (invertase), which hydrolyses sucrose to glucose + fructose.   However, many plant cells can take up sucrose intact, and possess an

intracellular enzyme, sucrose synthase, which can put the high-energy glycosidic bond of sucrose to good use in the direct formation of UDP-Glc:

$$\text{Sucrose} + \text{UDP} \leftrightarrow \text{UDP-Glc} + \text{Fru},$$

so it is unclear what benefit they would gain from invertase in the walls [237]. Walls also contain α-galactosidase, which would hydrolyse raffinose (and higher homologues) to galactose and sucrose. In cell suspension cultures, a wide variety of oligosaccharides will act as carbon source for growth; most of these are hydrolysed extracellularly and taken up as mono-saccharides. Coniferin is hydrolysed to glucose + coniferyl alcohol (the precursor of lignin) in the walls of young xylem vessel elements. Glyco-sidases also contribute to the mobilisation of polysaccharide-bound sugar residues, e.g. during germination [140,339,417] and senescence [567].

A further proposal is that wall-bound glycosidases serve to hydrolyse biologically-active oligosaccharides to inactive products (Ch 9), thereby controlling their activity.

6.1.2: Endoglycanases

Wall enzymes have been reported that catalyse the hydrolysis of mid-chain sugar residues in the following polysaccharides [173]:

cellulose [β-(1→4)-glucan]
dextran [α-(1→6)-glucan]
callose [β-(1→3)-glucan]
β-(1→3),(1→4)-glucan, including an interesting enzyme that cleaves this poly-
starch [α-(1→4)-glucan]   saccharide only at every 60–90 Glc
xyloglucan (Glc residues)   residues [235]
xylans [including arabinoxylans] (Xyl residues)
pectin (GalA residues)
arabinan
galactan
chitin

These enzymes can have a great impact on the chain-length of a polysaccha-ride: a single action could halve the M_r:

$$\text{Glc—Glc—Glc—Glc—Glc—Glc}\underset{(1)}{\text{—}}\text{Glc—Glc—Glc—Glc—Glc—Glc....}$$

Since it is easy to envisage such an effect influencing wall extensibility, endoglycanases have received much attention in the context of the physiology of growth [194]. The work is discussed further in Ch 10.

In view of the central importance of cellulose in the plant cell wall, cellulase is of special interest. It is clear from microscopy that walls are subject to erosion during the normal course of development. For instance, perforation plates appear in the end walls of xylem vessel elements; the juxtaposed walls of stock and scion become thinner during successful grafting; and a portion of the mother cell wall is excised during cell division (Fig. 1.1.4 b-d) [265]: all these effects seem likely to involve the action of cellulase.

An elegant approach used to show that cellulase plays a rôle in the loosening of the abscission zone prior to leaf fall is the application of anti-cellulase antibodies that specifically inhibit the enzyme's catalytic site. These antibodies delay abscission; pre-immune serum does not [445].

A further rôle for wall-bound endoglycanases is in the mobilisation of reserve polysaccharides. The cotyledons of certain seeds have thick walls that contain vast quantities of particular polysaccharides e.g. xyloglucan in Tropaeolum, arabinogalactan in lupin [417]. These walls thin dramatically during germination, owing to hydrolysis of the polysaccharide: the sugars are used by the seedling before it becomes photosynthetically self-sufficient. Tropaeolum seeds contain an endoglucanase which presumably takes part in hydrolysing the xyloglucan: it is unusual as it appears to be specific for xyloglucan [141]. The same (or very similar) reaction is also catalysed in vitro by normal cellulases (Panel 9.2.2).

Endo-acting enzymes, especially pectolytic ones, appear to be responsible for the cell-separation that softens fruits during ripening [567,206].

Another proposed rôle is the hydrolysis of polysaccharides in the walls of invading pathogens, especially fungi. Thus, the chitinase present in some plant cell walls (often at increased levels after infection) may breach the walls of fungal hyphae [397]. Similarly the endo-$(1 \rightarrow 3)$-glucanase may cleave oligosaccharide fragments from the fungal wall, and these may be important 'danger signals' (Ch 9).

Finally, it has been suggested that endoglycanases act on the plant's own polysaccharides to release biologically-active oligosaccharides that play a rôle in cell signalling and recognition (Ch 9).

6.1.3: Transglycosylases

Many glycosidases and endoglycanases can catalyse transglycosylation as well as hydrolysis. The enzyme splits the glycosidic linkage and then, instead of transferring the sugar residue to H_2O (= hydrolysis: Sections 6.1.1 & 6.1.2), transfers it to an alcohol e.g. another sugar. For transglycosylation to predominate over hydrolysis, a high concentration of the alcohol is usually required. Since primary cell walls are <u>ca</u>. 40 % polymer and only 60 % H_2O, this situation may pertain in the walls of living cells.

Endo-transglycosylation could graft segments of one polysaccharide on to another. The total number of glycosidic bonds is conserved, so there is no change in mean M_r, as shown in the following hypothetical example:

Xyl→Ẋyl→Xyl→Xyl→Xyl→Xyl→Xyl→Xyl→Xyl→Xyl→Xyl

+   Glc→Glc→Glc→Glc→Glc→Glc→Glc→Glc→Glc→Glc→Glc

→   Glc→Glc→Glc→Glc→Glc→Glc→Glc→Glc→Glc→Glc→Glc
                                 ↗
Xyl→Xyl→Xyl→Xyl→Xyl→Xyl

                                      +   Xyl→Xyl→Xyl→Xyl→Xyl

Cell walls have been reported to have a dextranase with a propensity to catalyse transglycosylation rather than hydrolysis [246]. However, its significance is unclear as dextran [= α-(1→6)-D-glucan] is not a known wall component. It has been hinted that the dextranase can also act on arabinan [246], which <u>is</u> a wall component.

Exo-transglycosylation would transfer single sugar residues from one complex carbohydrate molecule to another. A wall-bound β-glucosidase will do this <u>in vitro</u> [370]; its biological significance is unclear.

Transglycosylation of wall polysaccharides <u>in vivo</u> has not yet been reported; it would be very difficult to detect. However, endo-transglyco-sylation could play an important rôle in cell expansion, allowing creep to occur [101]; its demonstration would be very valuable.

6.1.4: Other hydrolases

Walls contain enzymes that hydrolyse carboxy-ester [571], phosphate-ester [111] and peptide [515] bonds. One carboxy-esterase, pectin-methyl-esterase, removes methyl groups from the GalA residues of pectins:

...methyl-galacturonate... + H_2O → ...galacturonate... + methanol.

In this way, the high methoxy pectin initially secreted into the cell wall can gradually be converted to a low-methoxy (highly acidic) pectin [570, 263]. Since the acidic form can take part in Ca^{2+}-bridges, de-esterifica-tion could have an important rôle in the control of wall properties.

Other carboxy-esterases of lower specificity are present in the wall [184] but their natural substrates are unknown. Possibilities include the acetyl, feruloyl and other acyl groups present on some polysaccharides.

Phosphatases are found in many plant cell walls. Their natural substrate is unknown since phosphate esters are normally assumed not to occur extracellularly. It is interesting, but unexplained, that wall phosphatase levels rise dramatically in response to phosphate starvation [509].

Protease activity has been reported from bean leaf cell walls, but its natural substrate is also unknown [515].

6.1.5: Peroxidases [518]

Peroxidases are haemoproteins which probably occur in all primary cell walls; they sometimes account for several % of the cell's total protein [188,472]. They are glycoproteins [100,338] and the complete amino acid sequences of horseradish [544] and turnip [338] peroxidases have been published. The general reactions that they catalyse are [188]:

$$2 \text{ AH} + \text{H}_2\text{O}_2 \rightarrow \text{A}_2 + 2 \text{ H}_2\text{O} \qquad \text{and}$$
$$\text{AH}_2 + \text{H}_2\text{O}_2 \rightarrow \text{A} + 2 \text{ H}_2\text{O},$$

where A is usually an aromatic ring, e.g. a phenol.

Peroxidases catalyse lignification e.g. in differentiating wood [228] (Section 5.1.8) and are also probably responsible for the dimerisation of phenolic side-chains on polysaccharides and glycoproteins in the growing cell wall to form isodityrosine [176], diferulate [335] and related products [171] (Section 3.3.2). The synthesis and secretion of peroxidases is under tight control, suggesting considerable biological significance. Thus, for example, the quantities and sub-cellular distribution of peroxidases change dramatically in response to a large number of stimuli, e.g. hormones [167], temperature stress, drought, infection [224] and Ca^{2+} levels [477,398]. Peroxidases tend to be unevenly distributed between specific cell types [220,461], and are often present in increased levels in genetically-dwarfed plants [274,519] and in slower-growing lines of tissue cultures [98]. It is possible that wall-bound peroxidases are involved in wall-tightening reactions, and thus in growth control (Section 10.1.4).

A number of peroxidase isozymes occur, differing greatly in isoelectric point [446]. These isozymes often differ in preferred substrate, although the specificity is far from absolute [288,176]. There are some reports of very large numbers (e.g. over 20) of isozymes occurring in a single tissue [190], but these may represent different post-translational modifications of the products of just a small number of genes [188].

6.1.6: Oxidases

Oxidases differ from peroxidases in using O_2 rather than H_2O_2 as electron acceptor. In some cases, a single enzyme protein can catalyse both the oxidatic and peroxidatic reactions: e.g. peroxidases can catalyse the oxidation by O_2 of IAA [368] and dihydroxyphenylalanine [210]. In some cases, additional compounds, e.g. malate, citrate or oxalate, + Mn^{2+}, + p-coumaric acid must also be present [430]. Similarly, peroxidase may be able to catalyse the oxidation of NADH [143,260]:

$$NADH + H^+ + O_2 \rightarrow NAD^+ + H_2O_2$$

Oxidases can catalyse two common general reactions:

$$2\ AH_2 + O_2 \rightarrow 2\ A + 2\ H_2O \qquad \text{and}$$
$$AH_2 + O_2 \rightarrow A + H_2O_2.$$

One wall-bound oxidase, ascorbate oxidase, is of the first type [AH_2 = ascorbate; A = dehydroascorbate], and another, polyamine oxidase, is of the second type [AH_2 = R—CH_2—NH_2; A (+ H_2O) = R—CHO + NH_3] [285]. The action of peroxidase on NADH is also of the second type. The second type is important as it could be the source of H_2O_2, which is required for the peroxidatic formation of lignin, isodityrosine etc. [143].

Cultured sycamore cells can secrete polyphenol oxidase into the apoplast in very large amounts (ca. 2 % of total cell protein), but the significance of this Cu-containing protein is unknown [62].

6.1.7: Malate dehydrogenase (MDH)

This enzyme, which is well known from Krebs' cycle, is also present in the cell wall [211], apparently as a different isozyme from that which occurs in the mitochondria. Its coenzyme, NAD, would not be expected to survive long in the presence of wall hydrolases, but it appears that the wall-bound MDH may carry its NAD in a permanently bound form, protected from hydrolases. This form is, however, apparently accessible to (per-) oxidase action (Section 6.1.6). Thus, when walls are incubated in the presence of malate, H_2O_2 can be generated:

$$Malate + NAD^+ \xleftarrow{\quad MDH \quad} Oxaloacetate + NADH + H^+.$$

$$NADH + H^+ + O_2 \xrightarrow{\text{'peroxidase'}} NAD^+ + H_2O_2.$$

This pair of reactions may play a crucial rôle in the generation of H_2O_2 needed for synthesis of lignin, isodityrosine etc.

It is unclear whether cells normally secrete malate, although the mesophyll cells of some C_4 plants can do so in bulk; it seems plausible that other cells should do so in the amounts required to make H_2O_2.

Many plant cells have an alternative and more direct way of 'secreting' reducing power across the plasma membrane, as monitored by reduction of extracellular ferricyanide [45,357].

6.2: EVIDENCE FOR LOCALISATION OF ENZYMES IN THE CELL WALL
6.2.1: Evidence from wall isolation

Some of the methods for isolation of cell walls (Sections 1.2.2 and 1.2.3), will retain the catalytic activity of wall enzymes. Methods involving homogenisation in glycerol seem particularly promising. If purified walls are found to exhibit catalytic activity, they are concluded to contain enzymes. Caution is required for the following reasons:

(a) Intracellular enzymes may bind ionically to acidic pectins (or possibly basic extensin) of the wall during homogenisation.

(b) If this is prevented by inclusion of salt, genuine wall enzymes may be leached off (Section 6.3.2).

(c) Intracellular enzymes may bind to the wall during isolation by tannin precipitation [236]. This can be suppressed by inclusion of bovine serum albumin in the homogenisation medium.

(d) Walls isolated under non-denaturing conditions are never totally free of contaminating membranes, chromatin and amyloplast fragments; enzymes present in these structures may thus be mistaken for wall enzymes.

6.2.2: Application of non-permeant substrates to living cells

This approach is particularly suitable for cell suspension cultures, but it could also be applied to intact plant parts by vacuum infiltration of the substrate (cf. Panel 6.3.1). The living cells are bathed in medium containing a substrate that will permeate the cell wall but not penetrate the plasma membrane, e.g. [403]:

p-nitrophenyl-glycosides	for glycosidases
feruloyl-oligosaccharides	for peroxidases; esterases
acetyl-oligosaccharides	for esterases
p-nitrophenyl-phosphate	for phosphatase

The potential problem that enzymes in <u>damaged</u> cells will have access to the substrate can be investigated by deliberate injury or permeabilisation. Thus, the rate of hydrolysis of extracellular p-nitrophenyl-α-D-

mannopyranoside is very low, but increases greatly on addition of 1 %
Triton X-100, which allows the substrate to penetrate the cell membranes
and reach the vacuole where most of the α-mannosidase is found. In contrast, hydrolysis of p-nitrophenyl-β-D-galactopyranoside is little affected
by Triton, showing that most of the β-galactosidase is extracellular.

6.2.3: Evidence from isolation of protoplasts

If an enzyme is absent from isolated, purified protoplasts, it was
probably located in the wall [285,331,514]. Even if enzyme activity is
found in the isolated protoplasts, some of it might still have been wall-
bound; since wall enzymes are synthesised intracellularly, there will
normally be a small pool of new enzyme molecules in transit to the wall.
Where there is no all-or-none effect of protoplast isolation, it will be
necessary to compare total enzyme activity per cell with enzyme activity
per isolated protoplast. The enzyme preparations used for isolation of
protoplasts are highly impure, and great care is needed to ensure removal
of the added enzymes from the isolated protoplasts.

6.3: MODE OF WALL-BINDING

Wall enzymes are either freely soluble in the water that permeates
the wall, or are bound to the wall polymers. Some bound enzymes can be
leached from the wall with salt, suggesting ionic bonding; others cannot
and are said to be covalently bound. The possibility that enzymes are held
in the wall by binding to lectins has not been substantiated.

6.3.1: Evidence for freely soluble wall enzymes

Enzymes accumulate in the media of suspension-cultures. Such enzymes
must have passed through the wall and some of them presumably remain there.
Enzymes in the medium will include some from cells that die during culture;
however, the enzymes detected differ from those detected in tissue homogenates, showing that there is a controlled secretion of selected enzymes.

The equivalent of the culture medium in an intact plant is the
surface film of fluid present on and in the cell walls. Enzymes dissolved
in this fluid can be sampled by vacuum-infiltration (Panel 6.3.1) of water
into the air-space system, followed by low-speed centrifugation [133,496,
513]. The water thus expelled represents washings of the cell surface.

```
┌─────────────────────────────────────────────────────────────────────────────┐
│ PANEL 6.3.1:    Sampling of water-soluble apoplastic enzymes and polysaccha- │
│ rides by vacuum-infiltration/centrifugation (modified after [496])           │
│                                                                               │
│ Sample:          Pea seedlings, grown in the dark in Vermiculite for 7 days.  │
│                                                                               │
│ Step 1:          Excise a 1.2-cm segment from each seedling, about 1 cm       │
│ below the apical hook.                                                        │
│                                                                               │
│ Step 2:          Pack the segments into a 5-ml syringe barrel.    Incubate    │
│ with the segments just submerged in water for 1 h.                            │
│                                                                               │
│ Step 3:          Wash copiously with solution A [20 mM K$_2$HPO$_4$, adjusted to│
│ pH 6.0 with HCl] for ½ h to remove materials leaking from the cut cells.   A  │
│ pump is useful for this washing.  Finally rinse with water at 0°C.            │
│                                                                               │
│ Step 4:          Submerge the segments, still packed in their barrel, in a    │
│ beaker of water (0°C) inside a vacuum desiccator.  Apply a vacuum for 3 min    │
│ — bubbles should exude from the tissue, which should remain submerged.[1]     │
│ Release the vacuum, and wait a further 3 min (water will enter the former     │
│ air spaces).                                                                  │
│                                                                               │
│ Step 5:          Gently expel surplus water from the tissue and barrel, then  │
│ centrifuge the barrel, as shown in Fig. 4.5.2 c, for 15 min.    With pea      │
│ stems, 1000 g can be used without appreciable cytoplasmic contamination       │
│ [496]; this must be established empirically for each tissue used (step 6).    │
│                                                                               │
│ Step 6:          Recover the eluate from the centrifuge tube, de-salt (Sec-   │
│ tion 3.5.1), and assay for (a) the enzyme(s) of interest (Section 6.5), and   │
│ (b) soluble cytoplasmic markers to demonstrate the extent of cytoplasmic      │
│ contamination.  A suitable marker is fumarase, measured as the increase in    │
│ A$_{250}$ of a solution containing 50 mM L-malic acid + 50 mM NaH$_2$PO$_4$, final pH│
│ adjusted to 7.3 with NaOH [250].  Malate dehydrogenase should not be used     │
│ as a cytoplasmic marker since it also occurs in cell walls (Section 6.1.7).   │
│                                                                               │
│ [1]If the segments fail to remain submerged, a piece of crumpled gauze can    │
│ be used to keep them at the bottom of the barrel.                             │
└─────────────────────────────────────────────────────────────────────────────┘
```

Small amounts of intracellular enzymes may also 'bleed' from damaged cells, but assay of marker enzymes suggests that this is often slight [496].

6.3.2: Evidence for ionically-bound wall enzymes

When a non-toxic salt is added to a suspension culture, more enzymes [538] and extensin [458] rapidly appear in the medium (Panel 6.3.2). This represents elution of enzymes from the cell surface by ion-exchange. The elution occurs at 0°C almost as fast as at 25°C, arguing against salt-induced active secretion. The effect is seen in cells killed by freezing/ thawing or sonication, again indicating passive leaching from the wall. In intact plants, vacuum-infiltration of salt solutions (cf. Section 6.3.1) can provide evidence for ionically wall-bound enzymes.

```
┌─────────────────────────────────────────────────────────────────────────┐
│ PANEL 6.3.2:   Salt-elution of enzymes from the walls of living cells      │
│                                                                            │
│ Sample:          Rapidly-growing suspension culture of spinach.            │
│                                                                            │
│ Step 1:          Collect the cells from 50 ml culture on 4 layers of muslin│
│ in a small funnel [sintered glass and glass fibre filters tend to clog].   │
│ Rinse the cells with 100 ml water.   [Pool the filtrates, de-salt, and     │
│ analyse for freely-soluble apoplastic enzymes — Section 6.3.1.]            │
│ Step 2:          Resuspend the cells in 25 ml 25 mM LaCl₃;[1] filter again.│
│ Step 3:          Repeat step 2. This can be done without removing the cells│
│ from the muslin.                                                           │
│ Step 4:          Pool the filtrates from steps 2 and 3, de-salt (Section   │
│ 3.5.1), and assay for the enzymes or proteins of interest (Section 6.5).   │
│ Major components in the salt-leachates of spinach cultures include extensin│
│ (Panel 3.5.3 b), peroxidase and several glycosidases.                      │
│                                                                            │
│ [1]Similar results are obtained with 100 mM CaCl₂ or 500 mM NaCl.          │
└─────────────────────────────────────────────────────────────────────────┘
```

An alternative approach is wall-isolation followed by salt-elution: this is not recommended because intracellular enzymes have the opportunity for binding ionically to the wall during homogenisation (Section 6.2.1). The possibility that this can be prevented by homogenisation in glycerol (Section 1.2.3) deserves careful investigation.

6.3.3: Evidence for covalently-bound wall enzymes

In some ways, this is the easiest class of wall enzyme to localise. Walls can be purified after vigorous homogenisation by rigorous washing in the presence of 1 M NaCl, 1 % Triton X-100 (a non-denaturing detergent) and 1 % bovine serum albumin without risk of losing the covalently-bound enzymes. A suspension of the final preparation is then assayed for enzyme activity [422]. Although there is common reference to 'covalently-bound' enzymes, the nature of the alleged covalent bond(s) is unknown.

6.4: EXTRACTION AND PURIFICATION OF WALL ENZYMES
6.4.1: Extraction of freely-soluble wall enzymes

Since cells contain more intracellular than extracellular enzymes, any source of apoplastic fluid that is free of cytoplasm is already highly purified relative to a tissue-homogenate. It is therefore best to avoid disruption of cells. For freely-soluble wall enzymes, an excellent starting material is the culture filtrate of a suspension culture, or the leachate obtained after vacuum-infiltration of intact plant parts which have a significant air-space system. Where there is no air space system, it may be necessary to resort to homogenisation.

6.4.2: Extraction of ionically-bound wall enzymes
Again it is a great advantage to avoid homogenisation. Suspension cultured cells are washed with water followed by a salt solution. The concentration used depends on valency, e.g. 25 mM $LaCl_3$, 100 mM $CaCl_2$ or 500 mM NaCl (Panel 6.3.2). In intact tissue with air spaces, vacuum-infiltration of the salt is suitable [515]. When there is no air space system, homogenisation may be necessary: the tissue is homogenised in a low-salt medium to remove freely soluble enzymes, and the residue is then washed in 25 mM $LaCl_3$, 100 mM $CaCl_2$ or 500 mM NaCl [370]. Some workers recommend the use of 3-4 M LiCl [235,322].

6.4.3: Extraction of covalently-bound wall enzymes
The only approach to the solubilisation of these wall enzymes is to isolate and extensively wash the walls (Section 6.3.3), and then to treat them with microbial enzymes of the type used for isolation of protoplasts. As the polysaccharides are digested, covalently-bound wall enzymes (e.g. peroxidases) are brought into solution [332,422,474]. It must be established that the microbial enzymes do not inactivate peroxidase. The wall peroxidases (glycoproteins) are likely to be stripped of carbohydrate side-chains, precluding the use of lectin affinity columns (Section 6.4.4). Another problem is that the microbial enzyme preparations are mixtures; no commercially available cellulase and pectinase is devoid of all other hyd-rolases. It is therefore essential to run a 'microbial-enzyme-only' control in all experiments of this type to check that the commercial enzyme is not contaminated with microbial peroxidase activity. 'Driselase' is a potent wall-digesting enzyme that appears to contain no peroxidase (or esterase) activity, nor does it inactivate horseradish peroxidase.

Certain wall enzymes can be solubilised by pectinase, and others by hemicellulase or DMSO [209]. Thus, purified wall-digesting hydrolases may provide a means of separating two (or more?) different groups of wall enzyme, simplifying subsequent purification.

230

6.4.4: Purification of wall enzymes

A valuable first step consists of addition of solid $(NH_4)_2SO_4$ (e.g. to 30, 50, 70 or 90 % saturation) to the enzyme solution and storage at 4°C. This precipitates glycoproteins, and thus separates the enzymes from polysaccharides, which are always a major contaminant of extracellular enzymes. It also concentrates the enzymes, which in the case of freely-soluble wall enzymes are often present in a large volume of culture medium.

The enzyme of interest can be purified by standard techniques [439, 124], usually including differential precipitation with $(NH_4)_2SO_4$, followed (after de-salting: Section 3.5.1) by gel-permeation chromatography, ion-exchange chromatography, 'Chromatofocusing', and affinity chromatography. Since wall enzymes are glycoproteins, affinity chromatography can be based on lectin binding: many different lectin-agarose conjugates are available commercially, and will bind wall proteins by their sugar side-chains. The enzymes can subsequently be eluted with an appropriate sugar, which is later removed by gel-permeation chromatography or dialysis. For instance, glycoproteins bearing α-Man residues will bind to immobilised Concanavalin A and can be eluted with 0.1 M methyl-α-D-mannopyranoside; glycoproteins with β-GlcNAc residues will bind to immobilised wheat-germ agglutinin and can be eluted with 0.1 M chitobiose [124]. Some wall proteins are themselves lectins, and can be bound by affinity chromatography on immobilised carbohydrates: for example the hydroxyproline-rich lectin of potato will bind to fetuin-agarose and can be eluted with chitobiose [390].

6.5: ASSAY OF WALL ENZYMES

Examples are given in Panel 6.5 of convenient assays for the main groups of enzymes mentioned in Section 6.1. All assays are conducted at 25°C. Buffer **A** is 50 mM acetic acid, pH adjusted to 4.7 with 1 M NaOH; buffer **B** is 50 mM acetic acid, adjusted to pH 4.7 with pyridine. Buffer **B** is used when the salts need to be volatile.

Assays for endoglycanases are particularly awkward because many plant-derived samples are much richer in glycosidases (exoglycanases) than endoglycanases. A commonly used approach is to monitor production of reducing groups during hydrolysis of a pure homopolysaccharide; however, this does not distinguish between exo- and endo-glycanases, and is not recommended. Panel 6.5, part **2**, provides various solutions to the problem:

(a) <u>Viscosity measurement.</u> Viscosity is closely dependent on DP, and thus distinguishes between mid-chain hydrolysis, where one act of hydrolysis is likely to cause an approximate halving of DP, and erosion from the non-reducing terminus, where each act of hydrolysis decreases the DP by only 1;

(b) <u>Assay of hydrolysis occurring specifically near the reducing terminus.</u> By labelling the reducing terminus, it is possible to monitor only those acts of hydrolysis that occur close to this end of the molecule. Since exohydrolases nibble from the non-reducing terminus, they take a long time to reach the other end.

(c) <u>Protection of the non-reducing terminus by a blocking group.</u> This prevents the action of exo-glycanases until an endo-glycanase has acted first. Naturally occurring feruloyl-polysaccharides are suitable substrates.

Most of the assays are designed for soluble or solubilised enzymes. However, many will also work with a suspension of cell walls in place of the enzyme solution. Some simplification in the assays may then be possible: the reaction can be stopped very conveniently by filtration or centrifugation, and removal of the filtrate or supernatant into a separate tube — see Panel 6.5, part **5(c)**. It should, however, be ascertained that the reaction product does not bind to the walls, as may be the case with peroxidase.

PANEL 6.5: **Assays of wall enzymes**

(1) GLYCOSIDASES.

Example (a): β-glucosidase [403]

Substrate: p-Nitrophenyl-β-D-glucopyranoside (pNP-Glc), which is commercially available.

Reaction: pNP-Glc + H_2O → Glc + pNP
(colourless) (yellow in alkali)

Incubation mix: 0.5 ml 5 mM pNP-Glc in buffer **A** + 0.5 ml enzyme solution.

Method: After t min, add 1 ml 1 M Na_2CO_3. Assay yellow colour within 5 min by measurement of A_{400}.

Others: Many other p-nitriphenyl-glycosides are available.

Example (b): α-Xylosidase [299]

Substrate: Xyloglucan oligosaccharides. [Wall-bound α-xylosidase is inactive on pNP-α-Xyl.] Extract xyloglucan from PAW-washed

232

walls (Panel 1.2.1 c) of **either** a Dicot cell culture [the assay is simplified if the walls were pre-labelled by culture of the cells in the presence of [^{3}H]arabinose or [^{14}C]glucose], **or** nasturtium (Tropaeolum) cotyledons, with 6 M NaOH in 1 % NaBH$_4$, de-salt, and digest with cellulase to yield xyloglucan hepta-, octa- and nonasaccharides (see Panel 9.2.2).

Reaction: Glc-Glc-Glc-Glc   + H$_2$O → Glc-Glc-Glc-Glc

              | | | | | + Xylose

              Xyl Xyl Xyl   Xyl Xyl

Incubation mix: 25 µl of 0.2 % (or 1 µCi/ml) xyloglucan oligosaccharide in buffer **B** + 25 µl enzyme solution.

Method: After t min, spot the entire reaction-mixture on to Whatman 3MM chromatography paper, dry in a stream of cold air, and run in BuOH/HOAc/H$_2$O (12:3:5) (Panel 4.5.2 a) overnight. Stain with aniline H-phthalate (Panel 4.5.2 f) or assay radioactivity in the xylose zone. Complete hydrolysis will yield about 15-18 µg xylose, which can be semi-quantified by visual comparison with a set of standard loading of pure xylose. [To make certain that the product is not arabinose, re-chromatograph a portion in EtOAc/pyridine/H$_2$O (8:2:1) (Panel 4.5.2 b).]

(2) ENDOGLYCANASES

Example (a):   Cellulase [9,347]

Substrate: High-viscosity carboxymethylcellulose (CMC), a soluble derivative of cellulose — available commercially.

Reaction: ...G-G*-G-G*-G*-G-G-G-G*-G*-G*-G-G*-G-G*-G*-G-G-G*-G-...
+ H$_2$O →
...G-G*-G-G*-G*-G-G + G-G*-G*-G*-G-G*-G-G*-G*-G-G-G*-G-...
(where G = glucose; G* = carboxymethylglucose).

Incubation mix: 2 ml 1 % CMC in buffer **B** + 0.1-1.0 ml enzyme solution. [CMC is slow to dissolve; use shaking, not stirring.]

Method: N.B. — it is important to control temperature; work in a controlled environment room. Immediately after mixing, draw the solution into a 1-ml pipette, and time the flow of the meniscus past two fixed points [flow-time should be ca. 10-60 s; cf. H$_2$O should have a flow-time of ca. 2-4 s]. Repeat the measurement of flow-time at intervals (e.g. up to 4 h). Hydrolysis of CMC causes loss of viscosity and decrease in flow-time. Absolute enzyme units can be calculated if necessary [9,347].

Others: The assay is applicable to other polysaccharides that form viscous solutions. [But pectin may **increase** in viscosity owing to a competing reaction with pectin methylesterase.]

Example (b):   Galactanase [170]

Substrate: Galactan, ^{3}H-labelled at the reducing terminus, prepared as follows. Extract arabinogalactan from de-fatted soya flour by heating in buffer **B** at 100°C for 2 h, dialyse, freeze-dry, partially hydrolyse in 30 mM oxalic acid at 100°C for 2 h to remove arabinose residues, dialyse again and freeze-dry. Re-dissolve the polysaccharide at 1 % in H_2O, add CTAB/Na_2SO_4 (Section 3.5.3) and remove the precipitated acidic polysaccharides by centrifugation. Recover the neutral galactan from the supernatant by precipitation with 5 volumes of ethanol, wash with 80 % ethanol, and dry. Dissolve a sample (40 mg) of the dried galactan in 4 ml of 0.5 M NaOH containing 10 mCi NaB^3H_4, incubate at 25°C for 2 h, acidify with HOAc **in a fume cupboard** (care: $^3H_2$ evolved), and dialyse. The product is β-(1→4)-galactan in which the single reducing terminus per molecule has been converted to [^3H]galactitol. This is the very last unit to be released by the action of β-galactosidase, which attacks from the **non**-reducing terminus; therefore endo-galactanase can be reliably assayed in the presence of a large excess of β-galactosidase.

Reaction: S-S-S-S-S-S-S-S-//-S-S-S-S-S-S-S-S-S-S-S-S-[^3H]galactitol
(ethanol-insoluble)
+ H_2O →

S-S-S-S-S-S-S-//-S-S-S-S-S-S + S-S-S-S-S-[^3H]galactitol
(ethanol-soluble)
[+ n S if β-galactosidase was also present]
(where S = galactose)

Incubation mix: 25 μl of [^3H]galactan (1 μCi) in buffer **A** + 25 μl enzyme solution.

Method: After t min, add 50 μl 1 % dextran (carrier polysaccharide) followed immediately by 1 ml 95 % ethanol containing 0.5 % ammonium formate. Mix. Stand at 25°C overnight in a capped tube. Centrifuge (2,500 **g**, 5 min). Assay exactly 0.75 ml of the clear supernatant for ^3H.

Others: In principle this method is applicable to any polysaccharide with a well-defined backbone.

Example (c):   **Arabinanase** [170]

Substrate: Feruloyl-oligoarabinan. This substrate has the fluorescent feruloyl group on the non-reducing terminus, and is therefore protected from arabinosidase. It can be used to assay endo-arabinanase in the presence of a large excess of arabinosidase. Feruloyl-oligoarabinans of DP ca. 4-10 can be prepared from spinach cell walls (suspension-cultures and probably also leaf tissue) by mild acid hydrolysis (Panel 4.4.2 b, scaled up with Whatman 3MM paper).

Reaction: Feruloyl-Ara-Ara-Ara-Ara-Ara-Ara-Ara-Ara (R_F ca. 0.1)
+ H_2O → Feruloyl-Ara-Ara (R_F = 0.76) + other fragments

Incubation mix: 25 μl of a solution of feruloyl-oligoarabinan (containing 0.5 μg equivalent of ferulic acid) in buffer **B** + 25 μl of enzyme solution. The concentration of the feruloyl-oligo-arabinan is estimated from A_{372} in a solution adjusted to pH 10 with dilute NH_3; the extinction coefficient of the feruloyl group is 28500 M^{-1} cm^{-1} [168], i.e. a 35 μM solution has a UV absorbance at 372 nm of 1.00.

Method: After t min, dry the whole sample on to Whatman 3MM paper and chromatograph in BuOH/HOAc/H$_2$O (12:3:5) overnight. Examine the dried paper under long-wavelength (366 nm) light in the presence of NH$_3$ vapour. An increase in the R$_F$ of the fluorescent spots relative the an enzyme-free control is evidence of endo-arabinanase activity. [Complete digestion of the substrate is acheived with 0.1 % crude Driselase, yielding Fer-Ara$_2$: this can then be used as a marker and the yield of Fer-Ara$_2$ (estimated from A$_{372}$ as above) taken as a quantitative estimate of arabinanase activity.]

Example (d)   **β-(1→3),(1→4)-Glucanase** [558]

Substrate: Oat β-(1→3),(1→4)-glucan [= ML-glucan] — commercially available.
Reaction: ML-Glucan   + $\underline{n}$ H$_2$O   →   ($\underline{n}$ + 1) Oligosaccharides.
                    (binds Congo Red)   (don't bind Congo Red)
Incubation mix: Prepare a mixture containing 1 % agar, 0.05 % oat ML-glucan, and 0.004 % Congo Red in buffer **A**, 'melt' at 100°C, pour into a Petri dish as a 5 mm deep layer, and allow to set. Punch holes in the gel: 5 mm diameter, spaced ca. 4 cm apart.
Method: Pipette 10 µl of enzyme solution into each well and incubate at 25°C for up to 18 h. A circle of decolorised dye is formed around the well; the diameter of the circle appears to be roughly proportional to the log of the enzyme concentration.
Others: The method is applicable to other dye-binding polysaccharides. E.g. cellulase and endo-β-(1→3)-glucanase can be assayed as above but with carboxymethylcellulose and carboxymethylpachyman respectively. An alternative protocol involves incubation of the plate without dye, followed by flooding with dye solution.

(3) TRANSGLYCOSYLASES

Example (a):   **Trans-β-glucosylase** [370]

Substrates: Cellobiose + p-nitrophenyl-β-D-glucopyranoside (pNP-Glc). It is not clear from [370] whether the pNP-Glc is needed.
Reaction(s):   pNP-Glc + Glc-Glc ↔ pNP-Glc-Glc-Glc + Glc   and/or
                    Glc-Glc + Glc-Glc ↔ Glc-Glc-Glc + Glc   and/or
                    pNP-Glc + Glc-Glc ↔ Glc-Glc-Glc + pNP
                    [The Substrates are also hydrolysed during the reaction:
                    pNP-Glc + H$_2$O → pNP + Glc
                    Glc-Glc + H$_2$O → 2 Glc.]
Incubation mix: 25 µl of 10? mM pNP-Glc and cellobiose (concentrations are not given in [370]) in buffer **B** + 25 µl enzyme solution.
Method: After $\underline{t}$ min, spot a portion of the reaction mixture on to a glass- or plastic-backed silica gel TLC plate and develop in BuOH/EtOH/H$_2$O (13:7:5) alongside markers (5 µg each) of pNP-Glc, glucose, cellobiose, raffinose and stachyose. Locate spots with AgNO$_3$/NaOH (Panel 4.5.2 f). Transglyco-

sylation is indicated by the formation of spots with lower R_F than the substrates. Hydrolysis is indicated by formation of glucose.

(4) OTHER HYDROLASES

Example (a):   **Pectinmethylesterase**

Substrate: Pectin

Reaction:

$$
\begin{array}{cccccc}
\text{Me} & \text{Me} & \text{Me} & \text{Me} & \text{Me} & \text{Me} \\
| & | & | & | & | & | \\
\end{array}
$$

GalA-GalA-GalA-GalA-GalA-GalA-... (soluble in 10 mM $CaCl_2$)
+ n H_2O →
GalA-GalA-GalA-GalA-GalA-GalA-... (gelled by 10 mM $CaCl_2$)
+ n MeOH

Incubation mix: 1 ml 1 % pectin in solution **C** [10 mM $CaCl_2$ containing 20 mM MES, pH adjusted to 6.0 with NaOH] + 0.5 ml enzyme solution. Set up several replicates in identical tubes.

Method: After t min, turn one of the replicates on its side. Record whether a gel has formed. Time taken for gel-formation is inversely related to enzyme activity. [As an alternative, make repeated measurements of viscosity as in the assay for cellulase.]
Note: this procedure is unreliable in the presence of pectinase, which reduces the DP of the pectin backbone. It should be ascertained that no increase in reducing power occurs during the reaction, e.g. by the PAHBAH test (Panel 3.7 a). If pectinase is present, an alternative pectin methylesterase assay can be used in which the pH decrease in very lightly buffered pectin solution is monitored (ester + H_2O → **acid** + alcohol) [571].

Example (b):   **Phosphatase** [509]

As for β-glucosidase [see this Panel, part **1(a)**], but using p-nitrophenylphosphate in place of pNP-Glc.

Example (c):   **Protease** [408,515]

Substrate: 'Azocoll' (Calbiochem), a red dye covalently linked to insoluble protein (powdered cow-hide).

Reaction:

$$
\begin{array}{ccccc}
\text{Dye} & & & & \text{Dye} \\
| & & & & | \\
\text{Protein} & + & \underline{n}\ H_2O & → & (\underline{n} + 1)\ \text{Oligopeptide} \\
(H_2O\text{-insol.}) & & & & (H_2O\text{-soluble})
\end{array}
$$

Incubation mix: 2 ml buffer **D** [25 mM Tris free base, pH adjusted to 9.0 with HCl] + 0.1 ml enzyme solution + 20 mg solid 'Azocoll'.

Method: Incubate with vigorous shaking for 2 h. Cool to 2°C for 15 min. Centrifuge (1500 **g**, 15 min). Read A_{520} of supernatant.

(5) OXIDOREDUCTASES

Example (a):   **Peroxidase** [167]

Substrate: o-Dianisidine.
Reaction: $2 AH + H_2O_2 \rightarrow A_2 + 2 H_2O$
Incubation mix: 1 ml of solution **G** [0.8 mM o-dianisidine, 1 mM H_2O_2 (= ca. 1/10 000 the concentration of commercial '100-vol' H_2O_2 solution) and 10 mM NaH_2PO_4 in buffer **A**] + 0.5 ml of enzyme solution.
Method: Take continuous colorimeter readings at 420 nm.
Others: Many alternative substrates have been used, and different peroxidases may differ in substrate preference. Examples are guaiacol (use 8 mM) and 3,3',5,5'-tetramethylbenzidine. Syringaldazine has been suggested to be specific for those isozymes of peroxidase involved in lignin synthesis [195].

Example (b):   **Spermine oxidase** [285] (polyamine oxidase)

Substrates: Spermine (or spermidine);   coupled reaction substrate = guaiacol.
Reaction: $R\text{---}CH_2\text{---}NH_2 + H_2O + O_2 \rightarrow R\text{---}CHO + NH_3 + H_2O_2$
                 then

$$2 AH + H_2O_2 \xrightarrow{\text{peroxidase}} A_2 + 2 H_2O$$
                 (colourless)   (brown)

Incubation mix: 1 ml buffer **E** [100 mM NaH_2PO_4, pH adjusted to 6.5 with NaOH] + 50 µl 25 mM guaiacol + 50 µl 0.1 % horseradish peroxidase + 100 µl oxidase solution.   Incubate 2 min. Then add 50 µl 4 mM spermine (or H_2O as a control).
Method: Take continuous colorimeter readings at 470 nm. Calibrate by addition of known amounts of H_2O_2 to an oxidase-free control.

Example (c):   **Malate dehydrogenase** (MDH) [211]

Substrates: Oxaloacetate + reduced nicotinamide adenine dinucleotide (NADH)
Reaction: Oxaloacetate + NADH $\leftrightarrow$ Malate + NAD^+
                 Conversion of NADH $\rightarrow NAD^+$ is monitored by measurement of A_{340}.
Incubation mix: 1 ml buffer **F** [50 mM KH_2PO_4, pH adjusted to 7.6 with KOH] + 50 µl 10 mM oxaloacetate + 50 µl 3 mM NADH + 4· mg cell walls.   [Note — wall-bound MDH cannot readily be solubilised.] Run a control without oxaloacetate.
Method: Shake for 2 min, then filter by rapid suction through a GF/C glass fibre filter paper disk. Read A_{340} of filtrate.
                 Note: Isolated cell walls (which contain both MDH and peroxidase) will also carry out the malate- and O_2-dependent production of H_2O_2, especially if an oxaloacetate-withdrawing system is included [143,212].

7 Wall architecture

7.1.1: Significance of wall cross-links

The individual polymers of the primary cell wall (except cellulose) are intrinsically **soluble**: for instance, during the early stages of wall regeneration by isolated protoplasts, matrix polymers are not retained on the cell surface but dissolve into the culture medium [225]. Also, once polymers have been extracted from an intact cell wall, even with the mildest extractants (Ch 3), they tend to remain H_2O-soluble. The water-solubility of the matrix polymers can be explained by their many hydrophilic —OH groups and the fact that most of them are poly-ionic.

Despite this, the wall is highly **coherent** [174]. The polysaccharides and glycoproteins are linked together into a fabric of great strength, coupled with elasticity and plasticity, resistance to digestion, and an ability to adhere to neighbouring walls. Above all, the intact wall matrix is highly recalcitrant to extraction in water. If water-soluble polymers are inextractable in water, the implication is that the molecules are **cross-linked.** This chapter explores the nature of the cross-linking.

An understanding of wall cross-links is essential for any attempt to explain the control of wall extensibility and plant growth (Ch 10): the composition of the wall is usually little affected by growth hormones, and this draws attention to interpolymeric bonds rather than primary structure. Cross-links in the wall matrix are also likely to govern wall digestibility (Ch 11). Cell/cell bonding, which plays a vital rôle in morphogenesis, is also dependent on cross-links formed between polymers of juxtaposed walls.

7.1.2: Non-covalent cross-links

The polymers of the cell wall could be cross-linked by covalent or non-covalent bonds, or both. They could also be held together by molecular intertwining, as in catenanes and rotaxanes, without any direct interpolymeric bonding. The major non-covalent cross-links that have been considered are the following [407]:

238

Hydrogen bonds: Concerted H-bonds are believed to bind the hemicelluloses to the cellulosic microfibrils of the wall [381,241]. Hemicelluloses with many side-chains (e.g. heavily arabinosylated zones of arabinoxylan) are less able to form H-bonds with cellulose than are linear glucans or xylans. Inter-polymeric H-bonds can be disrupted by chaotropic agents, which are low-M_r substances that can themselves form strong H-bonds with the polymer molecules (Ch 3). Weaker (and/or less frequent) H-bonds are also probably present between the pectic and hemicellulosic polysaccharides of the matrix. The extent of this, and its importance in wall architecture, is difficult to assess.

Ionic bonds: Positively-charged extensin is ionically bound to the negatively-charged polysaccharides of the wall [458,174]. These bonds can be broken by low-M_r salts, in a process of ion-exchange. Newly-deposited extensin (distinguished from older molecules by pulse-labelling) can thereby be eluted from the walls of living cells by treatment with salts (cf. Section 6.3.2). Later, covalent bonds are added which make this impossible [169,105] (Section 7.1.3).

Ca^{2+}-bridges: Ca^{2+}-bridges are a special case of ionic bond. Ca^{2+}, being divalent, can bind ionically to two pectin molecules, tending to hold them together. In addition, the Ca^{2+} ion can co-ordinate with neighbouring —OH groups in the pectins. Runs of contiguous acidic GalA⁻ residues will bind in a concerted way to a number of Ca^{2+} ions. Thus, strong cross-links can be formed between pectin molecules, probably in a well-defined arrangement known as the 'egg-box' model [263,406,415]. Ca^{2+}-Bridges can be broken by chelating agents (Ch 3).

The formation of runs of consecutive acidic (non-methylesterified) GalA⁻ residues is achieved by the action of pectin methylesterase, which, according to some workers [508], attacks contiguous GalA-Me residues of a pectin chain sequentially; in contrast, partial hydrolysis with NaOH removes the methyl groups randomly, and leads to a product with much reduced ability to form Ca^{2+}-bridges.

7.1.3: Covalent cross-links

Oxidative phenolic coupling products: Since some matrix polysaccharides possess phenolic side-chains, which can be cross-linked in vitro by the action of peroxidase + H_2O_2 [191], it can be suggested that a similar reaction occurs in vivo [166,174]. Evidence for this has come from the isolation of the putative cross-links from cell walls [92,171,231,335]. The generalised mechanism for the formation of this type of cross-link (see Section 5.1.8) is:

$$2 \text{ Ar—H} + H_2O_2 \rightarrow \text{Ar—Ar} + 2 H_2O$$

where **Ar** = phenolic ring. The two major types of cross-link formed are diphenyl-ethers and biphenyls (Fig. 3.3.2 d,e). The major phenols (other than lignin) thought to participate in this kind of reaction are the feruloyl side-chains of certain polysaccharides, which form diferulate and probably other products (Sections 3.3.2, 4.1.3, 4.1.4, 5.1.8 and 7.6), and the tyrosine residues of the glycoprotein extensin (Section 4.1.5), which form isodityrosine [105,169,147,174]. In addition, it appears likely that some of the initially-formed dimers, e.g. of ferulate, can be grafted via quinone-methide intermediates on to neighbouring polysaccharide molecules to form ether cross-links (Section 5.1.8).

The formation of phenolic cross-links is dependent on at least three things: peroxidase, a supply of H_2O_2 (or an equivalent oxidising agent), and the synthesis of the phenol-bearing polysaccharide or glycoprotein; any of these three could be rate-limiting for wall cross-linking [50].

The cross-links once formed seem likely to be permanent: plants contain no known enzymes that would cleave an oxidatively coupled phenolic dimer. Thus, reversal of phenolic cross-linking, if it occurs at all, would depend on cleavage of other bonds — in the case of isodityrosine-cross-linked extensin this could be hydrolysis of the peptide bonds of the protein backbone, and in the case of diferuloyl-polysaccharides it could be hydrolysis of either one of the ester bonds (**e**) or a glycosidic bond in the polysaccharide backbone (**g**):

Sugar—Sugar—Sugar—**g**—Sugar—Sugar—**g**—Sugar—Sugar—Sugar...
|
0—**e**—C=0
|
Diferulate
|
0=C---0
|
Sugar—Sugar—Sugar—Sugar—Sugar—Sugar—Sugar—Sugar...

Glycosidic bonds: It is possible that some wall polysaccharides are covalently linked to others via glycosidic bonds. For chemical reasons

(Section 3.1.2) this could only occur via the reducing terminus, and therefore any given polysaccharide can only be directly linked to <u>one</u> other,
this means that it is impossible to form a 'network', although an extremely
large and highly complex 'tree' could be formed. Hybrid polysaccharides
whose existence has been proposed include one where the reducing terminus
of xyloglucan is glycosidically linked to a side-chain of pectin [289].
This could take the following form:

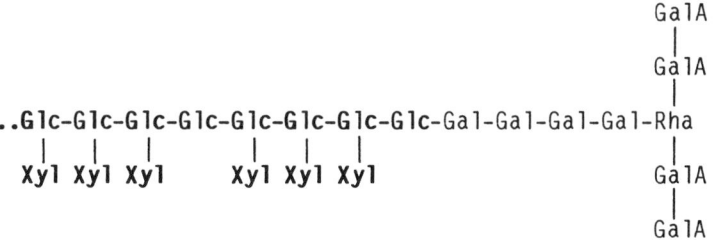

←----------**xyloglucan**----------→ ←---------pectin---------→

Some evidence for this type of hybrid polysaccharide has been obtained by
extraction, under mild conditions, of wall polymers in which it proved impossible to separate by chromatography the material composed of xyloglucan
building-blocks from that composed of pectin building blocks [442,444]. It
cannot, however, be stated categorically that such an extract contains
covalently-linked polysaccharides and not a mixture of two inseparable but
discrete polysaccharides; and this approach certainly does not permit the
conclusion that any putative cross-links are glycosidic bonds.

Glycosidic cross-links could theoretically be formed intracellularly,
for example if a xyloglucan synthase complex were to build a xyloglucan
chain on a pectic primer (Section 5.1.3), and the hybrid polysaccharide
would then be secreted, pre-formed, into the wall. Alternatively, hybrids
could form extracellularly by the action of wall-bound transglycosylases
(Section 6.1.3). Glycosidic cross-linking could be reversible by the action
of endo-glycanases.

Ester cross-links: The possibility of oxidative coupling of ester-
linked ferulate has already been mentioned. In addition, it is conceivable
that polysaccharides and/or glycoproteins could be directly cross-linked by
ester bonds. Since many of the GalA residues of the wall pectins are
methylesterified, it is theoretically possible that these could, by transesterification, become linked to other alcohols than methanol [173,174,

333]. Candidates include the —OH groups of sugars, Ser, Thr and Hyp:

$$\begin{array}{ccccccc} \text{Me} & & & & \text{R} & & \\ | & + & \text{R---OH} & \rightarrow & | & + & \text{MeOH} \\ \text{...GalA...} & & & & \text{...GalA...} & & \end{array}$$

In similar vein, <u>amide</u> cross-links could be formed by transamidation:

Glutamine + Lysine → N$^{\epsilon}$-(y-Glutamyl)-lysine + NH$_3$,

a reaction known to occur <u>in vivo</u> in animals [326]. Many other potential polypeptide cross-links, for which there is an equal lack of evidence in plants, are described in <u>Methods in Enzymology</u> Vols 106 & 107.

7.2: DETECTION OF NON-COVALENT CROSS-LINKS

Many of the extractants in Section 3.4 are based on attempts to break non-covalent cross-links. The fact that these mild extractants do not remove much polymer from the intact wall suggests that non-covalent cross-links provide only part of the basis of wall architecture.

Standard chaotropic agents (e.g. 8 M urea [39] or 5.5 M guanidinium thiocyanate [444]) extract small amounts of hemicellulose, indicating H-bonding to microfibrils. That only a small percentage is solubilised tends to suggest that other bonds also hold hemicelluloses in the wall.

The stronger chaotropic agent, MMNO-monohydrate at 120°C, dissolves almost the whole wall; however, it causes some depolymerisation of poly-saccharides [272] and extensive cleavage of esters [174]. Solubilisation by MMNO thus does not prove H-bonding. NaOH also dissolves hemicellulose, but the chaotropic effect cannot be achieved without hydrolysis of esters.

Salts can break ionic bonds, and thereby solubilise any polymer held in the wall only by such bonds, e.g. some enzymes (section 6.3.2) and lectins [215] and newly-secreted extensin (Section 6.4.2). However, salts extract very little polysaccharide from the wall.

Chelating agents extract any polymers held only by Ca^{2+}-bridges. As they do extract much of the pectin from **some** walls, Ca^{2+}-bridges seem to be involved. It can be shown by sensitive inorganic analysis that EGTA or CDTA at pH 7.5 and 25°C quickly remove all Ca^{2+} from isolated walls [262]. Therefore, when chelating agents fail to extract much pectin, other cross-links as well as (or instead of) Ca^{2+}-bridges must be invoked.

It would appear possible that some polymers are held in the wall by co-operation of two or more different types of non-covalent bond. This could be investigated by application of two or more different extractants at the same time or sequentially, although initial studies of a chelating agent (CDTA) plus high salt [(NH$_4$)$_2$SO$_4$] indicated little more extraction of pectin than would have occurred with the chelating agent alone [262].

7.3: DETECTION OF COVALENT CROSS-LINKS

Four types of evidence have been used as evidence for particular covalent cross-links between wall polymers (Fig. 7.3): **(a)** isolation of a molecular 'fragment' containing just enough of each polymer to permit identification, plus the covalent cross-link; **(b)** solubilisation (or decrease in M$_r$) of the cross-linked polymers by treatment with a reagent that specifically cleaves the cross-link without affecting the backbone; **(c)** demonstration that blocking the formation of a proposed cross-link abolishes the ability of the newly synthesised polymers to bind to the wall; and **(d)** formation of the cross-link between purified polymers in vitro. These four approaches are described in Sections 7.3.1 to 7.3.4 respectively.

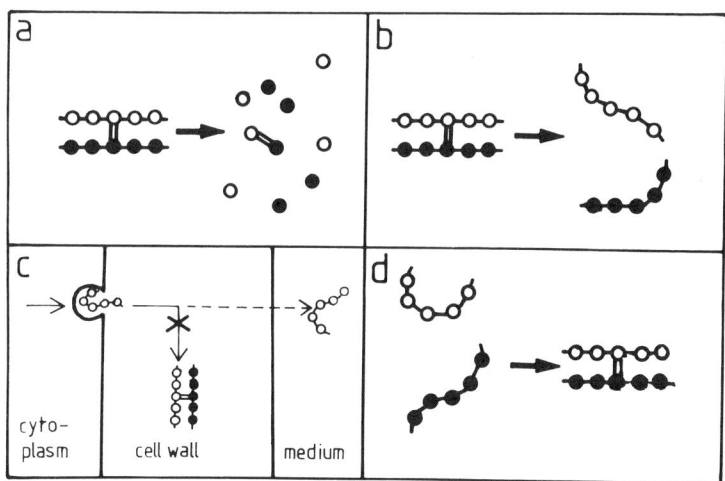

Fig. 7.3: **Types of evidence for cross-links (=) between polymers (o-o-o-o)**
(Reproduced with permission from Annual Review of Plant Physiology, Vol 37*,©1986 by Annual Reviews Inc.)
(a) Chemical isolation of cross-linked fragment.
(b) Chemical liberation of polymers by cleaving of cross-link.
(c) Blocking of wall-binding of newly-secreted polymer in vivo [———→, normal pathway; - - - - →, pathway in presence of inhibitor; x, site of action of inhibitor].
(d) Formation of cross-link in vitro. *[174]

7.3.1: Isolation of cross-links

For the isolation of a cross-linked fragment, an agent is required that will cleave the backbones of the polymers without affecting the cross-link itself. Thus isodityrosine, the diphenyl-ether dimer of tyrosine, can be isolated from extensin by hydrolysis of peptide bonds in 6 M HCl at 110°C for 20 h (Panel 4.3.3 a). The simplest way to detect isodityrosine [172] is by paper electrophoresis at pH 2 (Section 4.5.4), followed by staining with Folin & Ciocalteu's phenol reagent (Panel 4.5.2 f); isodityrosine and tyrosine are among the very few positively charged phenols likely to be encountered, and can thus be detected in the presence of a large excess of non-phenolic amino acids and neutral or acidic phenols. Alternative techniques include 2-dimensional paper chromatography [propan-2-ol/conc NH_3/H_2O (8:1:1) is a useful solvent since isodityrosine has a very low R_F (ca. 0.02) in it; for other solvents see Panels 4.5.2 a-e]; TLC [e.g. on silica-gel in PrOH/conc NH_3/H_2O (7:2:1) or phenol/H_2O (4:1, w/w)]; and HPLC of the PTC-derivative (Panel 4.3.3 b), which elutes on the HPLC system described in Section 4.5.7 between Leu and Phe. The PC and TLC methods can be made quantitative if the isodityrosine is eluted with dilute NH_3 and assayed by any convenient method, e.g. ninhydrin (Panel 3.7 a, part 14), Folin & Ciocalteu's phenol reagent (Panel 3.7 a, part 11) or UV absorbance (isodityrosine's molar extinction coefficient at 297 nm is ca. $4\,300$ M^{-1} cm^{-1}, measured in alkali [147]). Authentic isodityrosine is not commercially available but can easily be prepared (Panel 7.3.1) [172].

Diferulate, the biphenyl dimer of ferulic acid, survives a treatment with NaOH that completely hydrolyses the ester bonds that held the diferulate to the polysaccharides. Thus, diferulate can be detected as a minor component of the NaOH-hydrolysis products of certain walls (Panel 4.4.1). Authentic diferulic acid can be synthesised by oxidation of ferulic acid [92]. In order to identify the polysaccharides that may be cross-linked via diferulate bridges, it would be useful to be able to isolate fragments consisting of diferulate plus the immediately adjacent sugar residues of the two polysaccharide molecules to which it was linked. Unfortunately this is impossible with NaOH since the alkali hydrolyses the ester bonds. An alternative approach is to digest the wall polysaccharides with a hydrolase preparation (e.g. Driselase — Panel 4.4.2 a) that lacks esterase activity. Preliminary tests of this approach have yielded diferuloyl-carbohydrates [449], but their structure remains to be elucidated; the sort of reaction probably occurring can be represented as:

Sugar—Sugar—Sugar—**Sugar**—**Sugar**—Sugar—Sugar—...
|
Diferulate
|
Sugar—Sugar—Sugar—**Sugar**—**Sugar**—Sugar—Sugar—...

$$\xrightarrow{\text{Driselase}}$$

Sugar—Sugar
|
Diferulate   + $\underline{n}$ Sugar
|
Sugar—Sugar

Study of the oxidative coupling products of ferulate is simplified if the cells are pre-grown in the presence of **(a)** [^{14}C]cinnamate, an excellent precursor for ferulate and related phenolics that does not significantly

---

PANEL 7.3.1:   Synthesis of authentic isodityrosine

Step 1:   Dissolve 0.4 g L-tyrosine in 10 ml 17 % (w/v) NH$_3$.

Step 2:   Add 0.67 ml 0.75 M potassium ferricyanide with stirring. Incubate in a closed vessel at 25°C for 3 h.

Step 3:   Pour the solution into a shallow dish and stand it in a fume cupboard to allow some of the ammonia to evaporate.   Crystals of unreacted tyrosine will precipitate, and should be filtered off.

Step 4:   Load the filtrate on to Whatman 3MM paper as a 40-cm streak and run for 20 h in BuOH/HOAc/H$_2$O (12:3:5) [Panel 4.5.2 a].

Step 5:   Examine the chromatogram under long-wavelength (366 nm) UV light: dityrosine shows a strong blue fluorescence (intensified by NH$_3$ vapour);   isodityrosine does not fluoresce but can be assumed to be in the same zone of the chromatogram as the dityrosine.

Step 6:   Cut out the (iso)dityrosine band and elute (Panel 4.5.2 h) in 1 M NH$_3$.   Re-apply to a fresh sheet of chromatography paper and run overnight in BuOH/pyridine/H$_2$O (4:3:4).   This time the dityrosine (R$_F$ = 0.22) separates from isodityrosine (R$_F$ = 0.35).   Stain a narrow strip (cut in the direction of solvent flow) with Folin & Ciocalteu's phenol reagent [Panel 4.5.2 f] and hence, by comparison with the fluorescent dityrosine, locate the isodityrosine band.

Step 7:   Further purification is possible, by repeated chromato-graphy, e.g. in phenol/H$_2$O (4:1, w/v) — with this solvent, ninhydrin should be used in place of Folin & Ciocalteu's phenol reagent for staining the guide strip.

label unrelated metabolites [171], and **(b)** [1-^{3}H]glucose, which labels all carbohydrates. Spinach cell walls dual-labelled in this way and then digested with Driselase yield [^{14}C,^{3}H]-labelled products that, on treatment with NaOH, produce [^{3}H]sugars and many new ^{14}C,^{3}H-labelled compounds. The latter have R_F = 0.00 on silica-gel TLC in benzene/HOAc (9:1), which is used for analysis of ferulate, diferulate and related compounds. These R_F-0.00 substances are probably oligomers of ferulate (or related phenolics), probably with some carbohydrate still attached via NaOH-stable (ether?) bonds; their structure is under investigation in the author's laboratory.

7.3.2: Liberation of polymers by cleavage of cross-links

If covalent cross-links hold a particular polymer in the cell wall, their cleavage will facilitate the solubilisation of that polymer and/or reduce its apparent M_r. **Isodityrosine** is cleaved by warm, mildly-acidified sodium chlorite (NaClO$_2$), giving tyrosine as an initial product, showing that NaClO$_2$ splits the isodityrosine molecule in two [169]. Treatment of cell walls with NaClO$_2$ solubilises a proportion of the extensin, and this provides evidence that isodityrosine is an inter-extensin cross-link [389]. The difficulty with this approach lies in proof of specificity: it can be shown by viscometric measurements on carboxymethylcellulose [analogous to the cellulase assay described in Panel 6.5, part **2(a)**] that a few glycosidic bonds are broken during NaClO$_2$-treatment, although this probably does not account for the solubilisation of extensin because anhydrous HF (which completely de-polymerises the wall polysaccharides, including cellulose) fails to extract extensin from cell walls [365]. More seriously, NaClO$_2$ cleaves a small proportion of peptide bonds. The extent of this has been assessed by GPC of a standard protein (bovine serum albumin, BSA) after treatment under various conditions: the main determinant of peptide bond cleavage was the NaClO$_2$:peptide ratio. With 1 mol of NaClO$_2$ per mol of amino acid residues, peptide bond cleavage is unmeasurably low, and it is therefore recommended that this amount of BSA be added during NaClO$_2$-treatment of cell walls. BSA at this level only slightly retards the breakdown of isodityrosine by NaClO$_2$ [K.J. Biggs, University of Edinburgh; unpublished]. A method for the extraction of cross-linked extensin from cell walls with minimal breakdown is given in Panel 7.3.2

PANEL 7.3.2: Extraction of covalently-bound extensin from cell walls by cleavage of isodityrosine cross-links

Sample: Cell walls from suspension cultures, e.g. of Capsicum annuum, thoroughly homogenised and sonicated (Section 1.2.2) and washed in PAW (Panel 1.2.1 c).

Step 1:   Mix cell walls (1 g dry weight) to a paste with 10 ml of solution A [2.5 % BSA, adjusted to pH 3.4 with glacial HOAc] and add 50 ml of freshly-prepared soln B [0.5 % $NaClO_2$, adjusted to pH 3.4 with HOAc].

Step 2:   Incubate at 60°C with constant stirring for 15 min. This should be in an open beaker in a fume cupboard because the poisonous and explosive gas ClO_2 is evolved and must not be allowed to accumulate.

Step 3:   Filter on sintered glass, cool the filtrate, and adjust its pH to 8.5 with NH_3. Dialyse against H_2O and freeze-dry.

Next steps: Characterise the product by ion-exchange chromatography (Panel 3.5.3 b); GPC on Sepharose CL-6B (Panel 3.5.2 b); assay of total amino acids, hydroxyproline and total sugars (Panel 3.7 a); acid hydro-lysis to sugars (Panel 4.2.1 a) and amino acids (Panel 4.3.3 a); $Ba(OH)_2$-hydrolysis to yield the hydroxyproline-oligoarabinosides (Panel 4.3.2).

Ester bonds are cleaved fairly specifically by cold dilute aqueous alkali (e.g. 0.5 M Na_2CO_3 at 0°C for 24 h), sodium methoxide (NaOMe) in methanol, and in some cases by esterases. Potential side-effects of alkali are the elimination reactions (Section 4.2.4) and the rupture of hydrogen-bonds (Section 7.2). Therefore, extraction of a wall polymer by aqueous NaOH does not prove the existence of ester cross-links. The problem of H-bond breakage has been tackled in an interesting approach [363]: grass cell walls were treated with methanolic NaOMe to methanolyse any ester bonds without extracting the polysaccharides (polysaccharides are insoluble in methanol, and H-bonding is actually strengthened). The NaOMe was then washed out with pure methanol. When the methanol was later replaced by neutral water a substantial proportion of the hemicellulose was extracted, showing that a change had occurred in it during exposure to NaOMe. It is inferred that ester bonds had been holding the hemicelluloses in the walls but it is unknown whether these bonds were of the sugar—diferulate—sugar type or of the hypothetical uronic acid—sugar type (see Section 7.1.3).

7.3.3: Blocking of wall-binding of newly-secreted polymer

If a cross-link is responsible for the wall-binding of a particular polymer, then blocking the formation of that cross-link should stop the newly-synthesised polymer binding to the wall. Unfortunately there are few agents that will specifically inhibit cross-linking. One example of this approach is the use of peroxidase inhibitors to suppress the formation of phenolic dimers (Panel 7.3.3). Dithiothreitol, ascorbate, cyanide and other peroxidase inhibitors block the oxidative coupling of phenolics (e.g. tyrosine → isodityrosine), and concomitantly prevent the covalent binding of extensin to the wall. Thus the newly-synthesised and -secreted extensin remains in its initial, salt-extractable form (Section 6.3.2).

A possibility for future investigation is that **antibodies** directed against wall enzymes may inhibit wall cross-linking reactions. However, it is not certain that an exogenous antibody will reach many of its target enzyme molecules embedded within the wall matrix of a living cell [518].

7.3.4: Formation of cross-links in vitro

If wall polymers undergo cross-linking in vivo, they should also cross-link in vitro (in the presence of enzyme if needed). Cross-linking in vitro confirms that it **could** occur in vivo, but not that it does.

PANEL 7.3.3:   Inhibition of extensin cross-linking by ascorbate [105,169]

Sample: Cultured Capsicum annuum cells (2-5 days after sub-culture)

Step 1:   Incubate 2 samples of the cell suspension (each 5 ml containing ca. 0.5 ml PCV) in 25-ml flasks with gentle shaking for 1-2 h.

Step 2:   To sample **A**, add 0.5 ml of 0.1 M ascorbic acid [pH pre-adjusted to that of the culture medium (e.g. 5.0) with 10 M NaOH]. To sample **B**, add 0.5 ml of 0.1 M NaCl as a control.

Step 3:   To both samples, add 5 µCi of L-[U-^{14}C]proline. Incubate with gentle shaking for 2 h.

Step 4:   Kill the cells and extract non-covalently bound protein in PAW as in Panel 3.4.2 a [omitting the scintillation-counting in step 3, omitting step 4 altogether, and repeating step 6 several times — until the supernatant is no longer radioactive].

Step 5:   Suspend the dried samples from steps 3 & 7 of Panel 3.4.2 a in 6 M HCl/10 mM phenol (at least 0.25 ml per mg of dried sample), and hydrolyse the protein (Panel 4.3.3 a).

Step 6:   Load the hydrolysate on Whatman 3MM paper and run for 16 h in BuOH/HOAc/H$_2$O (12:3:5) (Panel 4.5.2 a) to separate Pro and Hyp. Assay ^{14}C in both zones (Section 2.8.2).   The [^{14}C]Hyp is a measure of the extensin (PAW-insoluble = covalently bound; PAW-soluble = unbound or only ionically-bound);   the [^{14}C]Pro is a measure of total protein, and allows detection of any effect of the ascorbate on general protein metabolism.

Attempts have been made to couple the tyrosine residues of salt-extracted extensin by in vitro treatments with peroxidase + H_2O_2, but the main product detected is dityrosine [105], which does not occur in plants, rather than isodityrosine, the naturally-occurring dimer. (Dityrosine is found in insect cuticle protein [14].) Only when present in whole, isolated walls does extensin form isodityrosine in vitro [105]. Free tyrosine does form some isodityrosine when incubated with peroxidase + H_2O_2 at low pH, but only dityrosine at high pH. The wall-factor directing isodityrosine formation in extensin is unlikely to be a specific peroxidase isozyme, as sharply contrasting isozymes do not differ in the dityrosine : isodityrosine ratio generated from free tyrosine; and the factor is also unlikely to be low pH, since this alone never completely suppresses dityrosine formation [176]. Rather, the presence of other wall polymers (perhaps acidic pectins) would appear to be what steers the exclusive formation of isodityrosine during assembly of the cell wall [176].

Feruloylated polysaccharides will also couple in vitro, with the formation of diferulic acid and possibly other dimers and higher oligomers [191,428]. This results in the gelation of formerly soluble polysaccharides as the individual molecules are cross-linked (Panel 7.3.4).

PANEL 7.3.4: Oxidative cross-linking of feruloyl-polysaccharides in vitro

Sample: Sugar beet root (50 g).

Step 1:   Isolate walls from beet as in Panel 1.2.2, omitting steps 4-6 if necessary.

Step 2:   Solubilise pectin by stirring the walls in solution **A** [0.25 M oxalic acid, pH adjusted to 4 with NH_3] at 100°C for 30 min.

Step 3:   Filter on sintered glass, de-salt the filtrate (Section 3.5.1) and freeze-dry.

Step 4:   Dissolve the pectin (0.15 g) in 5 ml of soln **B** [0.1 M HOAc, pH adjusted to 4.7 with 1 M NaOH].

Step 5:   Take four 1-ml portions of the solution [(a)-(d)]. Add:

Portion	H_2O	5 mM H_2O_2	0.1 % HRP[1]
(a)	0.2 ml	–	–
(b)	0.1 ml	0.1 ml	–
(c)	0.1 ml	–	0.1 ml
(d)	–	0.1 ml	0.1 ml

Step 6:   Incubate at 25°C without shaking. Gelation occurs only in sample **(d)**. [Alternatively, measure viscosity at intervals (Panel 6.5.2).]

[1]HRP = Horseradish peroxidase (e.g. Sigma P 8250).

8 Wall turnover and sloughing

8.1: OCCURRENCE AND SIGNIFICANCE OF WALL TURNOVER

'Turnover' means the simultaneous synthesis and breakdown of a polymer. Unfortunately, the term has also been applied to the simultaneous deposition of polysaccharides in the wall and sloughing into the medium in suspension cultures [489]. Sloughing does not necessarily involve breakdown of the polymer, but rather of the cross-links that held it in the cell wall. Although the distinction between turnover and sloughing is sometimes hard to draw, it is an important one.

Wall breakdown may play a number of important biological rôles, e.g. loosening the cell wall to permit expansion growth (Ch 10), production of biologically-active oligosaccharides (Ch 9), the execution of various lytic steps in morphogenesis (Ch 1 and Sections 6.1.1 to 6.1.3) and the mobilisation of food reserves from the thickened walls of certain seeds (Section 11.1.1).

The extent of breakdown varies enormously [306]. In thickened seed cell walls, virtually the whole matrix may be liquified during germination [319,417]; in developing xylem, the formation of perforation plates involves complete wall breakdown but only at highly localised sites [114]; in ripening fruit, a reduction in M_r of the bulk of the pectin may occur [156,213,293]; and in expanding cell walls, only a small proportion of the hemicellulose is broken down, although the bulk of it may undergo a reduction in mean M_r [377,378]. In many cells, very little wall turnover occurs.

Several biochemical (as well as histological) approaches are available to detect turnover [303,306]. One simple way is to show that the amount of a given wall polymer per cell decreases during a particular phase of development [306]; this approach depends on breakdown exceeding synthesis, a situation which is only likely to occur during ripening, senes-cence and germination. A second approach is to show that the mean M_r of a given wall polymer decreases during a particular phase of development [497,243]; here the difficulty is to establish that previously deposited wall molecules underwent a decrease in M_r rather than the cell starting to synthesise new polymers of lower M_r: a strictly quantitative analysis is required to demonstrate turnover by this method.

The third and most widely adopted approach consists of very carefully designed pulse-chase experiments: for example, Franz [163] injected [14C]glucose into soyabean hypocotyls and then at intervals thereafter analysed the [14C]polysaccharides in a defined zone of tissue, marked with Indian ink at the time of injection. The [14C]glucose was quickly incorporated into polysaccharide in this zone, and the majority of the samples were therefore taken during what was effectively a 'chase' period. In early samples, a [14C]glucan (probably callose) was heavily labelled, but over the next two weeks more than half of this [14C]glucan, in the marked zone, disappeared. This suggests that the glucan turned over (presumably to free glucose) in vivo. The experiment could underestimate turnover if the liberated [14C]glucose were re-incorporated into glucan. It could possibly be argued that the [14C]glucan was translocated out of the marked zone, rather than turned over, although there is little evidence for polysaccharide-translocation in plants. More seriously, it is possible that the [14C]callose was formed transiently as a wound-response to the trauma of injection and that turnover of normal wall polysaccharides might not occur. Turnover of callose has recently been confirmed [85].

Evidence that turnover of a normal wall component does indeed occur, and might be biologically significant, was provided by Labavitch and Ray [304]. [14C]Glucose was fed to pea stem segments in order to label the wall polysaccharides; in a subsequent chase period a small proportion of the [14C]xyloglucan disappeared. Xyloglucan turnover was promoted by auxin or H^+ (Ch 10). It was later found that, although only a small proportion of the xyloglucan was actually lost, the bulk of the wall xyloglucan decreased in mean M_r (assayed by GPC after solubilisation in alkali) [243, 378].

Similar experiments were performed on maize coleoptiles, with and without growth-promoting levels of auxin [119]. The 14C-labelled walls sampled during the chase period were examined structurally by methylation-analysis (Section 4.2.8), and the methyl-[14C]sugars obtained were identified by GC (of alditol-acetates) and assayed for 14C. It was found that auxin caused loss of non-reducing terminal [14C]arabinose residues and

of mid-chain [14C]xylose residues substituted on positions 2 **and** 4;   and
auxin simultaneously increased the levels of mid-chain [14C]xylose residues
substituted only on position 4.   It was concluded that auxin promoted the
loss of arabinose residues from arabinoxylan:

Xyl-(1→4)-**Xyl**-(1→4)-**Xyl**-(1→4)-**Xyl**-(1→4)-**Xyl**-(1→4)-**Xyl**-(1→4)-...
| (1→2) | (1→2) | (1→2) | (1→2)
Ara   **Ara**   **Ara**   **Ara**

Xyl-(1→4)-**Xyl**-(1→4)-**Xyl**-(1→4)-**Xyl**-(1→4)-**Xyl**-(1→4)-**Xyl**-(1→4)-...
auxin → | (1→2) | (1→2)
Ara   **Ara**

The enzymes involved have recently been studied [380]. Other workers have
failed to observe similar effects on the arabinoxylans of other plants than
maize. However, there are many reports of the turnover of β-(1→3),(1→4)-
glucan in grasses, in response to auxin. Oligoglucan breakdown products
have not been detected, suggesting that the principal product is the
monosaccharide, glucose [256].

It should be stressed that, during normal cell expansion, the great
majority of the polymer molecules of the plant cell wall are not turned
over, but remain as stable components of the wall. The extent of turnover
can be measured in cell suspension cultures by means of pulse-chase
experiments. The effect of various factors, e.g. hormones and pH of the
medium, on wall turnover can then be quantitatively compared (Panel 8.1).

8.2: WALL POLYMER SLOUGHING

It is difficult to be certain whether the loss of a polysaccharide
from the wall represents turnover or sloughing. There is a grey area
between the two, and it is possible that partial turnover (e.g. a reduction
in M_r) may sometimes underline cause sloughing. Evidence for sloughing is the
detection of soluble polysaccharides in the apoplast. In the case of cell
suspension cultures the apoplast is the culture medium, which is readily
accessible;   in intact plants the apoplast can be sampled by vacuum-
infiltration/centrifugation (Panel 6.3.1). As an example of the latter
technique, it was observed that during the promotion of growth by auxin in
legume stem segments an increase occurred in the amount of water-
extractable xyloglucan detected in the apoplastic fluid [497].

PANEL 8.1: Assay of wall turnover in cultured spinach cells

Sample: Suspension-cultured spinach cells, grown with 1 % glucose as sole carbon source, ca. 14 days after subculture.

Step 1: Inoculate the cells (ca. 1 ml PCV) aseptically into a 250-ml flask containing 50 ml of fresh medium with 1 % glucose and 10 μCi D-[U-^{14}C]glucose as sole carbon source. Incubate under normal conditions for 7 to 14 days [N.B. $^{14}CO_2$ will be evolved.] By this time, over 90 % of the [^{14}C]glucose will have been removed from the medium [this can be checked by scintillation-counting of a sample removed aseptically with a sterile pipette].

Step 2: Working aseptically, let the cells sediment, and pour off the spent medium. Resuspend the cells in 250 ml of fresh medium containing 1 % non-radioactive glucose, and let the cells sediment again. Resuspend them in 170 ml of fresh medium containing 1 % non-radioactive glucose, and aseptically dispense (e.g. via an autoclaved ARH pipetting unit) 14 10-ml portions into 50-ml flasks. Aseptically add the sterilised (see Panel 2.2 b) test solutions [e.g. buffer (see Panel 2.7.1 b), auxin, gibberellin, abscisic acid, oligosaccharide (Ch 9) etc.].

Step 3: Incubate under normal conditions. Each day for a week, harvest two 10-ml cultures.

Step 4: Pass each harvested culture through fine nylon gauze or Whatman GF/C glass fibre filter paper, and wash the cells with 2 10-ml portions of water. Pool the filtrates, dialyse against H$_2$O, freeze-dry.

Step 5: Treat the collected cells with PAW as in Panel 1.2.1 c.

Step 6: Assay [^{14}C]polysaccharides in the walls and medium e.g.:
    (a) Total ^{14}C by simple scintillation-counting. This measures polymer sloughed from the walls. It also gives (if total [^{14}C]polymer per 10-ml culture decreases) a **minimum** estimate of turnover. As Fig. 8.1 shows, this is only a minimum estimate since some of the sugars ('**s**') released by turnover will be re-incorpotated into new polysaccharides.
    (b) [^{14}C]Sugars after acid hydrolysis (Panels 4.2.1 a & b) and paper chromatography (e.g. Panel 4.5.2 c), to give a preliminary indication of the nature of the polysaccharides sloughed and/or turned over.
    (c) [^{14}C]Sugars after Driselase-digestion (Panel 4.4.2 a): this is particularly useful for study of xyloglucan turnover since Driselase converts xyloglucan into the diagnostic disaccharide, xylosyl-α-(1→6)-glucose [isoprimeverose] (Section 4.2.5).

It is important to ask whether the polysaccharides sloughed into the apoplast have been recently synthesised and secreted, simply having passed through the wall transiently, or whether they are molecules that, after a substantial period of residence in the wall, are subsequently liberated. There is evidence for both types of behaviour. It is easy to show that some of the polysaccharides liberated into the media of suspension-cultured cells are newly synthesised (Panel 8.2). For example, it was shown that,

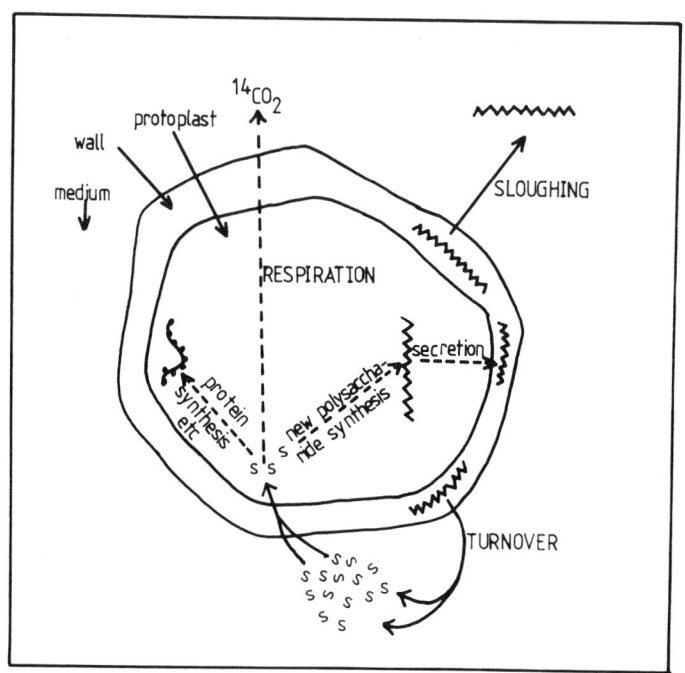

Fig. 8.1:   Sloughing and turn-over of wall carbohydrates. Wall-bound
polymers (**zigzags**) are subject to sloughing into the medium in more-or-less
intact form, and to degradation (turnover) to yield free sugars (**s**).   The
latter may be taken back up into the protoplast and then either respired,
or incorporated into new wall polymers, or [their carbon skeletons] used in
other biosynthetic processes.

during feeding of [14C]glucose or [14C]formaldehyde to cultured sycamore
cells, the sloughed polysaccharides had a higher specific activity than the
firmly wall-bound polysaccharides, indicating that the bulk of the latter
could not have acted as obligatory intermediate for the former [44].   As a
second example it was found that, during pulse-labelling with [3H]arabin-
ose, [3H]polysaccharide started to appear in the cells after a lag period
of ca. 5 min, and in soluble form in the medium after a lag period of ca.
25 min [177].   Thus, within 20 min of their synthesis, some polysaccharides
were being released into the medium, indicating that they were never
genuine structural components of the wall.   Importantly, the release of
[3H]polysaccharide reaches a _linear_ rate after about 30 minutes' labelling:
if the bulk of the released polysaccharide had first been bound firmly to
the wall and then liberated, the rate of release of [3H]polysaccharide

254

would have steadily risen over a period of several hours or days as the specific radioactivity of the bulk wall-bound polysaccharide increased. Furthermore, within about 20 min of the start of a chase period with non-radioactive arabinose, the rate of release of [^{3}H]polysaccharide fell off to a much lower level. From this type of experiment, it appears that, in cultured spinach cells, newly secreted polysaccharide molecules fall into two classes: (a) about 80 % of them bind firmly to the wall, and remain there for a long time, and (b) about 20 % never bind to the wall, but are quickly sloughed. Similar results have been obtained with other cultures than spinach, though the figures obtained vary widely from 80:20.

The polysaccharides in the medium resemble very closely those of the cell wall in sugar composition and linkage. All the polysaccharides known from the cell wall matrix have been detected in water-soluble form in the culture medium [476]. Some cultures accumulate in their media substantial clots of slime [361], which may contain lignin and cellulose as well as matrix-type polysaccharides. The formation of these clots, which occurs extracellularly in free solution, may be a valuable model for study of wall assembly. In addition, AGPs and a number of enzymes accumulate in the medium. Extensin does not normally appear in the medium, because it is firmly bound to the wall via ionic bonds and isodityrosine cross-links (Ch 7), but in suspension cultures of spinach, which form very little isodityrosine, extensin is sometimes sloughed into the medium when the culture enters stationary phase [Fry, unpublished].

It is unclear what distinguishes those polymer molecules that are destined to bind to the cell wall from those that are destined for rapid release. It seems possible that there is no chemical difference, but that any newly-secreted polysaccharide molecule has a 4:1 chance of binding to the wall and, if it misses this chance in the first few minutes after secretion, it has missed it forever.

In addition to this lottery at or soon after the moment of secretion, it is clear from experiments similar to that described in Panel 8.1 that some hemicellulose that had been firmly bound to the wall, even for several weeks, can subsequently be liberated [489].

Pectic polysaccharides are sloughed into the medium of cultured cells in response to growth-promoting treatments with gibberellic acid; it is uncertain whether such polysaccharides were newly synthesised [167,366].

PANEL 8.2:   Demonstration of release newly-secreted polysaccharides

Sample: Suspension-cultured spinach, 2-5 days after sub-culture.

Step 1:   Incubate 50 ml of suspension culture (containing ca. 5 ml PCV) in a 250-ml flask with gentle shaking for 1-2 h.

Step 2:   Add 0.1 mCi [^{3}H]arabinose[1] of high specific activity, with a minimum of disturbance, and continue the incubation.

Step 3:   At intervals, remove a 1-ml sample of the suspension, and quickly filter though a 2.5-cm disk of Whatman GF/C glass fibre paper to separate cells from medium.

Step 4:   Quickly transfer the filter + cells into 10 ml of 90 % ethanol containing 1 % non-radioactive arabinose. Incubate with stirring or shaking at 25°C for 1-16 h.   Arabinose and low-M_r intermediates are extracted; polymers are not. Filter again, wash the residue with 4 x 10 ml of 90 % ethanol, and transfer the final cell residue (+ both filters) into a scintillation vial. Dry, and scintillation-count **[= intracellular + wall-bound polymers]**.

Step 5:   Spot 0.1 ml of the filtrate from step 3 on to Whatman 3MM paper as a 4-cm streak.   Run by descending chromatography for 16 h in EtOAc/HOAc/H$_2$O (10:5:6).   Cut out and scintillation-count the origins **[= total released polymer]**.   The polymer in the filtrate can be analysed qualitatively, as described in Panel 8.1.

Next step: Qualitative analysis of the labelled polymers (see Panel 8.1, step 6).

[1]Other sugars give similar results —— see Panel 2.1.1.

9 Biologically-active wall oligosaccharides

9.1: BACKGROUND

9.1.1: Introduction

One of the most exciting developments in plant biology in the last decade has been the discovery of biologically-active oligosaccharides: certain 'fragments', excised from wall polysaccharides by partial breakdown in vitro, exert specific and potent biological effects when added to living plant cells. The effects are diverse, depending on the chemical structure of the oligosaccharide and on the type of cell tested; they include:

elicitation of phytoalexin* synthesis [8,121,122,385],
induction of phenylalanine ammonia-lyase [139],
evocation of synthesis of proteinase-inhibitors [54,432,536],
antagonism of auxin-stimulated growth [340,575],
decrease in general cell vitality [178,573],
induction of ethylene synthesis [505], and
induction of certain developmental processes e.g. flowering [507].

Wall oligosaccharides may thus play a rôle in disease resistance and in the control of growth and differentiation. The first biological activities were reported for oligosaccharides derived from fungal cell walls [8,547, 447]. Subsequently it was found that oligosaccharides derived from the walls of higher plants are also active. It has recently been established that certain biologically-active oligosaccharides are formed in vivo [175].

The hypothesis has been developed (the 'oligosaccharin' hypothesis) that wall polysaccharides serve a dual rôle: besides their structural rôle as the bricks and mortar of the wall, some of them also act as locked-up regulatory molecules which can be activated by partial enzymic hydrolysis [345]. In this way, wall-derived oligosaccharides may act as second messengers mediating and perhaps amplifying an initial stimulus perceived by the cell. Specific examples are cited in the following Sections.

----
*Phytoalexins are low-M_r antibiotics, produced by plant cells in response to invasion by a pathogen. Each species of plant tends to have its own characteristic phytoalexin(s), which is/are produced during infection regardless of the species of attacking pathogen. Chemically, phytoalexins are phenolics, terpenoids or polyacetylenes [8,197].

257

9.1.2: Oligosaccharides from β-(1→3),(1→6)-glucan: fungal elicitors

Soyabeans can often detect invading hyphae of the pathogenic fungus _Phytophthora megasperma_, and respond by synthesising phenolic phytoalexins that arrest further fungal growth. To identify the fungal component that the plant recognises as foreign, a bioassay was devised in which various components of the (killed) fungus were applied to soyabean cotyledons and the production of phytoalexin was measured. By such experiments, it was shown that the plant could detect fungal **walls** [8]. The active principle in the walls could be solubilised by autoclaving or by mild acid-hydrolysis. GPC of a mild acid-hydrolysate on Bio-Gel P-2 (Panel 3.5.2 b) showed that the smallest fragments of the fungal wall that could elicit phytoalexin synthesis co-chromatographed with heptasaccharides. Fractionation of the many different heptasaccharides present in this sample by HPLC revealed that only one was active, indicating that a very precise receptor for this oligosaccharide exists in soyabeans. The structure of the active heptasaccharide was deduced, and shown to be typical of β-(1→3),(1→6)-glucan, a major component of the fungal wall [447]. The heptasaccharide was synthesised chemically and shown to have activity identical with that isolated from fungal walls [448]. As little as 1 ng of heptasaccharide per cotyledon was enough to elicit the synthesis of phytoalexins, indicating that the receptor is very sensitive. Not only cotyledons but also many other parts of the plant and also cell suspension cultures respond to the oligo-β-glucans. Many other species than soyabean also respond, although by synthesising different phytoalexins. Plant cells respond to the oligosaccharides by switching on the transcription of the genes for certain key enzymes on the pathway to phytoalaxin synthesis, e.g. phenylalanine ammonia-lyase in the case of species with phenolic phytoalexins (Section 5.1.8) [139]. Although only one of the many heptasaccharide fragments that can be obtained from the fungal wall by random acid-hydrolysis was active, it seems possible that many larger oligosaccharides will prove to be active, perhaps so long as they possess the core heptasaccharide as part of their structure.

This phenomenon may be a warning system by which plant cells can detect the invading fungus and activate a defence response. The polysaccharide from which the active oligosaccharides arise is present in the fungal wall, and, during infection, may be released from it in a soluble form which the plant can detect. The release could be by spontaneous

258

sloughing on the part of the fungus (cf. Section 8.2) or by partial hydrolysis catalysed by endo-β-glucanases in the plant cell wall, or both [30].
There appears to be as yet no direct evidence that elicitor-active oligo-β-
glucans are present in infected plants in vivo.

9.1.3: Oligosaccharides of homogalacturonan: endogenous elicitors

Fungi secrete heat-labile elicitors (of phytoalexin synthesis) in addition to the heat-stable oligosaccharides mentioned above. Heat-lability
suggests an enzyme, and it was indeed shown that two of the major enzymes
secreted by many pathogens, pectinase [547] and pectate lyase [122], act as
elicitors. Since the substrates of these enzymes, pectins, are components
of the plant cell wall, and by analogy with the work described in Section
9.1.2, it seemed plausible that the enzymes act on the plant's own cell
wall to solubilise elicitor-active oligosaccharides. It was confirmed that
factitious oligosaccharides of homogalacturonan can elicit phytoalexins,
whether the oligosaccharides were made by acid-hydrolysis, by enzymic
hydrolysis (with pectinase) or by enzymic β-elimination (with pectate
lyase). The most active oligosaccharides of homogalacturonan have DP 10-14
[385]; they can act synergistically with oligoglucan elicitors [121].

Again, it is possible to argue for an advance warning system: many
phytopathogens produce pectinolytic enzymes in order to breach the host
cell walls; in so doing, they liberate pectic oligosaccharides; and the
plant has evolved a means to recognise these as alarm calls and respond to
them by switching on a defence response (phytoalexin synthesis). Although
it is likely that oligogalacturonans are solubilised during infection, this
solubilisation has not yet been demonstrated in vivo.

9.1.4: Pectic oligosaccharides: activators of protein synthesis

Mechanical injury of tomato plants switches on the synthesis of
certain proteins in the leaf blade. These include proteinase-inhibitors,
which may give some protection against insect attack by interfering with
the insect's ability to digest leaf protein. Soluble pectins and low-M_r
fragments of pectic polysaccharides, when supplied to tomato leaves via the
transpiration-stream, evoke the same response [53,54,432]. It is suggested
that pectins are fragmented upon wounding of the tissue, e.g. by an insect's
mandibles, releasing oligosaccharides that mediate the effect of the

primary stimulus (biting) on the final response (protein synthesis). However, there is as yet no clear evidence for the proposed fragmentation of pectins in response to mechanical injury, and the mechanism of any fragmentation is unknown. Nevertheless, the hypothesis is supported by the fact that very small injuries, which would not normally evoke the synthesis of proteinase-inhibitors, can be made to do so by application of pectinase to the wound site [537]. The possible rôle of pectic oligosaccharides as long-distance wound hormones is discussed in Section 9.4.1.

9.1.5: Xyloglucan nonasaccharide (XG9): an anti-auxin

The major nonasaccharide (XG9) released from xyloglucan by the action of cellulase (Section 4.1.4) shows anti-auxin properties in the classical pea-stem segment elongation bioassay [340,575]. Auxins, e.g. 2,4-D at 10^{-6} M, promote the elongation of segments cuts from young pea stems. XG9 at 10^{-8}-10^{-9} M blocks this effect of 2,4-D. Higher concentrations of XG9 are inactive, i.e. they restore the ability of 2,4-D to promote growth. The corresponding heptasaccharide (XG7), which lacks the fucose and galactose groups, is inactive, and an incompletely-characterised decasaccharide that appears to have one additional galactose residue, is also inactive [340]. XG9 is active whether it bears 0, 1 or 2 acetyl groups [575,576], but this could be because the acetyl groups are removed by wall-bound esterases. XG9 is thus another biologically-active, wall-related oligosaccharide that must act via a very sensitive and discriminating receptor.

It is attractive to postulate a feedback control loop in which XG9 production causes the inhibition of growth seen in response to supra-optimal concentrations of auxin: high auxin induces cellulase synthesis; cellulase catalyses the hydrolysis of xyloglucan to XG9; and XG9 blocks the action of auxin on growth [173]. It may be related that grasses, which are relatively resistant to the herbicidal effects of high auxin, have much less xyloglucan in their walls than do Dicotyledons, which are susceptible.

9.1.6: Unidentified wall fragments: hypersensitivity factors?

One highly effective mechanism by which all plants resist attack by almost all micro-organisms in their environment is the hypersensitive response: when a fungal cell gains entry to plant tissue, a small number of the plant cells at the site of attack quickly die, and, perhaps by

260

releasing toxins, arrest further growth of the fungus [197]. This hyper-sensitive death appears to be an active process, is usually accompanied by browning, and may depend on the plant cell being capable of respiration and protein synthesis. Death usually occurs at about the time the fungus penetrates its first plant cell. The plant must have detected the fungus somewhat earlier, probably while the fungus was still in the plant cell wall. It can therefore be speculated that as the fungus enzymically digests the plant cell wall, fragments are solubilised which the plant can recognise as danger signals and respond to by localised hypersensitive death. As a preliminary test of this hypothesis, random fragments of sycamore cell walls were prepared by partial acid-hydrolysis, and supplied to healthy cells. These cells were rapidly killed [573]. It remains to be seen whether this death is comparable with hypersensitivity, but the suggestion seems worth further investigation.

The cytotoxic wall fragments appeared to arise from the pectic poly-saccharides (or material with similar solubility properties) but has not been identified [178]. Acid-hydrolysis of homogalacturonan failed to release cytotoxic fragments, but the pectins of primary cell walls are much more complicated than homogalacturonan. It is interesting that highly purified pectinase and pectate lyase are also cytotoxic [37]: this has generally been ascribed to their weakening effect on the wall, rendering it osmotically fragile, but the alternative that they solubilise cytotoxic oligosaccharides deserves consideration.

9.1.7: Unidentified wall fragments: regulators of morphogenesis?

A number of preparations solubilised from plant cell walls by mild acid, alkali or enzymes have been reported to influence morphogenesis. For example, pectic fragments from sycamore at 1 μg/ml inhibited flowering in Lemna while promoting vegetative growth [345]. Other fragments, which were generated from sycamore cell walls, influenced the ratio of the various organised structures (roots, vegetative shoots, flowers, or callus) that developed on explants of tobacco [507].

9.1.8: Unidentified wall fragments: grafting compatibility factors?

Grafting is only possible between certain types of plant. A partial compatibility table for the Solanaceae can be drawn up as follows (where

+ = compatible; - = incompatible:

		Stock		
S		Lycopersicon	Datura	Nicandra
c	Lycopersicon	+	+	-
i	Datura	+	+	+
o	Nicandra	-	+	+
n				

In an investigation of the factors that prevent the grafting of Nicandra to Lycopersicon, it was found that cell walls isolated from Nicandra and placed between a Lycopersicon stock and a Lycopersicon scion would prevent the normally successful formation of a Lycopersicon/Lycopersicon homograft. On the other hand, Nicandra cell walls did not interfere in Nicandra/Nicandra homografts. It thus seems that incompatibility may be dictated by genus-specific components of the cell wall. There is preliminary evidence that these components may be pectic polysaccharides [266].

9.2: PREPARATION OF BIOLOGICALLY-ACTIVE WALL OLIGOSACCHARIDES
9.2.1: Mild acid-hydrolysis

The principle method for production of active oligosaccharides is hydrolysis of walls or polysaccharides with mild acid or specific enzymes. In addition, extraction of walls with strong base has sometimes been used. Mild acid-hydrolysis was introduced in Section 4.2.2. In the present context, it is usual to aim for higher-M_r products than the di- or trisaccharides often sought in structural studies. This is achieved by use of lower acid concentrations, lower temperatures or shorter times of hydrolysis. Precise conditions will depend on the chemical nature of the polysaccharide. One standard set of conditions is given in Panel 9.2.1. This method is suitable for preparation of fungal elicitors (e.g. from yeast glucan, which is available from Sigma), endogenous elicitors and inducers of the synthesis of proteinase-inhibitors (from commercial pectin), and the unidentified fragments that are reported to be cytotoxic or influence morphogenesis and flowering (e.g. from the walls of cultured plant cells).

9.2.2: Enzymic hydrolysis

Enzymic hydrolysis is more specific than acid-hydrolysis. Thus, while it is possible that acid could solubilise fragments from other wall components than polysaccharides or even from contaminants of the wall (e.g. tannins), it is more likely that any fragment generated by treatment with a

PANEL 9.2.1:   Isolation of wall fragments by mild acid-hydrolysis [573]

Sample: Isolated walls of cultured sycamore cells (Panel 1.2.2).

Step 1:   Suspend 1 g dry wt of walls in 100 ml 2 M trifluoroacetic acid in a capped bottle, and incubate in an 85°C water-bath for 2 h.

Step 2:   Cool. Filter through Whatman GF/C glass fibre paper, collect the filtrate and dry in vacuo (e.g. in a rotary evaporator, below 40°C). Re-dry from a small volume of 50 % MeOH three times.

Step 3:   Re-dissolve in 10 ml H_2O, adjust to pH 6.5 with a small volume of 1 M NaOH, and dialyse (Panel 3.5.1) in 'Spectra-Por 6' (1000-M_r cut-off tubing, washed thoroughly to remove preservative) against several changes of H_2O to remove mono- and very small oligosaccharides. Freeze-dry the contents of the dialysis sac. [As an alternative to dialysis, de-salt the sample on Bio-Gel P-2 (Panel 3.5.1 b, scaled up).]

Next step: Bio-assay (Section 9.3).

PANEL 9.2.2:   Preparation of XG9, an anti-auxin oligosaccharide [175]

Sample: PAW-washed (Panel 1.2.1 c) cell walls of a Dicotyledonous cell suspension culture, e.g. rose.

Step 1:   Extract hemicellulose from the cell walls (2 g) by stirring overnight at 25°C in 100 ml of 6 M NaOH containing 1 % $NaBH_4$ (Panel 3.4.1 a). De-salt by dialysis (Panel 3.5.1 a). Freeze-dry.

Step 2:   Suspend 0.1 g of the hemicellulose in 10 ml buffer A [50 mM HOAc, pH adjusted to 4.7 with 1 M NaOH]. Add 10 mg of Trichoderma viride cellulase (Sigma C 2274), and incubate at 25°C for 3-4 h.

Step 3:   Centrifuge (2,500 **g**, 5 min). Load the supernatant (10 ml) on to a column of Bio-Gel P-2 (bed volume 200 ml) pre-equilibrated with buffer B [50 mM HOAc, pH adjusted to 4.7 with pyridine], and elute with buffer B, collecting 50 4-ml fractions.

Step 4:   Assay 10 µl of each fraction for hexose by the anthrone test [10 µl sample + 490 µl H_2O + 1 ml 0.2 % anthrone in H_2SO_4 — see Panel 3.7 a]. XG9 and XG7 elute as peaks at $k_{av.}$ ca. 0.42 and 0.51 respectively; these values can be compared with those of markers run previously: dextran = 0.00, maltoheptaose = 0.54, glucose = 1.00.

Step 5:   Pool and freeze-dry the fractions of interest [this removes the pyridine and HOAc]. Determine sugar composition by acid-hydrolysis of a small portion (Panel 4.2.1 a) followed by paper chromatography (Panel 4.5.2 c) or TLC (Section 4.5.3).

Next steps: Further purification, if required, by preparative PC [175]:

Solvent	$R_{maltoheptaose}$ of		see Panel
	XG7	XG9	
EtOAc/HOAc/H_2O (10:5:6)	0.49	0.95	4.5.2 d
BuOH/pyridine/H_2O (4:3:4)	1.00	0.92	4.5.2 e

pure endo-glycanase originated from polysaccharides. The enzymic approach would appear to be necessary for isolation of fragments that have essential residues that are highly acid-labile (e.g. the fucose residue of XG9). As mentioned in Section 4.2.5, the main problem is the lack of commercially-available pure enzymes. An example of the sucessful use of enzymic hydrolysis is in the preparation of XG9 from xyloglucan (Panel 9.2.2).

9.3: BIOASSAY OF BIOLOGICALLY-ACTIVE OLIGOSACCHARIDES

Bioassay is the only way of discovering new biological activities and of purifying the molecules responsible for known activities. Oligosaccharides are supplied to a plant tissue, whose response is taken as evidence for the presence of the molecules of interest. Unlike chemical assays, bioassays do not detect inactive oligosaccharides. One drawback to the use of bioassays is that the biological properties of mixtures of oligosaccharides may be different from those of pure oligosaccharides; therefore certain activities may vanish during the purification scheme. In

---

PANEL 9.3 a: Bioassay of cytotoxic oligosaccharides [573]

Sample: Oligosaccharide mixture prepared from cell walls of a suspension culture by mild acid-hydrolysis (Panel 9.2 a).

Step 1:   Transfer 0.5 ml of a suspension of cultured cells (e.g. sycamore, PCV = 1 %, in normal culture medium supplemented with 30 mM MES, adjusted to pH 6.0 with 1 M NaOH) into a flat-bottom glass vial (2 cm internal diameter). Add 10 μl of a 2 % solution of the oligosaccharides [pH pre-adjusted to about 6.0 with 0.1 M NaOH or 0.1 M HCl (not HOAc as acetate is highly toxic to cultured cells)], or 10 μl of a control solution (e.g. 2 % starch-hydrolysate). Incubate with gentle shaking for 2 h.

Step 2:   Add exactly 10 μl of an aqueous solution of L-[U-^{14}C]-leucine (0.5 μCi, specific activity 0.3 Ci/mmol) to each vial. Incubate for a further 3 h.

Step 3:   Add 5 ml of ice-cold 10 % trichloroacetic acid (TCA) to kill the cells, to extract the unused [^{14}C]leucine and to precipitate the [^{14}C]protein that has been synthesised. Incubate at 0°C for at least 30 min.

Step 4:   Filter the suspension through a 2.5-cm disk of Whatman GF/C glass fibre paper, and wash with 3 x 10 ml 10 % TCA. Finally wash with ethanol, dry the filter, and assay ^{14}C by scintillation-counting in non-Triton scintillant or by a Geiger-Müller tube.

Interpretation: Healthy cells rapidly incorporate [^{14}C]leucine into TCA-insoluble protein, which is collected on the filter. This process is reduced or prevented in cells treated with cytotoxic wall-fragments.

---

particular, large amounts of contaminating inactive oligosaccharides may protect trace quantities of active ones from digestion by the enzymes of the plant cell wall. A possible solution to this problem would be the deliberate addition of appropriate inactive oligosaccharides to the bioassay solution. Two bioassays are outlined in Panels 9.3 a and b.

PANEL 9.3 b: Anti-auxin activity of xyloglucan nonasaccharide [340]

Sample: Xyloglucan nonasaccharide, prepared as in Panel 9.2.2.

Step 1:   Soak 100 g of pea seeds (variety 'Alaska') overnight in a jar of running tap-water. The water should be delivered to the very bottom of the jar via a rubber tube so that the seeds do not become anaerobic.

Step 2:   Germinate the soaked seeds in moist Vermiculite at 25°C in the dark for 7-8 days.

Step 3:   Examine the seedlings in dim red light, and check that the 3rd internode (counting the cotyledons as the 1st node) is 1-3 cm long.

Step 4:   Working under dim red light, cut a 0.6-cm length from the 3rd internode, 0.3-0.9 cm from the 4th node. Place the cut segments in a beaker of H_2O, and incubate with gentle shaking at 25°C for 30 min. Then transfer the segments into a beaker of solution **A** [5 mM KH_2PO_4, 0.02 % potassium benzyl-penicillin, 1 % sucrose; pH adjusted to 6.1 with 1 M KOH] and incubate with gentle shaking at 25°C for a further 1½ h.

Step 5:   Still working under dim red light, transfer 8 of the segments into each of six 5.5-cm Petri dishes containing 5 ml solution **A**. To dishes 2-6, add 50 µl of a 10^{-4} M solution of 2,4-dichlorophenoxyacetic acid (2,4-D) made up in solution **A**. To dishes 3-6, add 50 µl of 10^{-8}, 10^{-7}, 10^{-6} and 10^{-5} M XG9 respectively[1].

Step 6:   Incubate the dishes for 16 h in the dark at 25°C with gentle orbital shaking (100 r.p.m.).

Step 7:   Measure the lengths of the segments: place the Petri dishes on an overhead projector and measure the magnified images of the segments on a screen.

Interpretation: Typical results obtained in this laboratory have have been: Initial length of stem segments: 6.0 mm.   Final length of segments:

dish no.:	1	2	3	4	5	6
[2,4-D] (µM):	0	1	1	1	1	1
[XG9] (nM):	0	0	0.1	1	10	100
length (mm):	7.5	9.2	9.2	8.0	8.3	8.3

These results confirm that 2,4-D promotes growth, but 1 nM XG9 antagonised this effect. Higher concentrations of XG9 had little influence, indicating that the effect was not simple cytotoxicity of XG9.

[1]Concentration of XG9 estimated by assay in the anthrone test, allowing for the fact that 1 mg of XG9 (M_r = 1371.2) yields 0.657 mg of hexose.

9.4: FATE OF OLIGOSACCHARIDES IN VIVO

Very little work has been done on the fate of wall-derived oligosaccharides in vivo, and further studies are urgently required.

9.4.1: Transport

It has been suggested that oligosaccharides can act as hormones (defined as substances produced in one part of an organism and exerting potent effects on a distant part of the same organism). For example, a signalling molecule must be present to account for the fact that, when one tomato leaf is injured, other leaves on the same plant respond by synthesising proteinase-inhibitors [203]. Since exogenous pectic oligosaccharides evoke the synthesis of proteinase-inhibitors (Section 9.1.4), the idea evolved that such oligosaccharides, solubilised at the site of injury, are hormones that move to distant uninjured leaves to tell them to produce proteinase-inhibitors. However, observation of highly radioactive oligosaccharides applied to injury sites showed that such translocation did not occur [41]. This does not detract from the potential importance of pectic oligosaccharides as signalling molecules involved in the sensing of injury, but it indicates that they act at or very near their site of production, perhaps by initiating the dispatch of a second (long-distance) messenger.

9.4.2: Binding

Animal peptide hormones, which circulate in the blood, interact with cells via receptors in the plasma membrane. Plant oligosaccharides might act in a similar way: they are thought to be produced extracellularly by partial enzymic hydrolysis of cell walls; they are hydrophilic molecules, incapable of penetrating a lipoprotein membrane; and yet they have effects on intracellular processes such as protein synthesis. Very little is known of the binding of oligosaccharides to plasma membranes. Fungal oligoglucans may bind to isolated plasma membranes, but the specificity of the receptor and its mode of coupling to intracellular processes will require careful investigation.

9.4.3: Degradation

Plant cell walls contain enzymes with specificities suggesting that they would degrade (as well as generate) biologically-active oligosaccharides. However, little work has been done to explore the metabolic fate of such oligosaccharides. Fungal oligo-β-glucans (Section 9.1.2) are probably broken down by plant β-glucanases [103] Radioactive xyloglucan nonasaccharide (Section 9.1.5) has been fed to suspension-cultured spinach cells [Baydoun & Fry, unpublished]: it underwent slow breakdown ($t_{\frac{1}{2}}$ ca. 1 day) by at least two different processes — **(a)** loss of the fucose residue, which must occur by the action of α-fucosidase, and **(b)** loss of xylose and glucose residues, which probably occurs by the action of a specific α-xylosidase [299] and a β-glucosidase. In addition, when nonasaccharide labelled with ^{3}H in its xylose residues was fed to the cultures, ^{3}H appeared not only in low-M_r products but also in high-M_r polysaccharide. This may represent a novel 'degradation' method; extracellular incorporation of a nonasaccharide into a polysaccharide is also of interest in providing possible evidence for transglycosylation (cf. Section 6.1.3).

9.5: NATURAL OCCURRENCE OF BIOLOGICALLY-ACTIVE OLIGOSACCHARIDES

If oligosaccharides are to have a regulatory rôle in the living plant, they must occur naturally. Since they are active at (and therefore likely to be present in vivo at) exceedingly low concentrations, sensitive methods are required for their detection. As shown in Section 2.1.1, in vivo radioactive labelling is the method of choice. The relevant oligosaccharides are likely to be found extracellularly, i.e. in the apoplast. The task of isolating a specific oligosaccharide, perhaps present at less than 10^{-7} M, from cells, is greatly facilitated if a purified source of apoplastic fluid can be used as starting material. Potential sources include the fluid obtained by vacuum-infiltration/centrifugation (Panel 6.3.1), which has apparently not yet been tested, and the spent medium of cultured cells. The latter source has been shown to contain physiologically significant quantities of xyloglucan nonasaccharide [175]. It also contains similar amounts of many other, incompletely characterised oligosaccharides; it will be of interest to test these for possible biological activity.

Oligosaccharides are probably overlooked in many searches for biologically-active substances. They are hydrophilic and would thus be lost during the solvent-partitioning which is common practice in the isolation of plant hormones (auxins, gibberellins etc.). They also have M_r ca. 1000 to 5000, a range usually ignored in GPC and dialysis. These two properties — hydrophilicity and intermediate M_r — however, can be used to advantage in the isolation of naturally-occurring oligosaccharides. Thus, these molecules remain in the aqueous phase upon partitioning with even the most polar solvents, e.g. butanol; and they elute between V_0 and V_i during GPC on gels such as Sephadex G25 and Bio-Gel P-2 (Panel 3.5.2 b).

PANEL 9.5:   Isolation of naturally-occurring xyloglucan nonasaccharide and other oligosaccharides from cultured spinach[1] cells

Step 1:   Dry 0.5 mCi of L-[3H]arabinose or L-[3H]fucose (ca. 10 Ci/mmol) or D-[14C]glucose (ca. 0.3 Ci/mmol) into a 50-ml flask, add 10 ml standard culture medium (Panel 2.7.1 a) containing 1 % glucose, and autoclave. Inoculate with stationary-phase spinach[1] cells, and incubate under aseptic conditions for ca. 7 days.

Step 2:   Filter the culture through GF/C glass fibre paper and retain the filtrate. Add 15 mg dextran (e.g. M_r 10,000), 5 mg malto-heptaose and 15 mg glucose to the filtrate as internal markers.

Step 3:   Load the filtrate (ca. 8 ml) on to a 200-ml bed volume column of Bio-Gel P-2 [pre-equilibrated with HOAc/pyridine/H_2O (3:2:245)], Elute with the same solution, collecting fifty 4-ml fractions.

Step 4:   Assay 30 µl of each fraction for internal marker hexoses by the anthrone test (30 µl sample + 470 µl H_2O + 1 ml 0.2 % anthrone in H_2SO_4). This should reveal 3 major peaks, at $k_{av.}$ 0.00, 0.54 and 1.00.

Step 5:   Assay 0.5 ml of each fraction for 3H or 14C, and look for any peak of radioactivity at $k_{av.}$ 0.42 (xyloglucan nonasaccharide) or 0.51 (xyloglucan heptasaccharide).

Step 6:   Dry each fraction containing radioactive material of interest (a 'SpeedVac' is very useful for this), re-dissolve in 100 µl H_2O and apply to Whatman 3MM chromatography paper. Run overnight in EtOAc/HOAc/H_2O (10:5:6) by the descending method.[2] Locate radioactive spots. This can effectively give a 2-dimensional separation [1st = Bio-Gel; 2nd = paper chromatography].

Step 7:   Elute fractions of interest, and re-run on paper chromatography in BuOH/pyridine/H_2O (4:3:4) overnight by the descending method.[2]

Next steps: Further characterisation of the separated oligosaccharides e.g. by acid-hydrolysis, Driselase-hydrolysis, electrophoresis etc. (Ch 4).

[1]Preliminary evidence suggests that similar results are obtained with other cultures e.g. of 'Paul's Scarlet' rose.

[2]R_F values of xyloglucan oligosaccharides are given in Panel 9.2.2.

---

10 Wall tightening and loosening: growth

10.1: BACKGROUND

10.1.1: The biophysical basis of cell growth [101,109,110,120]

The osmotic pressure of the cell contents is usually higher than that of the fluid bathing the cell (whether this is natural apoplastic fluid or an artificial culture medium). The gradient in osmotic pressure drives water into the cell, which thus develops an internal hydrostatic pressure (**turgor**) that tends to make the cell increase in volume and stretch the wall. Part of the stretching is **elastic**, i.e. the wall shrinks when turgor is decreased (e.g. if the cell is experimentally bathed in a solution of high osmotic pressure); part, however, is **plastic** (irreversible) and this is what is meant by **growth** (defined as irreversible increase in volume).

Plant growth is subject to strict spatial and temporal controls, imposed both from within the plant (its developmental programme) and from the external environment (light, temperature, etc.). Hormones (e.g. auxins, gibberellins, cytokinins and abscisic acid) and H^+ play a major rôle in mediating both the internal and external control of growth, and a great deal of thought has been applied to understanding their modes of action. For a hormone to promote growth, that hormone **must** cause existing wall material to expand in area; the expansion may or may not be accompanied by cell division. This can be clarified by consideration of a resting filamentous alga capable of growing one-dimensionally: a hormone that induced cell division alone would not initiate growth (as defined), since it merely results in an increased number of cross-walls —

before:
after:

For the hormone to initiate growth, it would need to induce the elongation of side-wall material —

before:
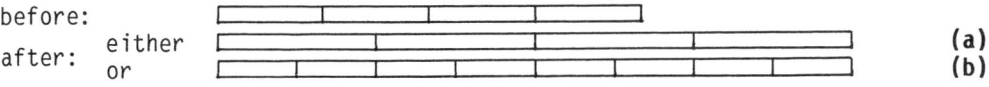
after: either
or

(a)
(b)

either without **(a)** or with **(b)** cell division. The same argument can be applied to non-resting higher plant tissues growing three-dimensionally, where the hormone increases the rate of growth. Of course, for growth to be maintained for a long period of time (weeks), cell division would need to keep pace [the alternative is infinitely long cells], but this does not detract from the argument the the cause of growth must be wall expansion.

In the simple case where cell division is absent, a hormone (e.g. GA_3 or IAA) that promotes growth could act either by raising turgor pressure or by making the wall more susceptible to turgor-driven plastic extension. A growth-retarding hormone (e.g. ABA) could act by either decreasing turgor or reducing the susceptibility of the wall to plastic extension. Direct and indirect measurements of turgor are almost unanimous in indicating that hormones do not act via changes in turgor.[*] Any effects of hormones on turgor are usually in the wrong direction to account for the hormone's effect on growth [101], and are thus probably <u>caused</u> by the change in growth rate [growth dilutes the cell sap and thus reduces its osmotic pressure].

10.1.2: Physiological mode of action of growth hormones [101,110,120,486]

If growth hormones do not affect turgor, they must affect the susceptibility of the wall to plastic extension. Strong evidence that this is the case comes from direct studies of the physical properties of cell walls in the absence of natural turgor. Pieces of individual walls large enough to be stretched <u>in vitro</u> can be isolated from the giant internodal cells of certain algae, e.g. <u>Nitella</u> [354]. No higher plant cells are large enough to allow this, and the alternative has been adopted of using whole organs (e.g. stems) or tissues (e.g. epidermal strips) in which the contribution of turgor has been eliminated by plasmolysis, freezing/thawing or treatment with boiling methanol. Such specimens can be held in a tensile tester (e.g. an 'Instron') and stretched, and their behaviour observed (for details, see [101,102,521]). Observations made in this type of experiment clearly demonstrate changes in the physical properties of the wall in response to hormone treatment, and the changes are very often in such a direction as to suggest that they cause the change in growth rate.

Increases and decreases in the susceptibility of walls to turgor-driven stretching can be described in general terms as wall 'loosening' and

----
[*]An exception to this generalisation is the cucumber hypocotyl, where GA_3 seems to promote growth by increasing turgor, although IAA promotes growth in the same tissue without increasing the turgor [283].

'tightening' respectively. The chemical basis of these physical changes is unknown and has proved very hard to investigate satisfactorily. The major problem is our lack of a reliable molecular model of the cell wall (Ch 7).

Consider wall-loosening. It can be argued that a cell that has been fully turgid for some time (a common situation in vivo) will have stretched its wall to the point where little further plastic extension is possible, and would thus be unlikely to show any substantial loosening without the cleavage of load-bearing bonds. This does not specify whether the bonds broken are covalent or non-covalent, nor does it specify whether simple breakage occurs or whether new wall material is necessarily inserted (like new links into a broken chain). Deposition of new wall material may or may not accompany the cleavage of load-bearing bonds in the short term (although wall deposition is clearly needed for long term growth, just as the de novo synthesis of all essential cell components is required for continued growth). It is hard to see how wall deposition alone could loosen the wall: only bond-breaking could do it alone. Progress awaits a better definition of the molecular architecture and workings of the cell wall.

Secretion of H^+ ions is necessary (but not sufficient) for the action of IAA etc. on wall-loosening [217,520,550]. The salient pieces of evidence for this 'acid growth hypothesis' are: (a) H^+ ions, measured with a pH meter, are very rapidly secreted by auxin-sensitive cells in response to IAA-treatment, (b) exogenous H^+ ions mimic the action of IAA on growth, both in vivo and in isolated, stretched cell walls in vitro, and (c) applied neutral buffers (which would abolish the wall-acidifying effect of secreted H^+ ions) block the action of IAA on growth in vivo. It seems likely that, in cereal coleoptiles, H^+ ions loosen walls indirectly by activating wall enzymes rather than by a direct action of low pH on the structural polymers. Evidence for this is that oat coleoptile walls in which the enzymes have been denatured or removed are not loosened by H^+ ions [486]. However, this may not be the case in algae and Dicots [110]. In addition, detectable H^+-induced wall loosening is dependent on the walls being stretched during the H^+-treatment: this may indicate that acid induces reversible bond breakage, and that the broken bonds simply re-form in the same place when the pH is restored to neutrality if there was no stretching force to move the polymer molecules during the low-pH period. In addition, it seems likely that covalent bonds which are being stretched by turgor-driven expansion are more susceptible to acid- and enzyme-catalysed hydrolysis [295]. One type of enzyme that could account for 'acid growth' would be a transglycosylase with a low pH optimum (Sections 6.1.3 & 10.1.3). However, it should be stressed that there is no direct evidence that transglycosylation occurs in the walls of living cells.

10.1.3: Chemical basis of wall loosening

The cell wall is a network composed of cellulose and a variety of soluble polymers that are rendered insoluble by cross-links (Ch. 7). Which bonds are broken to allow wall loosening? There is little evidence for erosion of cellulose microfibrils — the main 'skeleton' of the cell wall — during the action of growth-promoting hormones. This leaves, as the only alternative, bond breakage in the network of matrix polymers. A network can be broken either by cutting the strings (polymer backbones) or undoing the knots (cross-links). Either of these strategies would seem able to explain the wall-loosening action of hormones. A potential method of identifying bonds whose breakage would <u>not</u> cause wall-loosening is to subject a wall sample (e.g. a strip of methanol-killed tissue) to experimental treatments that would be expected to break a particular bond, and then to examine the effect of this treatment on the behaviour of the specimen in a tensile tester. In experiments like this a chaotropic agent, 8 M urea, did not loosen oat coleoptile cell walls, suggesting that the wall's H-bonds are too abundant to be targets of wall-loosening. Similarly, chelating agents had little effect on wall extensibility in oat coleoptiles, arguing against Ca^{2+}-bridges as targets [494]. However, Ca^{2+}-chelators did loosen walls isolated from algae and Dicots [110], and promoted growth in living Dicots [356]. It is hard to see how a mechanism reliant on the breakage of H-bonds or Ca^{2+}-bridges could be enzyme-mediated. A flaw in this type of experiment is that the aqueous chaotropic and chelating agents may have been unable to penetrate to the growth-limiting wall zone, which may well be the cuticularised (and relatively hydrophobic) outer wall of the epidermis (Section 10.1.5).

Approaches that have been adopted to gain evidence for the bonds and enzymes that <u>are</u> involved in wall loosening include:

(a) <u>Detection of enzymes whose levels increase in response to IAA-treatment or whose activities are enhanced by low pH.</u> Auxin induces the <u>de</u>

novo synthesis of cellulase in pea stems [526], and existing cellulase is activated by low pH. Cellulase alone has little effect on wall-bound cellulose, but does hydrolyse xyloglucan [240], and this may be part of the mechanism of wall loosening [see (d) below]. Cellulase would also attack, in grass cell walls, those domains of β-$(1\rightarrow3),(1\rightarrow4)$-glucan that consist of ca. 10 contiguous $(1\rightarrow4)$-linked glucose residues (see Section 4.1.4). Auxin also induces β-$(1\rightarrow3)$-glucanase, an enzyme that might also attack the β-$(1\rightarrow3),(1\rightarrow4)$-glucan component of grass hemicellulose [337].

In addition, many other wall enzymes have low pH optima, and it has been speculated that β-galactosidase and β-glucosidase mediate auxin-induced wall-loosening. However, this appears unlikely since galactono-lactone and gluconolactone inhibit these enzymes respectively, without blocking the action of auxin on growth [151].

(b) Detection of enzyme inhibitors that block auxin action. Nojiri-mycin is an inhibitor of β-$(1\rightarrow3)$-glucanase that simultaneously blocks the action of auxin on growth in grasses [434]. This evidence is compatible with a rôle of β-$(1\rightarrow3)$-glucanase in wall loosening; however, the specificity of nojirimycin has been questioned [134]. Antibodies have been raised against a number of wall proteins, and there is evidence that some of these antibodies, when applied to the walls of living cells, can block the action of auxin. The identity of the proteins (enzymes?) inhibited by these antibodies is an area of great interest [110].

(c) Detection of effects of exogenous enzymes on growth. If a wall enzyme is responsible for auxin-induced wall-loosening, application of that enzyme to living cells might be expected to mimic auxin. However, failure of this experiment does not negate the proposed rôle of the enzyme since it is quite likely that some exogenous enzymes cannot gain access to certain wall components (the 'exclusion limit' of the cell wall matrix has been estimated by different methods to be 1 000 - 5 000 [94] and ca. 50 000 [495]). There have nevertheless been reports that added β-$(1\rightarrow3)$-glucanase mimics auxin in vivo in oat coleoptiles [569], again hinting that β-$(1\rightarrow3),(1\rightarrow4)$-glucan metabolism is at the heart of auxin action in grasses.

(d) Detection of chemical changes in structural wall polymers. The fourth approach is the direct detection of bond breakage during the action of auxin or H^+. Treatment of legume stems with auxin induces the nicking, sloughing and turnover of xyloglucan. Nicking is detected by a decrease in the mean M_r (measured by GPC) of the bulk xyloglucan in the wall, although it remains part of the wall [379,243]. Sloughing is the solubilisation of xyloglucan and its accumulation in solution in the apoplastic fluid [497]. Turnover is the complete hydrolysis of the polymer to monosaccharides and low-M_r oligosaccharides [193,304]. The effects are induced both by auxin and by H^+ ions, and probably occur rapidly enough to account for the action of these factors on wall loosening. Since xyloglucan is the major hemi-cellulose of the Dicot primary cell wall, and therefore presumably of great structural significance, it is easy to imagine how its hydrolysis could loosen the wall. Xyloglucan hydrolysis is thus of great interest in understanding the mode of action of auxin in Dicots.

An interesting feature of H^+-induced xyloglucan nicking is that the mean M_r seems to increase rapidly when the pH is restored to neutrality [378]. This observation, if confirmed, would be important as it might indicate that the xyloglucan participates in a transglycosylation reaction.

In grasses, where there is little xyloglucan and auxin does not induce cellulase synthesis, the widely-observed hydrolysis of β-$(1\rightarrow3)$, $(1\rightarrow4)$-glucan may be the mechanism of auxin-induced wall loosening [434]. There is also a report of the turnover of arabinoxylan in maize coleoptile cell walls in response to auxin treatment [119].

10.1.4: Chemical basis of wall tightening

Certain physiological treatments cause a pronounced decrease in growth rate, e.g. application of abscisic acid (ABA) to cereal coleoptiles [302], and exposure of cucumber stems to blue light [108]; the latter effect is detectable within 1-2 min. There is good evidence that these effects are due to changes in the physical properties of the cell wall, but a tightening rather than a loosening. An understanding of wall tightening reactions would clearly aid in the explanation of the mechanisms of growth inhibition; in addition, wall tightening reactions might be relevant to the action of growth-promoting hormones since a growth promoter could act by inhibiting an otherwise constitutive wall-tightening reaction [166].

Wall tightening could be due to de novo deposition of wall material, making the wall stronger. Taken to the extreme (secondary wall deposition, perhaps even with lignification), this would stop growth completely. However, the blue light effect is so rapid that it is difficult to imagine a 'stoichiometric' wall-thickening mechanism like this coming into play quickly enough. Rather, a tightening of existing wall material appears more plausible, and this view points to a rôle for cross-linking mechanisms (Ch 7). One type of cross-link, which could be formed very rapidly in response to a stimulus, would be a phenolic dimer e.g. isodityrosine or diferulic acid (Section 3.3.2), formed under the catalysis of peroxidase [174]. There is a suggestion that such cross-links are indeed involved [450]: when cucumber stem segments were exposed to blue light, there was a rapid secretion of peroxidase into the apoplastic fluid. Unfortunately, experimental limitations thwarted attempts to find out if this effect began before or after the deceleration of growth. Ascorbic acid, which inhibits phenolic dimerisation, abolished the effect of blue light, i.e. allowed growth to continue unabated in blue light. It remains to be seen whether the expected cross-linking products can be detected in blue light-treated stems. The negative correlation between wall-bound extensin and growth rate [181,182,186,433,552,] supports this mechanism of wall tightening.

Gibberellic acid, which promotes growth, has the opposite effect: it suppresses the secretion of peroxidase, and permits the accumulation of extracellular phenolics in uncross-linked form [166,167]. Certain growth retardants, e.g. chlorocholine chloride, which may antagonise the synthesis of gibberellins, generally increase peroxidase levels, and dwarf mutants generally possess higher peroxidase levels than tall varieties of the same plant [50]. All these examples are compatible with the view that extracellular peroxidase-catalysed reactions tighten the cell wall.

Other types of cross-link could also conceivably cause wall tightening, such as Ca^{2+}-bridges (e.g. by secretion of Ca^{2+} from within the cell, or by raising the pH of the wall [40]), or the hypothetical uronoyl-ester bonds (e.g. catalysed by transesterases), or inter-polysaccharide glycosidic bonds (e.g. catalysed by aglycone-specific transglycosylases) [see Section 7.1.3]. This is an area requiring considerable further work.

Phenolic dimers (isodityrosine, diferulic acid etc.) may act as irreversible cross-links, since plants possess no known enzymes for their cleavage. It may thus be questioned how growth can be restored after tightening via such bonds. One possibility is that the backbones of the polymers to which the phenolics are attached (extensin, pectin etc.) can be broken. The ester bonds that link diferulate to the polysaccharides could also be hydrolysed. However, it is also relevant that only the innermost portion of the wall is considered to be limiting for growth [420] and so as a given wall layer is displaced towards the exterior and becomes architecturally irrelevant, the cross-links it contained could be ignored.

10.1.5: Special rôle of the epidermis

The walls most directly involved in the control of growth (at least in stems and petioles) seem to be those of the epidermis [82,159,300,301, 499,530]. These walls are very different from those of the rest of the stem. Epidermal walls are ca. 10 times thicker, and they possess waxes, cutin and cutin—polysaccharide complexes; it appears possible that the cutin—polysaccharide zone is load-bearing. This zone is autofluorescent [461,232] and clearly possesses abundant phenolics — some probably associated with the cutin, but others perhaps linked via ester bonds to the polysaccharides. It is thus particularly likely that this wall layer is capable of tightening by the formation of phenolic dimers. There is an urgent need to study the chemistry of the cutin—polysaccharide complex, and to relate its chemical properties to its physical properties.

10.2: EXPERIMENTAL APPROACH TO STUDY OF WALL TIGHTENING AND LOOSENING
10.2.1: Important unanswered questions

(a) What is the molecular architecture of the primary cell wall? A knowledge of this is essential to any clear understanding of the control of wall loosening and tightening, but unfortunately we do not yet have a reliable model of the cell wall.

(b) Following on from this first question is the problem of identifying those load-bearing bonds which resist turgor-driven expansion and whose breakage could account for wall loosening. Even more importantly, we need to discover the proportion of the total load borne per individual bond of any given type when the wall is stretched. For example, consider the following scheme, which shows two cellulose (GGGGGGGG) molecules on different microfibrils being prised apart by turgor-driven expansion (▽▽▽, △△△) but held together by a hydrogen-bonded (∷∷∷∷) xylan (XXXXXXXX) molecule. Cleavage of a few hydrogen-bonds would have little effect, but cleavage of just one (well-chosen) xylose→xylose glyco-sidic bond would have a dramatic effect on the ability of the two microfibrils to move apart and allow growth; this is despite the fact that both hydrogen-bonds and xylosyl linkages are load-bearing.

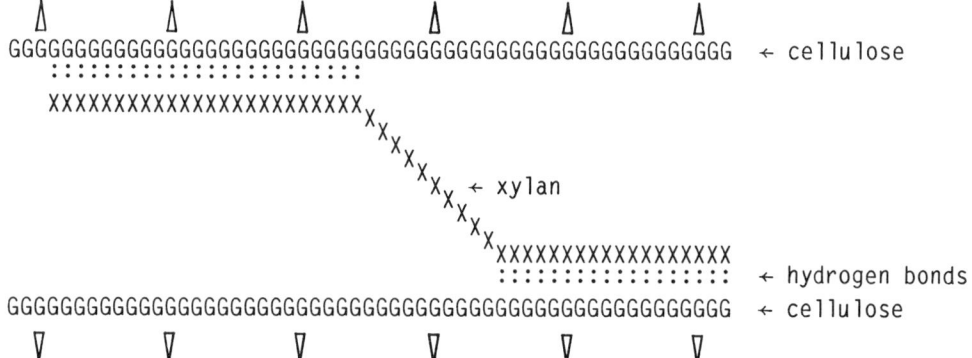

It is possible that different wall-loosening agents act by cleaving different bonds; this too will need defining.

(c) Which bonds form during wall tightening upon application of growth-inhibitory treatments such as ABA and blue light?

(d) Does the breakage of covalent bonds involve transglycosylation, transesterification, simple hydrolysis of glycosidic bonds, or simple hydrolysis of ester bonds?

(e) What are the special structural features of the epidermal cell wall that allow it to control the growth of the whole organ?

10.2.2: Choice of plant material

The principal decision to be taken is whether to use cell suspension cultures or intact organs/tissues. Some of their salient advantages and disadvantages are summarised in Panel 10.2.2.

10.2.3: Analytical techniques

Chapters 1-4, 6 and 8 describe techniques that have been or could be applied to the study of cell wall metabolism during growth. The approach chosen will depend on the biological question being posed. Many of the studies reported to date, using these approaches, have revealed only disappointingly small changes in the chemical composition of walls during dramatic changes in wall extensibility. It may be that more progress will be made when sufficient attention is paid (a) to interpolymeric cross-links

[174] rather than simple polymer composition, and **(b)** to epidermal walls — probably the target of many plant growth hormones [300] — rather than total walls from whole organs.

PANEL 10.2.2: Suspension cultures versus organs and tissues as objects for the investigation of hormone action

Feature	Suspension cultures	Organs/tissues
Biochemical analysis	Excellent: all cells similar and participate in growth regulation	Questionable unless relevant tissue (e.g. epidermis) is isolated for analysis (see Section 10.1.5)
Radioactive labelling of wall in vivo	Excellent (see Ch 2)	Difficult
In-vivo application of hormones, antibodies, etc.	Simple; all cells receive same concentration instantaneously	Internal tissues shielded
Response to auxins & cytokinins	Abnormal; often required simply to keep cells alive; no specific promotion of cell expansion [154]	Normal response; often dramatic promotion of cell expansion [120]
Response to gibberellins	Rarely reported; depends on cells being in carefully controlled physiological condition [180,450]	Often dramatic promotion of cell expansion [120]
Response to ABA	Rarely reported; little effect at physiological concentrations [555]	Often dramatic inhibition of cell expansion [120]
Response to H^+	Strong promotion of growth; little studied	Strong promotion of growth; extensively characterised [486]
Short-term measurements of growth	No reliable method	Good methods available for detection of changes in rate of cell expansion over 1 min [152]
Measurements of wall extensibility in vitro	Impossible?	Several methods available [101]

11 Wall degradation

11.1: BACKGROUND

The plant cell wall is chemically stable and physically strong, but it is nevertheless breached in a number of different situations. These are worth studying because **(a)** they are biologically important, **(b)** they are economically important, and **(c)** they give insight into wall architecture.

11.1.1: Biologically important situations where the wall is attacked

Walls are attacked in the healthy plant during the normal course of growth and development (Section 6.1.2). Besides this, walls may be degraded in the living but diseased plant [38,197]. The penetration of a pathogenic fungus, e.g. on a leaf surface, may require a hole to be made (by digestion) in the epidermis. In addition, many pathogenic fungi obtain their nutrition from the contents of their host cells rather than from the apoplast, and must therefore penetrate the wall of each cell they attack.

Cell walls are also degraded after the plant (or organ) has died. Since walls constitute a high proportion of the plant's total biomass, it is essential for the global carbon cycle that the walls of dead plant cells are rotted. This important task is carried out by soil microbes.

Another biologically important situation where walls are digested is in the guts of animals. Herbivores may derive much of their carbon from plant cell wall polymers. In the human diet, cell walls (= dietary fibre) are important to the proper functioning of the bowel, and probably also to the prevention of 'Western diseases'; the cell walls are quite extensively hydrolysed by microbes in the large intestine [113,145,473,539]. Different types of fibre are hydrolysed to different extents, with important consequences for the contribution of a given fibre to our health.

11.1.2: Economically important situations where the wall is attacked

Cellulose is the most abundant organic chemical on earth and lignin is the second most abundant; hemicelluloses may be the third. These are cell wall components, and it would therefore seem potentially profitable to exploit walls commercially. We already exploit cell wall-rich materials in a number of ways e.g. as building materials (timber), fuels (firewood, peat), textiles (cotton, jute), fodder (straw, hay) and food (dietary fibre from cereals, fruit and vegetables). But much wall material is wasted. The U.K. wastes ca. 10^7 tonnes of straw a year. It would be a cheap raw material, and this annual amount could, on hydrolysis of its polysaccharides, yield over 5 x 10^6 tonnes of sugar [value over £500 000 000]. The difficulty in exploiting waste cell walls (other than as firewood etc.) is their indigestibility [10,78,324].

On a totally different scale, plant biotechnology makes frequent use of isolated protoplasts: these are prepared by enzymic hydrolysis of walls (Section 2.7.6). Not all tissues readily yield protoplasts: some have indigestible walls. Better methods of wall digestion are needed [161].

11.1.3: Information from digestibility about cell wall architecture

Although the walls of diverse plant cells differ greatly in their digestibility, they generally show only small differences in polymer composition [though the grasses are rather exceptional]. Compositional differences do not seem big enough to explain the huge differences in observed digestibility, which may therefore reflect the superstructural make-up of the wall — how polymer molecules are held together (Ch 7). Research into digestibility could thus give information on wall architecture [427], which would be valuable in understanding how the wall carries out its botanical functions (Section 10.2.1).

11.2: ATTACKING THE PLANT CELL WALL

Since cell walls are largely composed of polysaccharides, the enzymes most involved in wall breakdown are likely to be polysaccharide **hydrolases.** Both **glycosidases,** which remove sugars 1-by-1 [or occasionally 2-by-2] from the non-reducing end(s) of polysaccharides, and **endoglycanases,** which chop polysaccharides in mid-chain, are involved (Section 4.2.5). Enzymes exist that hydrólyse all the glycosidic linkages of the plant cell wall; this is ecologically vital since even a very minor wall component is produced in huge amounts on a global scale. In general, attack of polysaccharides is initiated by endoglycanases and the glycosidases hydrolyse the resultant fragments down to monosaccharides. Enzymes of both classes are produced in all systems where appreciable wall breakdown occurs. Assays for some of these enzymes are outlined in Panel 6.5.

Endoglycanases and glycosidases act synergistically: this is illustrated by enzyme systems that attack cellulose. As cellulose is semi-crystalline it is difficult for an enzyme to attack. Yet, as the world's most abundant organic compound, cellulose is an important target. The fungus <u>Trichoderma</u> produces large amounts of endo-β-(1$\rightarrow$4)-glucanase (**cellulase**), whose activity is easily demonstrated <u>in vitro</u> by digestion of CMC (Panel 6.5, part 2a). But the purified enzyme has almost no detectable effect on native cellulose (e.g. cotton), despite the fact that the fungus can utilise cotton as sole carbon source. Synergism of two or more enzymes was suspected. A factor (C_1) was found which, when added to pure cellulase, would digest native cellulose. Pure C_1 very slowly releases cellobiose from the non-reducing terminus of cellulose, and is therefore called cellobiohydrolase [559]. It is thought that cellulase acts first, nicking in mid-chain a cellulose molecule on the exposed surface of the microfibril, and that cellobiohydrolase then attacks the newly-created non-reducing terminus, and nibbles along, releasing glucose units 2-by-2 (i.e. as cellobiose) [81,577]. Complete digestion needs a third enzyme, β-glucosidase (cellobiase), to hydrolyse cellobiose to glucose.

11.3: DEFENDING THE PLANT CELL WALL

It is often in the plant's interests to resist wall degradation. An obvious case is during attack by pathogenic fungi. It may be less obvious that a plant derives much benefit from resisting digestion once consumed by a herbivore since vegetative plant cells are unlikely to survive whether digested or not. However, animals which are smaller than the plant (consider a newly-hatched caterpillar on an indigestible tree) would die or move on to another plant leaving the indigestible plant better able to reproduce. Larger animals might come to recognise particular plant populations (or species) as failing to sustain and therefore avoid eating them.

In many cases the reason for indigestibility of a particular tissue is unknown, but three general mechanisms can be considered: **(a)** the wall contains non-polysaccharide matter that protects the otherwise digestible polysaccharides; **(b)** the wall's polysaccharides are tightly cross-linked, forming a network that physically excludes the enzymes; or **(c)** the wall contains inhibitors of the attacking enzyme. These three possibilities will be discussed in turn.

11.3.1: Enzyme-proofing agents

Lignin, cutin and suberin tend to make wood, epidermis and cork cell walls indigestible. These substances are hydrophobic and are therefore effective at excluding enzymes that only work in an aqueous environment. Evidence that lignin serves this rôle is two-fold: **(a)** when xylem is de-lignified, it becomes much more susceptible to decay [compare wood and paper], and **(b)** when pure cellulose is artificially re-lignified, e.g. by incubation with coniferyl alcohol + H_2O_2 + peroxidase, it becomes more resistant to enzymic hydrolysis [204]. Deposition of lignin and related compounds is a common defence response shown by plants to fungal infection [165,421].

Extensin is another highly indigestible substance, deposited in many cell walls. It resists hydrolysis by most proteases, especially while it still has its oligoarabinose side-chains attached. Extensin-rich walls appear in general to be very resistant to digestion, e.g. with the enzyme mixtures normally used for the isolation of protoplasts (Section 2.7.6). Cell suspension cultures of Capsicum and Acer are highly resistant to 'Driselase' [Fry, unpublished], and rich in isodityrosine-crosslinked extensin. Extensin is also often induced in response to pathogen attack, e.g. in the Colletotrichum lindemuthianum/melon interaction, where extensin deposition has been suggested to confer disease-resistance [150].

Silica is another indigestible accretion found in some cell walls; for example it constitutes about 40% of the dry weight of the epidermal hairs of some grasses [399]. Deposition of this too may be evoked by infection [46].

11.3.2: Cross-linking of structural polymers

A small percentage of the sugars in wall polysaccharides carry ferulic acid and related phenolic side-chains, which may be cross-linked by the action of peroxidase + H_2O_2 to form inter-polysaccharide bridges e.g. of diferulate, thus tying polysaccharide molecules together in a tight network (section 7.1.3). A rôle for this process in excluding hydrolytic enzymes from wall may be proposed [171]. The tyrosine units of extensin undergo a similar cross-linking reaction to form isodityrosine cross-links [169,174]; the resultant network is probably one of the factors that renders extensin so indigestible. It is of interest that many plant tissues, including those that are non-lignified, respond to pathogen attack by enhancing their peroxidase levels [224].

11.3.3: Production of enzyme inhibitors

Some cells secrete into their walls a protein that inhibits pectinase (usually the first enzyme produced by a pathogenic fungus) [345]. Tannins may also be present in plants, often compartmentalised in the vacuole or in specialised tanniniferous cells, from where they are released and come into contact with the walls upon damage, e.g. during mastication: tannins are relatively non-specific enzyme inhibitors [234,401]. The brown compounds, formed for example when an apple is cut open, might play a similar rôle.

11.4: THE COUNTER-ATTACK

Despite the defence mechanisms mentioned above, it is unfortunately true that not all plants always resist disease. Since normal plant morpho-genesis depends on limited wall-hydrolysis, the wall cannot afford to be completely indigestible. This Achilles heel allows infection; it can also suggest some clues to help in the biotechnologist's search for efficient ways of digesting plant cell walls. Possible counterattacks on lignin-fortified walls include the following.

11.4.1: Enzymic degradation of the lignin

Lignin is difficult to digest because **(a)** it is a hydrophobic polymer, held together by bonds (C—C and C—O—C) that are not readily hydrolysed; and **(b)** lignin is a random structure, with a huge diversity of shapes. Enzymes act by a 'lock/key' mechanism, which fails if the lock has no definite shape. However, as already argued, anything produced by Nature must also be broken down by it, and certain fungi, especially white rots, e.g. Phanerochaete chrysosporium, can live on lignin. There is evidence that a 'ligninase' is involved [233,457], which degrades lignin by a process culminating in a non-enzymic free radical reaction: it is hence unimportant that the lock has no fixed shape. Further information on 'ligninase' might help in the exploitation of waste cell walls.

11.4.2: Prevention of lignin formation

Possible ways of preventing plant cells making lignin include **(a)** inhibition of the first committed step of phenolic synthesis (catalysed by phenylalanine ammonia-lyase) e.g. by the application of amino-oxyphenyl-propionic acid (AOPP) [12], and **(b)** inhibition of the enzyme that catalyses the polymerisation of phenolic alcohols to lignin (peroxidase). The latter approach would also block the cross-linking of ferulic acid and tyrosine groups. Peroxidase can indeed be inhibited by use of reducing agents e.g. ascorbate, dithiothreitol or cysteine [104,169]. Pretreatment of cultured cells with such compounds may increase protoplast yield by blocking the action of extracellular peroxidase.

11.4.3: Chemical de-lignification

Certain oxidising agents, e.g. $NaClO_2$, destroy lignin and at the same time make the remaining walls more susceptible to polysaccharide-hydrolysing enzymes. Treatment of plant tissue with alkali may also remove some phenolic material — lignin in the case of grasses, and possibly phenolic esters in these and other plants.

11.4.4: Chemical degradation of the polysaccharides

Treatments that degrade wall polysaccharides by non-enzymic methods are not necessarily inhibited by lignin. One promising approach is the use of anhydrous hydrogen fluoride, one of the few treatments that efficiently degrade cellulose. The initial products are glycosyl fluorides [227].

Perhaps the most promising approaches to the hydrolysis of waste cell walls are enzymic, augmented by pre-treatment of the walls with a chemical, heat or mechanical treatment to increase the accessibility of the polysac-charides [324]. A smaller-scale alternative, perhaps more appropriate to tissue cultres, is pre-treatment of the living plants with agents that suppress lignification.

11.5: METHODS

Chapters 1-4, 6 and 8 give techniques applicable to the study of wall degradation. The approach used will depend on the biological questions under investigation. However, an approach that can be particularly recommended is radioactive labelling of the walls of interest (Ch 2). This is especially helpful in studies of complex natural digestion systems because it allows ready discrimination between the substrate (radioactive cell walls) and the organism(s) or enzymes (or their contaminants) carrying out digestion. Thus, radioactive cell walls can be fed to micro-organisms or animals and the rate and type of degradation studied. Theoretically it could be difficult in in vivo systems to distinguish between radioactive substrate and the radioactive metabolites synthesised by the animal or microbe from the wall-hydrolysis products, but in practice this is unlikely to be a problem because the major animal and microbial metabolites are so different from components of the plant cell wall. For example, if large amounts of [14C]glucosamine are found among the products, microbial metabolism can be inferred; if most of the xylose is recovered as [14C]-xylosyl-α-(1 → 6)-glucose upon Driselase-digestion, it was in the form of xyloglucan, a polysaccharide only found in plants (see Fig. 4.2.5).

It must be decided whether to label the walls in specific sugar residues or uniformly, and whether to fractionate the walls into component polymers prior to digestion or to investigate the digestion of the whole walls. Useful labelled substrates include:

(a) Uniformly labelled whole walls, e.g. prepared by growth of a tissue culture for several cell cycles in [U-14C]glucose (Section 2.1.3). Such walls are useful for drawing up a general balance sheet of the fate of all the major components, but would be difficult to use for study of the detailed fate of any given component.

(b) A specific [U-14C]polymer extracted and purified [in soluble form — see Ch 3] from substrate (a). This overcomes the problem of studying the detailed fate of a particular polymer, but raises the new problem that the polymer is not presented to the digestive system in a life-like physical state. The solubilised polymer may thus be readily digested whereas the same polymer would have been protected to some extent by the three-dimensional structure of the intact wall.

(c) Whole walls labelled only in specific residues. Careful choice of radioactive precursor is required (consult Panel 2.1.1 and Fig. 2.5). For example, if pectin digestion is of interest, plant cells could be fed [6-14C]glucuronic acid [464] so that uronic acids would be the only wall components labelled. In this approach, the problem of (b) is overcome.

Further reading

ASPINALL, G.O. (ed.) (1982) The Polysaccharides, Vols I-III. Academic Press, New York.

BRETT, C.T., HILLMAN, J.R. (eds) (1985) Biochemistry of Plant Cell Walls. Cambridge University Press.

CUTTER, E.G. (1978) Plant Anatomy: Experiment and Interpretation. Part I: Cells and Tissues. (2nd edn.) Edward Arnold, London.

FREY-WYSSLING, A. (1976) The Plant Cell Wall. Gebr. Borntraeger, Berlin.

FRY, S.C. (1985) Primary cell wall metabolism. Oxford Surv. Plant mol. Cell Biol. **2**, 1-42. [Miflin, B.J., ed., Oxford University Press].

HALL, J.L., FLOWERS, T.J., ROBERTS, R.M. (1982) Plant Cell Structure and Metabolism. (2nd edn.) Longman, London.

LOEWUS, F. (ed.) (1973) Biogenesis of Plant Cell Wall Polysaccharides. Academic Press, New York.

PILET, P.-E. (1971) Les Parois Cellulaires. Doin, Paris.

PREISS, J. (ed.) (1980) The Biochemistry of Plants, a Comprehensive Treatise, Vol. 3. Academic Press, New York.

PRESTON, R.D. (1974) The Physical Biology of Plant Cell Walls. Chapman & Hall, London.

ROBERTS, K., JOHNSTON, A.W.B., LLOYD, C.W., SHAW, P., WOOLHOUSE, H.W. (eds) (1985) The Cell Surface in Plant Growth and Development. J. Cell Sci., suppl. **2**. [The Company of Biologists Ltd, Cambridge].

ROELOFSEN, P.A. (1959) The Plant Cell-Wall. Gebr. Borntraeger, Berlin.

SIEGEL, S.M. (1962) The Plant Cell Wall. Pergamon, Oxford.

TANNER, W., LOEWUS, F.A. (eds) (1981) Plant Carbohydrates. II: Extracellular Carbohydrates. [Encyclopedia of Plant Physiology New Series **13B**.] Springer, Berlin.

WHISTLER, R.L. and others (1962-) Methods in Carbohydrate Chemistry. Vols I- . Academic Press, New York.

Journals frequently reporting original work on primary cell walls
    Planta, Plant Physiology, Carbohydrate Research, Physiologia Plantarum, Journal of experimental Botany, Phytochemistry, Journal of Plant Physiology [Zeitschrift für Pflanzenphysiologie], Plant & Cell Physiology, Plant Science, Australian Journal of Plant Physiology, Plant Physiology and Biochemistry [Physiologie Végétale], Biochemical Journal, Journal of biological Chemistry, Plant Cell and Environment, Agricultural and biological Chemistry, Physiological and molecular Plant Pathology

Journals publishing review articles of relevance to the primary cell wall
    Annual Review of Plant Physiology, Advances in Carbohydrate Chemistry and Biochemistry, Methods in Enzymology.

References

1 ADAMS., G.A. (1965) Complete acid hydrolysis. In: Methods in Carbo-hydrate Chemistry, Vol. 5, ed. R.L. Whistler. Academic Press, New York, pp. 269-75.

2 ADLER, E. (1957) Newer views of lignin formation. Tech. Assoc. Pulp Paper Ind. **40**: 294-311.

3 AHLUWALIA, B., FRY, S.C. (1986) Barley endosperm cell walls contain a feruloylated arabinoxylan and a non-feruloylated β-glucan. J. Cer. Sci. **4**: 287-95.

4 AITCHISON, P.A., MACLEOD, A.J., YEOMAN, M.M. (1977) Growth patterns in tissue (callus) cultures. In: Plant Tissue and Cell Culture, ed. H.E. Street. Blackwell, Oxford, pp. 267-306.

5 AKIYAMA, Y., MORI, M., KATÔ, K. (1980) ^{13}C-NMR Analysis of hydroxy-proline arabinosides from Nicotiana tabacum. Agric. biol. Chem. **44**: 2487-9.

6 ALBERSHEIM, P. (1963) Hormonal control of myo-inositol incorporation into pectin. J. biol. Chem. **238**: 1608-10.

7 ALBERSHEIM, P., NEVINS, P.D., ENGLISH, P.D., KARR, A. (1967) A method for the analysis of sugars in plant cell wall polysaccharides by gas-liquid chromatography. Carbohydr. Res. **5**: 340-5.

8 ALBERSHEIM, P., VALENT, B.S. (1978) Host—pathogen interactions in plants: plants, when exposed to oligosacchrides of fungal origin, defend themselves by accumulating antibiotics. J. Cell Biol. **78**: 627-43.

9 ALMIN, K.E., ERIKSSON, K.E., JANSON, C. (1967) Enzymic degradation of polymers. II Viscometric determination of cellulase activity in absolute terms. Biochim. Biophys. Acta **139**: 248-53.

10 ÅMAN, P., NORDKVIST, E. (1983) Chemical composition and in vitro degradability of botanical fractions of cereal straw. Swed. J. agric. Res. **13**: 61-7.

11 AMINO, S., TAKEUCHI, Y., KOMAMINE, A. (1984) Changes in cell wall constituents during the cell cycle in a synchronous culture of Catharanthus roseus. Physiol. Plant. **60**: 326-32.

12 AMRHEIN, N., FRANK, G., LEMM, G., LUHMANN, H-B. (1983) Inhibition of lignin formation by L-α-aminooxy-β-phenylpropionic acid, an inhibitor of phenylalanine ammonia-lyase. Eur. J. Cell Biol. **29**: 139-44.

13 AMRHEIN, N., TOPP, H., JOOP, O. (1984) The pathway of gallic acid biosynthesis in higher plants. Plant Physiol. **75 suppl**: 18.

14 ANDERSEN, S.O. (1964) The cross-links in resilin identified as dityrosine and trityrosine. Biochim. Biophys. Acta **93**: 213-15.

15 ANDERSON, R.L., CLARKE, B.A., JERMYN, M.A., KNOX, R.B., STONE, B.A.

(1977) A carbohydrate-binding arabinogalactan-protein from liquid suspension cultures of endosperm from Lolium multiflorum. Aust. J. Plant Physiol. **4**: 143-58.

16 ANDREWARTHA, K.A., PHILLIPS, D.R., STONE, B.A. (1979) Solution properties of wheat flour arabinoxylans and enzymically modified arabinoxylans. Carbohydr. Res. **77**: 191-204.

17 ANDREWS, P., JONES, J.K.N. (1954) The isolation of oligosaccharides from gums and mucilages. III Golden apple gum. J. Chem. Soc. **1954**: 4134-8.

18 ANDRULIS, I.L., ARFIN, S.M. (1979) Methods for determining the extent of tRNA aminoacylation in vivo in cultured mammalian cells. Meth. Enzymol. **59**: 268-71.

19 ASAMIZU, T., NISHI, A. (1979) Biosynthesis of cell-wall polysaccharides in cultured carrot cells. Planta **146**: 49-54.

20 ASHFORD, D., DESAI, N.N., ALLEN, A.K., NEUBERGER, A., O'NEILL, M.A., SELVENDRAN, R.R. (1982) Structural studies of the carbohydrate moieties of lectins from potato (Solanum tuberosum) tubers and thorn-apple (Datura stramonium) seeds. Biochem. J. **210**: 199-208.

21 ASIKIN, N., KOEPPE, R.E. (1979) Mannose-6-P and mannose-1-P in rat brain, kidney and liver. Biochem. biophys. Res. Comm. **89**: 279-85.

22 ASPINALL, G.O. (1965) Reduction of uronic acids in polysaccharides. In: Methods in Carbohydrate Research Vol. 5, ed. R.L. Whistler. Academic Press, New York, pp.397-400.

23 ASPINALL, G.O. (1973) Degradation of polysaccharides. In: Elucidation of Organic Structures by Physical and Chemical Methods, 2nd edn., Part II, eds. K.W. Bentley and G.W. Kirby. Wiley, New York, pp. 379-450.

24 ASPINALL, G.O. (1980) Chemistry of cell wall polysaccharides. In: The Biochemistry of Plants, a Comprehensive Treatise, Vol. 3, ed. J. Preiss. Academic Press, New York, pp. 473-500.

25 ASPINALL, G.O. (1982) Chemical characterization and structure determination of polysaccharides. In: The Polysaccharides, Vol. 1, ed. G.O. Aspinall. Academic Press, New York, pp. 35-131.

26 ASPINALL, G.O., BEGBIE, R., HAMILTON, A., WHYTE, J.N.C. (1967) Poly-saccharides of soy-beans. III. Extraction and fractionation of poly-saccharides from cotyledon meal. J. Chem. Soc. (C) **1967**: 1065-7.

27 ASPINALL, G.O., COTTRELL, I.W., MATHESON, K.W. (1971) Synthesis of uridine 5'-(L-arabinofuranosyl pyrophosphate) and the structure of the UDP-L-arabinose formed in plants from UDP-α-D-xylopyranose. Can. J. Biochem. **50**: 574-80.

28 ASPINALL, G.O., JIANG, K.-S. (1974) Rapeseed hull pectin. Carbo-hydr. Res. **38**: 247-55.

29 ASPINALL, G.O., MOLLOY, J.A., CRAIG, J.W.T. (1969) Extracellular polysaccharides from suspension-cultured sycamore cells. Can. J. Biochem. **47**: 1063-70.

30 AYERS, A.R., EBEL, J., FINELLI, F., BERGER, N., ALBERSHEIM, P. (1976) Host — pathogen interactions. IX Quantitative assays of elicitor

activity and characterization of the elicitor present in the extra-cellular medium of cultures of Phytophthora megasperma var. sojae. Plant Physiol. **57:** 751-59.

31 AZUMA, J., TAKAHASHI, N., KOSHIJIMA, T. (1981) Isolation and characterisation of lignin—carbohydrate complexes from the milled-wood lignin fraction of Pinus densiflora Sieb. et Zucc. Carbohydr. Res. **93:** 91-104.

32 BACIC, A., CHURMS, S.C., STEPHEN, A.M., COHEN, P.B., FINCHER, G.B. (1987) Fine structure of the arabinogalactan-protein from Lolium multiflorum. Carbohydr. Res. **162:** 85-93.

33 BAGDASARIAN, M., MATHESON, N.A., SYNGE, R.L.M., YOUNGSON, M.A. (1964) New procedures for isolating polypeptides and proteins from tissues. Metabolic incorporation of L-[^{14}C]valine into fractions of inter-mediate molecular weight in broad bean (Vicia faba) leaves. Biochem. J. **91:** 91-105.

34 BAILEY, R.W., PRIDHAM, J.B. (1962) The separation and identification of oligosaccharides. Chromatogr. Revs. **4:** 114-36.

35 BARBER, G.A. (1966) Formation of UDP-L-rhamnose from UDP-D-glucose. Meth. Enzymol. **8:** 307-9.

36 BARRETT, A.J., NORTHCOTE, D.H. (1965) Apple fruit pectic substances. Biochem. J. **94:** 617-27.

37 BASHAM, H.G., BATEMAN, D.F. (1975) Relationship of cell death in plant tissue treated with a homogeneous endopectate lyase to cell wall degradation. Physiol. Plant Pathol. **5:** 249-62.

38 BATEMAN, D.F., BASHAM, H.G. (1976) Degradation of plant cell walls and membranes by microbial enzymes. In: Physiological Plant Pathology [Encyclopedia of Plant Physiology, N.S., Vol. 4], ed. R. Heitefuss and P.H. Williams. Springer, Berlin, pp.316-55.

39 BAUER, W.D., TALMADGE, K.W., KEEGSTRA, K., ALBERSHEIM, P. (1973) The structure of plant cell walls. II. The hemicellulose of suspension-cultured sycamore cells. Plant Physiol. **51:** 174-84.

40 BAYDOUN, E. A.-H., BRETT, C.T. (1984) The effect of pH on the binding of calcium to pea epicotyl cell walls and its implications for the control of cell expansion. J. exp. Bot. **35:** 1820-31.

41 BAYDOUN, E.A-H., FRY, S.C. (1985) The immobility of pectic substances in injured tomato leaves and its bearing on the identity of the wound hormone. Planta **165:** 269-76.

42 BAYDOUN, E.A-H., FRY, S.C. (1988) [2-^{3}H]Mannose incorporation in cultured plant cells: investigation of L-galactose residues in the primary cell wall. J. Plant Physiol. (in press).

43 BAYDOUN, E.A-H., NORTHCOTE, D.H. (1981) The extraction from maize (Zea mays) root cells of membrane-bound protein with Ca^{2+}-dependent ATPase activity and its possible role in membrane fusion in vitro. Biochem. J. **193:** 781-92.

44 BECKER, G.E., HUI, P.A., ALBERSHEIM, P. (1964) Synthesis of an extra-cellular polysaccharide by suspension cultures of Acer pseudoplatanus cells. Plant Physiol. **39:** 913-20.

45 BELKOURA, M., RANJEVA, R., MARIGO, G. (1986) Cations stimulate proton pumping in Catharanthus roseus cells: implication of a redox system? Plant Cell Env 9: 653-6.

46 BELL, A.A. (1981) Biochemical mechanisms of disease resistance. Ann. Rev. Plant Physiol. 32: 21-81.

47 BEHRENS, N.H., PARODI, A.J., LELOIR, L.F. (1971) Glucose transfer from dolichol monophosphate glucose: the product formed with endogenous microsomal acceptor. Proc. natl Acad. Sci. (US) 68: 2857-60.

48 BENNETT, A.B., CHRISTOFFERSEN, R.E. (1986) Synthesis and processing of cellulase from ripening avocado fruit. Plant Physiol. 81: 830-35.

49 BERGH, M.L.E., HOOGHWINKEL, G.J.M., VAN DEN EIJNDEN, D.H. (1983) Biosynthesis of the O-glycosidically linked oligosaccharide chains of fetuin. J. biol. Chem. 258: 7430-6.

50 BIGGS, K.J, FRY, S.C. (1987) Phenolic cross-linking in the plant cell wall. In: Physiology of Cell Expansion During Plant Growth, eds. D.J. Cosgrove and D.P. Knievel. American Society of Plant Physiologists (in press.)

51 BIGGS, K.J., FRY, S.C. (1987) Unpublished observations.

52 BIRNIE, G.D., RICKWOOD, D. (eds.) (1978) Centrifugal Separations in Molecular and Cell Biology. Butterworths, London.

53 BISHOP, P.D., MAKUS, D.J., PIERCE, G., RYAN, C.A. (1981) Proteinase inhibitor-inducing factor activity in tomato leaves resides in oligosaccharides enzymically released from cell walls. Proc. natl Acad. Sci. (US) 78: 3536-40.

54 BISHOP, P.D., RYAN, C.A. (1987) Plant cell wall polysaccharides that activate natural plant defenses. Meth. Enzymol. 138: 715-24.

55 BLAKE, D.A., GOLDSTEIN, I.J. (1982) Resolution of carbohydrates by lectin affinity chromatography. Meth. Enzymol. 83: 127-32.

55a BLAKENEY, A.B., HARRIS, P.J., HENRY, R.J., STONE, B.A. (1983) A simple and rapid preparation of alditol acetates for monosaccharide analysis. Carbohydr. Res. 113: 291-9.

56 BLAKENEY, A.B., STONE, B.A. (1985) Methylation of carbohydrates with lithium methylsulphinyl carbanion. Carbohydr. Res. 140: 319-24.

57 BLASCHEK, W. (1983) Complete separation and quantification of neutral sugars from plant cell walls and mucilages by high-performance liquid chromatography. J. Chromatogr. 256: 157-63.

58 BLASCHEK, W., HAASS, D., KOEHLER, H., FRANZ, G. (1981) Cell wall regeneration by Nicotiana tabacum protoplasts: chemical and biochemical aspects. Plant Sci. Lett. 22: 47-57.

59 BLASCHEK, W., HAASS, D., KOEHLER, H., SEMLER, U., FRANZ, G. (1983) Demonstration of a β-1.4-primer glucan in cellulose-like glucan synthesized in vitro. Z. Pflanzenphysiol. 111: 357-64.

60 BLASCHEK, W., KOEHLER, H., SEMLER, U., FRANZ, G. (1982) Molecular weight distribution of cellulose in primary cell walls. Investigations with regenerating protoplasts, suspension cultured cells and mesophyll of tobacco. Planta 154: 550-5.

61   BLIGH, E.G., DYER, W.J. (1959) A rapid method of total lipid extraction and purification. Can. J. Biochem. Physiol. **37:** 911-17.

62   BLIGNY, R., DOUCE, R. (1983) Excretion of laccase by sycamore (<u>Acer pseudoplatanus</u> L.) cells. Purification and properties of the enzyme. Biochem. J. **209:** 489-96.

63   BLUMENKRANTZ, N., ASBOE-HANSEN, G. (1973) New method for quantitative determination of uronic acids. Anal. Biochem. **54:** 484-9.

64   BOBBITT, J.M. (1956) Periodate oxidation of carbohydrates. Adv. Carbohydr. Chem. **11:** 1-41.

65   BOFFEY, S.A., NORTHCOTE, D.H. (1975) Pectin synthesis during the regeneration of plasmolysed tobacco leaf cells. Biochem. J. **150:** 433-40.

66   BOLWELL, G.P. (1986) Microsomal arabinosylation of polysaccharide and elicitor-induced carbohydrate-binding glycoprotein in French bean. Phytochemistry **25:** 1807-13.

67   BOLWELL, G.P., NORTHCOTE, D.H. (1981) Control of hemicellulose and pectin synthesis during differentiation of vascular tissue in bean (Phaseolus <u>vulgaris</u>) callus and in bean hypocotyl. Planta **152:** 225-33.

68   BOLWELL, G.P., NORTHCOTE, D.H. (1984) Demonstration of a common anti-genic site on endomembrane proteins of Phaseolus <u>vulgaris</u> by a rat monoclonal antibody: tentative identification of arabinan synthase and consequences for its regulation. Planta **162:** 139-46.

69   BOUVENG, H.O., LINDBERG, B. (1965) Native acetylated wood polysaccha-rides. Extraction with dimethyl sulfoxide. In: Methods in Carbo-hydrate Chemistry, Vol. 5, ed. R.L. Whistler. Academic Press, New York, pp. 147-50.

70   BOUVENG, H.O., MEIER, H. (1959) Studies on a galactan from Norwegian spruce compression wood. Acta Chem Scand. **13:** 1884-9.

71   BOWER, C.A., WILCOX, L.V. (1965) Soluble salts. In: Methods in Soil Analysis, Part 2, ed. C.A. Black. American Society of Agronomy, Madison, Wisconsin, pp. 933-51.

72   BOWLES, D.J., NORTHCOTE, D.H. (1974) The amounts and rates of export of polysaccharides found within the membrane system of maize root cells. Biochem. J. **142:** 139-44.

73   BRANT, D.A. (1980) Conformation and behavior of polysaccharides in solution. In: The Biochemistry of Plants, a Comprehensive Treatise, Vol. 3, ed. J. Preiss. Academic Press, New York, pp. 425-72.

74   BRETT, C.T. (1978) Synthesis of β-(1→3)-glucan from extracellular uridine diphosphate glucose as a wound response in suspension cultured soybean cells. Plant Physiol. **62:** 377-82.

75   BRETT, C.T. (1980) The isolation and characterisation of polyprenyl-phosphate-sugars. Techniques in Carbohydrate Biochemistry, **B305:** 1-14. Elsevier.

76   BRETT, C.T. (1981) Polysaccharide synthesis from GDP-glucose in pea epicotyl slices. J. exp. Bot. **32:** 1067-77.

77   BRETT, C.T., NORTHCOTE, D.H. (1975) The formation of oligoglucans

linked to lipid during synthesis of β-glucan by characterized membrane fractions isolated from peas. Biochem. J. **148**: 107-17.

78 BRICE, R.E., MORRISON, I.M. (1982) The degradation of isolated hemicelluloses and lignin—hemicellulose complexes by cell free, rumen hemicellulases. Carbohydr. Res. **101**: 93-100.

79 BRILLOUET, J.-M. (1985) A rapid procedure for the purification of an endo-(1→4)-β-D-xylanase from Polyporus tulipiferae (Irpex lacteus). Carbohydr. Res. **144**: 346-9.

80 BRILLOUET, J.-M., MOULIN, J.-C., AGOSIN, E. (1985) Production, purification and properties of an α-L-arabinosidase from Dichomitus squalens. Carbohydr. Res. **144**: 113-26.

81 BROWN, R.M., Jr (ed.) (1982) Cellulose and Other Natural Polymer Systems. Plenum, New York.

82 BRUMMELL, D.A., HALL, J.L. (1980) The role of the epidermis in auxin-induced and fusicoccin-induced elongation of Pisum sativum stem segments. Planta **150**: 371-9.

83 BRUMMELL, D.A., HALL, J.L. (1983) Regulation of cell wall synthesis by auxin and fusicoccin in different tissues of pea stem segments. Physiol. Plant. **59**: 627-34.

84 BUCHANAN, B.B., CRAWFORD, N.A., WOLOSINK, R.A. (1979) Activation of plant acid phosphatases by oxidized glutathione and dehydroascorbate. Plant Sci. Lett. **14**: 245-51.

85 BUCHELI, P., BUCHALA, A.J., MEIER, H. (1987) Autolysis in vitro of cotton fibre cell walls. Physiol. Plant. **70**: 633-8.

86 BULIGA, G.S., BRANT, D.A., FINCHER, G.B. (1986) The sequence statistics and solution conformation of a barley (1→3,1→4)-β-D-glucan. Carbohydr. Res. **157**: 139-56.

87 CACAN, R., CECCHELLI, R., HOFLACK, B., VERBERT, A. (1984) Intra-lumenal pool and transport of CMP-N-acetylneuraminic acid, GDP-fucose and UDP-galactose (Study with plasma-membrane-permeabilized mouse thymocytes). Biochem. J. **224**: 277-84.

88 CAMIRAND, A., BRUMMELL, D., MACLACHLAN, G. (1987) Fucosylation of xyloglucan: localization of the transferase in dictyosomes of pea stem cells. Plant Physiol. **84**: 753-6.

89 CAMIRAND,A., MACLACHLAN, G. (1986) Biosynthesis of fucose-containing xyloglucan nonasaccharide by pea microsomal membranes. Plant Physiol. **82**: 379-83.

90 CARMINATTI, H., PASSERON, S. (1966) Chromatography of sugar nucleotides in morpholinium borate. Meth. Enzymol. **8**: 108-11.

91 CARPITA, N.C. (1984) Fractionation of hemicelluloses from maize cell walls with increasing concentrations of alkali. Phytochemistry **23**: 1089-93.

92 CARPITA, N.C. (1986) Incorporation of proline and amino acids into the cell walls of maize coleoptiles. Plant Physiol. **80**: 660-6.

93 CARPITA, N.C., DELMER, D.P. (1981) Concentration and metabolic turnover of UDP-glucose in developing cotton fibers. J. biol. Chem. **256**: 308-15.

94 CARPITA, N.C., MONTEZINOS, D., SABULARSE, D., DELMER, D.P. (1979) Determination of the pore size of cell walls of living plants. Science **205**: 1144-7.

95 CHANG, H.-M., ALLEN, G.G. (1971) Oxidation. In: Lignins. Occurrence, Formation, Structure and Reactions, ed. K.V. Sarkanen and C.H. Ludwig. Wiley, New York, pp. 433-85.

96 CHANZY, H., CHUMPITAZI, B., PEGUY, A. (1982) Solutions of polysaccharides in $\underline{N}$-methyl morpholine $\underline{N}$-oxide (MMNO). Carbohydr. Polym. **2**: 35-42.

97 CHEN, J., VARNER, J.E. (1985) An extracellular matrix protein in plants: characterization of a genomic clone for carrot extensin. EMBO J. **4**: 2145-51.

98 CHIBBAR, R.N., CELLA, R., ALBANI, D., VAN HUYSTEE, R.B. (1985) The growth of and peroxidase synthesis by two carrot cell lines. J. exp. Bot. **35**: 1846-52.

99 CHURMS, S.C. (ed.) (1982) CRC Handbook of Chromatography. Carbohydrates. Vol 1. CRC Press, Boca Raton, Florida.

100 CLARKE, J., SHANNON, L.M. (1976) The isolation and characterisation of the glycopeptides from horseradish peroxidase isoenzyme C. Biochim. Biophys. Acta **427**: 428-42.

101 CLELAND, R.E. (1981) Wall extensibility: hormones and wall extension. In: Plant Carbohydrates II Extracellular Carbohydrates [Encyclopedia of Plant Physiology, N.S., vol. **13B**], eds. W. Tanner and F.A. Loewus. Springer, Berlin, pp. 255-76.

102 CLELAND, R.E. (1984) The Instron technique as a measure of immediate-past wall extensibility. Planta **160**: 514-20.

103 CLINE, K., ALBERSHEIM, P. (1981) Host—pathogen interactions. XVII Hydrolysis of biologically-active fungal glucans by enzymes isolated from soybean cells. Plant Physiol. **68**: 221-8.

104 COOPER, J.B., VARNER, J.E. (1983) Insolubilization of hydroxyproline-rich cell wall glycoprotein in aerated carrot root slices. Biochem. Biophys. Res. Comm. **112**: 161-7.

105 COOPER, J.B., VARNER, J.E. (1984) Cross-linking of soluble extensin in isolated cell walls. Plant Physiol. **76**: 414-17.

106 COHEN, S.A., BIDLINGMEYER, B.A., TARVIN, T.L. (1986) PITC derivatives in amino acid analysis. Nature **320**: 769-70.

107 CONRAD, H.E. (1976) Radiochromatographic analysis of reducing carbohydrates by paper chromatography. In: Methods in Carbohydrate Research, Vol 7, eds. R.L. Whistler and J.N. BeMiller. Academic Press, New York, pp. 71-5.

108 COSGROVE, D.J. (1981) Rapid suppression of growth by blue light. Occurrence, time-course, and general characteristics. Plant Physiol. **67**: 584-90.

109 COSGROVE, D.J. (1987) Wall relaxation and the driving forces for cell expansive growth. Plant Physiol. **84**: 561-64.

110 COSGROVE, D.J., KNIEVEL, D.P. (eds.) (1987) Physiology of Cell

Expansion During Plant Growth. Am. Soc. Plant Physiol. (in press)

111 CRASNIER, M., NOAT, G., RICARD, J. (1980) Purification and molecular properties of acid phosphatase from sycamore cell walls. Plant Cell Env. **3**: 217-24.

112 CUMMING, D.F. (1970) Separation and identification of soluble nucleotides in cambial and young xylem tissues of Larix decidua Mill. Biochem. J. **116**: 189-98.

113 CUMMINGS, J.H. (1984) Cellulose and the human gut. Gut **25**: 805-10.

114 CUTTER, E.G. (1984) Plant Anatomy: Experiment and Interpretation. Part I: Cells and Tissues. 2nd edn. E. Arnold, London.0

115 DALESSANDRO, G., NORTHCOTE, D.H. (1981) Increase in xylan synthetase activity during xylem differentiation of the vascular cambium of sycamore and poplar trees. Planta **151**: 61-7.

116 DALESSANDRO, G., PIRO, G., NORTHCOTE, D.H. (1986) Glucomannansynthase activity in differentiating cells of Pinus sylvestris L. Planta **169**: 564-74.

117 DANIEL, P.F. (1987) Separation of benzoylated oligosaccharides by reversed-phase high-pressure liquid chromatography: application to high-mannose type oligosaccharides. Meth. Enzymol. **138**: 94-116.

118 DARVILL, A.G., McNEIL, M., ALBERSHEIM, P. (1978) Structure of plant cell walls VIII. A new pectic polysaccharide. Plant Physiol. **62**: 418-22.

119 DARVILL, A.G., SMITH, C.J., HALL, M.A. (1978) Cell wall structure and elongation growth in Zea mays coleoptile tissue. New Phytol. **80**: 503-16.

120 DAVIES, P.J. (ed.) (1987) Plant Hormones and their Role in Plant Growth and Development. M. Nijhoff, Amsterdam.

121 DAVIS, K.R., HAHLBROCK, K. (1987) Induction of defense responses in cultured parsley cells by plant cell wall fragments. Plant Physiol. **85**: 1286-90.

122 DAVIS, K.R., LYON, G.D., DARVILL, A.G., ALBERSHEIM, P. (1984) Host—pathogen interactions. XXV Endopolygalacturonic acid lyase from Erwinia carotovora elicits phytoalexin accumulation by releasing plant cell wall fragments. Plant Physiol. **74**: 52-60.

123 DAWSON, R.M.C., ELLIOTT, D.C., ELLIOTT, W.H., JONES, K.M. (1986) Data for Biochemical Research. 3rd edn. Clarendon, Oxford.

124 DEAN, P.D.G., JOHNSON, W.S., MIDDLE, F.A. (1985) Affinity Chromatography, a Practical Approach. IRL Press, Oxford.

125 DELMER, D.P. (1987) Cellulose biosynthesis. Ann. Rev. Plant Physiol. **38**: 259-90.

126 DELMER, D.P., ALBERSHEIM, P. (1970) The biosynthesis of sucrose and nucleoside diphosphate glucoses in Phaseolus aureus. Plant Physiol. **45**: 782-86.

127 DELMER, D.P., BEASLEY, C.A., ORDIN, L. (1974) Utilization of nucleoside diphosphate glucoses in developing cotton fibers. Plant Physiol. **53**: 149-53.

128 DELMER, D.P., COOPER, G. (1987) Identification of a receptor protein for the herbicide 2,6-dichlorobenzonitrile. Plant Physiol. (in press).

129   DESAI, N.N., ALLEN, A.K., NEUBERGER, A. (1983) The profiles of potato (Solanum tuberosum) lectin after deglycosylation by trifluoromethane-sulphonic acid. Biochem. J. **211**: 273-6.

130   deVRIES, J.A., den UIJL, C.H., VORAGEN, A.G.J., ROMBOUTS, F.M., PILNIK, W. (1983) Structural features of the neutral sugar side chains of apple pectic substances. Carbohydr. Polym. **3**: 193-205.

131   deVRIES, J.A., ROMBOUTS, F.M., VORAGEN, A.G.J., PILNIK, W. (1983) Distribution of methoxyl groups in apple pectic substances. Carbohydr. Polym. **3**: 245-58.

132   deVRIES, J.A., ROMBOUTS, F.M., VORAGEN, A.G.J., PILNIK, W. (1984) Comparison of the structural features of apple and citrus pectin substances. Carbohydr. Polym. **4**: 89-101.

133   deWIT, P.J.G.M., SPIKMAN, G. (1982) Evidence for the occurrence of race and cultivar-specific elicitors of necrosis in intercellular fluids of compatible interactions of Cladosporium fulvum and tomato. Physiol. Plant Pathol. **21**: 1-11.

134   DIGBY, J., FIRN, R.D. (1977) Some criticisms of the use of nojiri-mycin as a specific inhibitor of auxin-induced growth. Z. Pflanzen-physiol. **82**: 355-62.

135   DISCHE, Z. (1962) Color reactions of carbohydrates. In: Methods in Carbohydrate Chemistry Vol. 1, ed. R.L. Whistler and M.L. Wolfrom. Academic Press, New York, pp. 475-514.

136   DREYER, W.J., BYNUM, E. (1967) High voltage paper electrophoresis. Meth. Enzymol. **11**: 32-39.

137   DUBOIS, M., GILLES, K.A., HAMILTON, J.K., REBERS, P.A., SMITH, F. (1956) Colorimetric method for determination of sugars and related substances. Anal. Chem. **28**: 350-6.

138   DuPONT, M.S., SELVENDRAN, R.R. (1987) Hemicellulosic polymers from the cell walls of beeswing wheat bran. I. Polymers solubilised by alkali at 2°. Carbohydr. Res. **163**: 99-113.

139   EDWARDS, K., CRAMER, C.L., BOLWELL, G.P., DIXON, R.A., SCHUCH, W., LAMB, C.J. (1985) Rapid transient induction of phenylalanine ammonia-lyase mRNA in elicitor-treated bean cells. Proc. natl Acad. Sci. (US) **82**: 6731-5.

140   EDWARDS, M., DEA, I.C.M, BULPIN, P.V., REID, J.S.G. (1985) Xyloglucan (amyloid) mobilisation in the cotyledons of Tropaeolum majus L. seeds following germination. Planta **163**: 133-40.

141   EDWARDS, M., DEA, I.C.M, BULPIN, P.V., REID, J.S.G. (1986) Purification and properties of a novel xyloglucan-specific endo-$(1\rightarrow4)$-β-D-glucanase from germinated nasturtium seeds (Tropaeolum majus L.). J. biol. Chem. **261**: 9489-94.

142   ELBEIN, A.D. (1966) Microscale adaptation of the morpholidate proced-ure for the synthesis of sugar nucleotides. Meth. Enzymol. **8**: 142-5.

143   ELSTNER, E.F., HEUPEL, A. (1976) Formation of hydrogen peroxide by isolated cell walls from horseradish (Armoracia lapathifolia Gilib.). Planta **130**: 175-80.

144   ENGLISH, P.D., MAGLOTHIN, A., KEEGSTRA, K., ALBERSHEIM, P. (1972) A

cell wall-degrading endopolygalacturonase secreted by Colletotrichum lindemuthianum. Plant Physiol. **49:** 293-7.

145 ENGLYST, H.N., HAY, S., MacFARLANE, G.T. (1987) Polysaccharide breakdown by mixed population of human faecal bacteria. FEMS Microbiol. Ecol. **95:** 163-71.

146 ENOKI, A., YAKU, F., KOSHIJIMA, T. (1983) Synthesis of LCC model compounds and their chemical and enzymatic stabilities. Holzforschung **37:** 135-41.

147 EPSTEIN, L., LAMPORT, D.T.A. (1984) An intramolecular linkage involving isodityrosine in extensin. Phytochemistry **23:** 1241-6.

148 ERICKSON, M., LARSSON, S., MIKSCHE, G.E. (1973) Gaschromatographische Analyse von Ligninoxidationsprodukten. VII Ein verbessertes Verfahren zur Characterisierung von Ligninen durch Methylierung und oxydativen Abbau. Acta Chem Scand. **27:** 127-40.

149 ESHDAT, Y., MIRELMAN, D. (1972) An improved method for the recovery of compounds from paper chromatograms. J. Chromatogr. **65:** 458-9.

150 ESQUERRÉ-TUGAYÉ, M.-T., MAZAU, D. (1981) Les glycoprotéines à hydroxyproline de la paroi végétale. Physiol. Vég. **19:** 415-26.

151 EVANS, M.L. (1974) Evidence against the involvement of galactosidase or glucosidase in auxin- or acid-stimulated growth. Plant Physiol. **54:** 213-15.

152 EVANS, M.L. (1974) Rapid responses to plant hormones. Ann. Rev. Plant Physiol. **25:** 195-223.

153 EVANS, P.K., COCKING, E.C. (1977) Isolated plant protoplasts. In: Plant Tissue and Cell Culture, ed. H.E. Street. Blackwell, Oxford, pp. 103-36.

154 EVERETT, N.P., STREET, H.E. (1979) Studies on the growth in culture of plant cells. XXIV Effects of 2,4-D and light on the growth and metabolism of Acer pseudoplatanus L. suspension cultures. J. exp. Bot. **30:** 409-17.

155 FEINGOLD, D.S., AVIGAD, G. (1980) Sugar nucleotide transformations in plants. In: The Biochemistry of Plants, a Comprehensive Treatise, Vol. 3, ed. J. Preiss. Academic Press, New York, pp. 101-70.

156 FINCHER, G.B., STONE, B.A. (1981) Metabolism of noncellulosic polysaccharides. In: Plant Carbohydrates II [Encyclopedia of Plant Physiology, N.S., Vol. **13B**), eds. W. Tanner and F.A. Loewus. Springer, Berlin, pp. 68-132.

157 FINCHER, G.B., STONE, B.A., CLARKE, A.E. (1983) Arabinogalactanproteins: structure, biosynthesis and function. Ann. Rev. Plant Physiol. **34:** 47-70.

158 FINNE, J., KRUSIUS, T. (1976) O-Glycosidic carbohydrate units from glycoproteins of different tissues: demonstration of a brain-specific disaccharide, α-galactosyl-$(1\rightarrow3)$-N-acetylgalactosamine. FEBS Letts **66:** 94-7.

159 FIRN, R., DIGBY, J. (1977) The role of the peripheral cell layers in the geotropic curvature of sunflower hypocotyls: a new model of shoot geotropism. Aust. J. Plant Physiol. **4:** 337-47.

160 FISCHER, L. (1980) Gel Filtration Chromatography: Laboratory Techniques in Biochemistry and Molecular Biology, 2nd edn. Elsevier, Amsterdam.

161 FITZSIMONS, P.J., WEYERS, J.D.B. (1985) Properties of some enzymes used for protoplast isolation. In: The physiological properties of plant protoplasts, ed. P.-E. Pilet. Springer, Berlin, pp. 12-23.

162 FORREST, G.I., BENDALL, D.S. (1969) The distribution of polyphenols in the tea plant (Camellia sinensis L.) Biochem. J. 113: 741-55.

163 FRANZ, G. (1972) Polysaccharidmetabolismus in den Zellwänden wachsender Keimlinge von Phaseolus aureus. Planta 102: 334-47.

164 FREUDENBERG, K., NEISH, A.C. (1968) Constitution and Biosynthesis of Lignin. Springer, Berlin.

165 FRIEND, J. (1976) Lignification in infected tissue. In: Biochemical Aspects of Plant—Parasite relationships, eds. J. Friend and D.R. Threlfall. Academic Press, London, pp. 291-303.

166 FRY, S.C. (1979) Phenolic components of the primary cell wall and their possible rôle in the hormonal regulation of growth. Planta 146: 343-51.

167 FRY, S.C. (1980) Gibberellin-controlled pectinic acid and protein secretion in growing cells. Phytochemistry 19: 735-40.

168 FRY, S.C. (1982) Phenolic components of the primary cell wall: feruloylated disaccharides of D-galactose and L-arabinose from spinach polysaccharide. Biochem. J., 203: 439-504.

169 FRY, S.C. (1982) Isodityrosine, a new cross-linking amino acid from plant cell-wall glycoprotein. Biochem. J., 204: 449-55.

170 FRY, S.C. (1983) Feruloylated pectins from the primary cell wall: their structure and possible functions. Planta 157: 111-23.

171 FRY, S.C. (1984) Incorporation of [14C]cinnamate into hydrolase-resistant components of the primary cell wall. Phytochemistry 23: 59-64.

172 FRY, S.C. (1984) Isodityrosine — its detection, estimation and chemical synthesis. Meth. Enzymol. 107: 388-97.

173 FRY, S.C. (1985) Primary cell wall metabolism. Oxford Surv Plant mol. Cell Biol. 2: 1-42, ed. B.J. Miflin. Oxford University Press.

174 FRY, S.C. (1986) Cross-linking of matrix polymers in the growing cell walls of angiosperms. Ann. Rev. Plant Physiol. 37: 165-86.

175 FRY, S.C. (1986) In-vivo formation of xyloglucan nonasaccharide: a possible biologically active cell-wall fragment. Planta 169: 443-53.

176 FRY, S.C. (1987) Formation of isodityrosine by peroxidase isozymes. J. exp. Bot. 38: 853-62.

177 FRY, S.C. (1987) Intracellular feruloylation of pectic polysaccharides. Planta 171: 205-11.

178 FRY, S.C., DARVILL, A.G., ALBERSHEIM, P. (1983) Amino acid transport and protein synthesis, possible primary targets of biologically active cell wall fragments. In: Interactions between nitrogen and growth regulators in plant development, Monograph 9, ed. M.B.

Jackson. British Plant Growth Regulator Group, Wantage, Oxfordshire, pp. 33-44.

179 FRY, S.C., NORTHCOTE, D.H. (1983) Sugar-nucleotide precursors of the arabinofuranosyl, arabinopyranosyl and xylopyranosyl residues of spinach polysaccharides. Plant Physiol. **73**: 1055-61.

180 FRY, S.C., STREET, H.E. (1980) Gibberellin-sensitive suspension cultures. Plant Physiol. **65**: 472-7.

181 FUJII, T. (1978) Effects of IAA on oxygen-sensitive growth and on hydroxyproline protein level in cell wall. Plant Cell Physiol. **19**: 927-33.

182 FUJII, T., SUZUKI, T., KATO, R. (1981) Effect of a growth inhibitor on the hydroxyproline level in cell wall of Zea primary roots. Plant Cell Physiol. **22**: 1185-90.

183 FUKUDA, M., KONDO, T., OSAWA, T. (1976) Studies on the hydrolysis of glycoproteins. Core structures of oligosaccharides obtained from porcine thyroglobulin and pineapple stem bromelain. J. Biochem. **80**: 927-33.

184 GAHAN, P.B., BELLANI, L.M. (1984) Identification of shoot apical meristem cells committed to form vascular elements in Pisum sativum L. and Vicia faba L. Ann. Bot. **54**: 837-41.

185 GAILLARD, B.D.E. (1961) Separation of linear from branched polysaccharides by precipitation as iodine complexes. Nature **191**: 1295-6.

186 GAN, S., SHEN, Z., ZHANG, Z., YAN, J. (1986) The transformation of extensin in cucumber cell wall and its relations with hydroxyproline and isodityrosine. Acta Phytophysiol. Sin. **12**: 272-80.

187 GARCIA-TREJO, A., HADDOCK, J.W., CHITTENDEN, G.J.F., BADDILEY, J. (1971) The biosynthesis of galactofuranosyl residues in galactocarolose. Biochem. J. **122**: 49-57.

188 GASPAR, T., PENEL, C., THORPE, T., GREPPIN, H. (1982) Peroxidases 1970-1980: A Survey of the Biochemical and Physiological Roles in Higher Plants. Université de Genève.

189 GAUTHERET, R.J. (1959) La Culture des Tissus Végétaux, Techniques et Réalisations. Masson, Paris, pp. 585-94.

190 GEBHARDT, K., SCHMID, P.P.S., FEUCHT, W. (1982) Isoenzyme der Indol-3-essigsäure-Oxidase und Peroxidase in Geweben verschiedener Prunus-arten. Gartenbauwissenschaft **47**: 265-9.

191 GEISSMANN, T., NEUKOM, H., (1971) Vernetzung von Phenolcarbonsäure-estern von Polysacchariden durch oxydative phenolische Kupplung. Helv. Chim. Acta **54**: 1108-12.

192 GERWIG, G.P., KAMERLING, J.P., VLIEGENTHART, J.F.G. (1979) Determination of the absolute configuration of monosaccharides in complex carbohydrates by capillary G.L.C. Carbohydr. Res. **77**: 1-7.

193 GILKES, N.R., HALL, M.A. (1977) The hormonal control of cell wall turnover in Pisum sativum L. New Phytol. **78**: 1-15.

194 GOLDBERG, R. (1980) Cell wall polysaccharidase activities and growth processes: a possible relationship. Physiol. Plant. **50**: 261-64.

195 GOLDBERG, R. (1985) Cell-wall isolation, general growth aspects. In: Cell Components [Modern Methods of Plant Analysis, N.S., Vol. 1], eds. H.F. Linskens and J.F. Jackson. Springer, Berlin, pp. 1-30.

196 GOLDSTEIN, I.J., HAY, G.W., LEWIS, B.A., SMITH, F. (1965) Controlled degradation of polysaccharides by periodate oxidation, reduction, and hydrolysis. In: Methods in Carbohydrate Chemistry, Vol. 5, ed. R.L. Whistler. Academic Press, New York, pp. 361-70.

197 GOODMAN, R.N., KIRÁLY, Z., WOOD, K.R. (1986) The Biochemistry and Physiology of Plant Disease. Univ. Missouri Press, Columbia, MO.

198 GORDON, A.H., BACON, J.S.D. (1981) Fractionation of cell-wall preparations from grass leaves by centrifuging in non-aqueous density gradients. Biochem. J. **193**: 765-71.

199 GOULDING, K.H. (1986) Radioisotope techniques. In: A Biologist's Guide to Principles and Techniques of Practical Biochemistry, eds. K. Wilson and K.H. Goulding. E. Arnold, London, pp. 314-44.

200 GOWDA, D.C., SARATHY, C. (1987) Structure of an L-arabino-D-xylan from the bark of Cinnamomum zeylanicum. Carbohydr. Res. **166**: 263-9.

201 GREEN, J.R., NORTHCOTE, D.H. (1978) The structure and functions of glycoproteins synthesised during slime polysaccharide production by membranes of the root cap cells of maize (Zea mays). Biochem. J. **170**: 599-608.

202 GREEN, J.R., NORTHCOTE, D.H., (1979) Location of fucosyltransferases in the membrane system of maize root cells. J. Cell Sci. **40**: 235-44.

203 GREEN, T.R., RYAN, C.A. (1972) Wound-induced proteinase inhibitor in plant leaves: a possible defense mechanism against insects. Science **175**: 776-7.

204 GRESSEL, J., VERED, Y., BAR-LEV, S., MILSTEIN, O., FLOWERS, H.M. (1983) Partial suppression of cellulase action by artificial lignification of cellulose. Plant Sci. Lett. **32**: 349-53.

205 GREVE, L.C., ORDIN, L. (1977) Isolation and purification of an α-mannosidase from the coleoptiles of Avena sativa. Plant Physiol. **60**: 478-81.

206 GRIERSON, D., TUCKER, G.A. (1983) Timing of ethylene and polygalacturonase synthesis in relation to the control of tomato fruit ripening. Planta **157**: 174-9.

207 GRISEBACH, H. (1980) Branched-chain sugars: occurrence and biosynthesis. In: The Biochemistry of Plants, a Comprehensive Treatise, Vol. 3, ed. J. Preiss. Academic, New York, pp. 171-97.

208 GRISEBACH, H., BARON, D., SANDERMANN, H., WELLMANN, E. (1974) Formation of UDP-apiose from UDP-glucuronic acid. Meth. Enzymol. **28**: 439-46.

209 GRISON, R., PILET, P-E. (1985) Cytoplasmic and wall isoperoxidases in growing maize roots. J. Plant Physiol. **118**: 189-99.

210 GRISON, R., PILET, P-E. (1985) Maize root peroxidases: relationship with polyphenol oxidases. Phytochemistry **24**: 2519-21.

211 GROSS, G.G. (1977) Cell wall-bound malate dehydrogenase from horse-radish. Phytochemistry **16**: 319-21.

212 GROSS, G.G. (1980) The biochemistry of lignification. Adv. Bot. Res. **8**: 25-63.

213 GROSS, K.G. (1984) Fractionation and partial characterisation of cell walls from normal and non-ripening tomato fruit. Physiol. Plant. **62**: 25-32.

214 GUBLER, F., ASHFORD, A.E., BACIC, A., BLAKENEY, A.B., STONE, B.A. (1985) Release of ferulic acid esters from barley aleurone. II Characterization of feruloyl compounds released in response to GA$_3$. Aust. J. Plant Physiol. **12**: 307-17.

215 HAASS, D., FREY, R., THIESEN, M., KAUSS, H. (1981) Partial purification of a hemagglutinin associated with cell walls from hypocotyls of Vigna radiata. Planta **151**: 490-96.

216 HADDON, L.E., NORTHCOTE, D.H. (1975) Quantitative measurement of the course of bean callus differentiation. J. Cell Sci. **17**: 11-26.

217 HAGER, A., MENZEL, H., KRAUSS, A. (1971) Versuche und Hypothese zur Primärwirkung des Auxins beim Streckungswachstum. Planta **100**: 47-75.

218 HAHLBROCK, K., SCHRÖDER, J. (1975) Specific effects on enzyme activities upon dilution of Petroselinum hortense cell cultures into water. Arch. Biochem. Biophys. **171**: 500-6.

219 HAIS, I.M., MACEK, K. (eds) (1963) Paper Chromatography, a Comprehensive Treatise, 3rd edn, English Translation. Academic Press, New York and Nakladatelství Ceskoslovenské akademie věd, Prague.

220 HALL, J.L., SEXTON, R. (1972) Cytochemical localisation of peroxidase activity in root cells. Planta **108**: 103-20.

221 HALL, R.D., HOLDEN, M.A., YEOMAN, M.M. (1987) The accumulation of phenylpropanoid and capsaicinoid compounds in cell cultures and whole fruit of the chilli pepper, Capsicum frutescens Mill. Plant Cell, Tissue & Organ Culture **8**: 163-76.

222 HAMES, B.D., RICKWOOD, D. (eds.) (1981) Gel Electrophoresis of Proteins: a Practical Approach. IRL Press, Oxford.

223 HAMILTON, B., MORT, A. (1986) Sodium chlorite reveals cell wall architecture in soybean root tissue. Plant Physiol. **80** (suppl.): 77.

224 HAMMERSCHMIDT, R., NUCKLES, E.M., KUĆ, J. (1982) Association of enhanced peroxidase activity with induced systemic resistance of cucumber to Colletotrichum lagenarium. Physiol. Plant Pathol. **20**: 73-82.

225 HANKE, D.E., NORTHCOTE, D.H. (1974) Cell wall formation by soybean callus protoplasts. J. Cell Sci. **14**: 29-50.

226 HARBORNE, J.B. (1983) Phytochemical Methods. 2nd edn. Chapman & Hall, London.

227 HARDT, H., LAMPORT, D.T.A. (1982) Hydrogen fluoride saccharification of cellulose and xylan: isolation of α-D-glucopyranosyl fluoride and α-D-xylopyranosyl fluoride intermediates, and 1,6-anhydro-β-D-glucopyranoside. Phytochemistry **21**: 2301-3.

228 HARKIN, J.M., OBST, J.R. (1973) Lignification in trees: indication of exclusive peroxidase participation. Science **180**: 296-7.

300

229 HARRIS, P.J. (1983) Cell walls. In: Isolation of Membranes and Organelles from Plant Cells, eds. J.L. Hall and A.L. Moore. Academic Press, London, pp. 25-53.

230 HARRIS, P.J., ANDERSON, M.A., BACIC, A., CLARKE, A.E. (1984) Cell—cell recognition in plants with special reference to the pollen—stigma interaction. Oxford Surv. Plant mol. Cell Biol. **1:** 161-203, ed. B.J. Miflin. Clarendon, Oxford.

231 HARRIS, P.J., HARTLEY, R.D. (1980) Phenolic constituents of the cell walls of Monocotyledons. Biochem. Syst. Ecol. **8:** 153-60.

232 HARTLEY, R.D., HARRIS, P.J. (1981) Phenolic constituents of the cell walls of Dicotyledons. Biochem. Syst. Ecol. **9:** 189-203.

233 HARVEY, P.J., SHOEMAKER, H.E., BOWEN, R.M., PAINTER, J.M. (1985) Single-electron transfer processes and the reaction mechanism of enzymic degradation of lignin. FEBS Lett. **183:** 13-16.

234 HASLAM, E. (1981) Vegetable tannins. In: The Biochemistry of Plants, a Comprehensive Treatise, Vol. 7, ed. E.E. Conn. Academic Press, New York, pp. 527-56.

235 HATFIELD, R.D., NEVINS, D.J. (1987) Hydrolytic activity and substrate specificity of an endoglucanase from Zea mays seedling cell walls. Plant Physiol. **83:** 203-7.

236 HAWKER, J.S. (1969) Insoluble invertase from grapes: an artifact of extraction. Phytochemistry **8:** 337-44.

237 HAWKER, J.S. (1985) Sucrose. In: Biochemistry of Storage Carbohydrates in Green Plants, eds. P.M. Dey and R.A. Dixon. Academic, London, pp. 1-51.

238 HAY, G.W., LEWIS, B.A., SMITH, F. (1965) Periodate oxidation of polysaccharides: general procedures. In: Methods in Carbohydrate Research, Vol. 5, ed. R.L. Whistler. Academic Press, New York, pp. 357-61.

239 HAYASHI, T., MACLACHLAN, G. (1984) Biosynthesis of pentosyl lipids by pea membranes. Biochem. J. **217:** 791-803.

240 HAYASHI, T., MACLACHLAN, G. (1984) Pea xyloglucan and cellulose. I. Macromolecular organization. Plant Physiol. **75:** 596-604.

241 HAYASHI, T., MARSDEN, M.P.F., DELMER, D.P. (1987) Pea xyloglucan and cellulose. V. Xyloglucan—cellulose interactions in vitro and in vivo. Plant Physiol. **83:** 384-9.

242 HAYASHI, T., MATSUDA, K. (1981) Biosynthesis of xyloglucan in suspension-cultured soybean cells. Evidence that the enzyme system of xyloglucan biosynthesis does not contain $\beta(1 \rightarrow 4)$glucan 4-β-D-glucosyltransferase activity. Plant Cell Physiol. **22:** 1571-84.

243 HAYASHI, T., WONG, Y., MACLACHLAN, G. (1984) Pea xyloglucan and cellulose. II Partial hydrolysis by pea endo-1,4-β-glucanases. Plant Physiol. **75:** 605-10.

244 HENRY, R.J., STONE, B.A. (1982) Solubilization of β-glucan synthases from the membranes of cultured ryegrass endosperm cells. Biochem. J. **203:** 629-36.

245 HESLOP-HARRISON, J. (1978) Cellular Recognition Systems in Plants.

E. Arnold, London.

246 HEYN, A.N.J. (1981) Molecular basis of auxin-regulated extension growth and role of dextranase. Proc. natl Acad. Sci. (US) **78:** 6608-12.

247 HICKS, K.B., LIM, P.C., HAAS, M.J. (1985) Analysis of uronic acids, their lactones, and related compounds by high-performance liquid chromatography on cation-exchange resins. J. Chromatogr. **319:** 159-71.

248 HILDEBRAN, J.N., AIRHART, J., STIREWALT, W.S., LOW, R.B. (1981) Prolyl-tRNA-based rates of protein and collagen synthesis in human lung fibroblasts. Biochem. J. **198:** 249-58.

249 HILDEBRANDT, A.C., RIKER, A.J. (1949) The influence of various carbon compounds on the growth of marigold, Paris-daisy, periwinkle, sunflower and tobacco tissue in vitro. Am. J. Bot. **36:** 74-85.

250 HILL, R.L., BRADSHAW, R.A. (1969) Fumarase. Meth. Enzymol. **13:** 91-9.

251 HOLLOWAY, P.J., BROWN, G.A., WATTENDORFF, J. (1981) Ultrahistochemical detection of epoxides in plant cuticular membranes. J. exp. Bot. **32:** 1051-66.

252 HOLLOWAY, P.J. (1982) Structure and histochemistry of plant cuticular membranes: an overview. In: The Plant Cuticle, eds. D.F. Cutler, K.L. Alvin and C.E. Price. Academic Press, London, pp. 1-32.

253 HOLLOWAY, P.J. (1982) The chemical constitution of plant cutins. In: The Plant Cuticle, eds. D.F. Cutler, K.L. Alvin and C.E. Price. Academic Press, London, pp. 45-85.

254 HON, D.N.-S., SRINIVASAN, K.S.V. (1983) Mechanochemical process in cotton cellulose fiber. J. appl. Polym. Sci. **28:** 1-10.

255 HONDA, S. (1984) High-performance liquid chromatography of mono- and oligosaccharides. Anal. Biochem. **140:** 1-47.

256 HUBER, D.J., NEVINS, D.J. (1979) Autolysis of the cell wall β-D-glucan in corn coleoptiles. Plant Cell Physiol. **20:** 201-12.

257 HUGHES, R., STREET, H.E. (1974) Galactose as an inhibitor of the expansion of root cells. Ann. Bot., **38:** 555-64.

258 HUWYLER, H.R., FRANZ, G., MEIER, H. (1979) Changes in the composition of cotton fibre cell walls during development. Planta **146:** 635-42.

259 ILER, R.F. (1979) The Chemistry of Silica. Wiley, New York.

260 ISHIDA, A., OOKUBO, K., ONO, K. (1987) Formation of hydrogen peroxide by NAD(P)H oxidation with isolated cell wall-associated peroxidase from cultured liverwort cells, Marchantia polymorpha L. Plant Cell Physiol. **28:** 723-6.

261 ISHII, S. (1981) Isolation and characterization of cell-wall pectic substances from potato tuber. Phytochemistry **20:** 2329-33.

262 JARVIS, M.C. (1982) The proportion of calcium-bound pectin in plant cell walls. Planta **154:** 344-6.

263 JARVIS, M.C. (1984) Structure and properties of pectin gels in plant cell walls. Plant Cell Env. **7:** 153-64.

264 JARVIS, M.C., THRELFALL, D.R., FRIEND, J. (1977) Separation of macro-molecular components of plant cell walls: electrophoretic methods. Phytochemistry **16**: 849-52.

265 JEFFREE, C.E., DALE, J.E., FRY, S.C. (1986) The genesis of inter-cellular spaces in developing leaves of Phaseolus vulgaris L. Proto-plasma **132**: 90-8.

266 JEFFREE, C.E., YEOMAN, M.M., PARKINSON, M., HOLDEN, M.A. (1987) The chemical basis of cell to cell contact and its possible role in differentiation. In: Advances in the Chemical Manipulation of Plant Tissue Cultures, Monograph 16, ed. M.B. Jackson. British Plant Growth Regulator Group, Wantage, Oxfordshire (in press).

267 JOHNSON, D.B., MOORE, W.F., ZANK, L.C. (1961) The spectrophotometric determination of lignin in small wood samples. TAPPI **44**: 793-8.

268 JOHNSON, D.C., NICHOLSON, M.D., HAIGH, F.C. (1976) Dimethylsulfoxide/paraformaldehyde: a non-degrading solvent for cellulose. Appl. Polym. Symp. **28**: 931-43.

269 JONES, D.H. (1984) Phenylalanine ammonia-lyase: regulation of its induction, and its role in plant development. Phytochemistry **23**: 1349-59.

270 JONES, D.H., NORTHCOTE, D.H. (1981) Induction by hormones of phenyl-alanine ammonia-lyase in bean-cell suspension cultures. Eur. J. Biochem. **116**: 117-25.

271 JONES, J.K.N, STOODLEY, R.J. (1965) Fractionation using copper complexes. In: Methods in Carbohydrate Chemistry, Vol. 5, ed. R.L. Whistler. Academic Press, New York, pp. 36-8.

272 JOSELEAU, J.-P., CHAMBAT, G., CHUMPITAZI-HERMOZA, B. (1980) Solubilization of cellulose and other plant structural polysaccha-rides in 4-methylmorpholine-N-oxide: an improved method for the study of cell wall constituents. Carbohydr. Res. **90**: 339-44.

273 KAKES, P. (1985) Linamarase and other β-glucosides are present in the cell walls of Trifolium repens L. leaves. Planta **166**: 156-60.

274 KAMERBEEK, G.A. (1956) Peroxydase content of dwarf types and giant types of plants. Acta bot. Néerl. **5**: 257-63.

275 KAMSTEEG, J., VAN BREDERODE, J., VAN NIGTEVECHT, G. (1979) Properties and genetic control of UDP-L-rhamnose : anthocyanidin 3-0-glucoside, 6"-0-rhamnosyltransferase from petals of red campion, Silene dioica. Phytochemistry **18**: 659-60.

276 KANU, H. SHIBATA, S., SATSUMA, Y., SONE, Y., MISAKI, A., (1986) Interactions of α-L-arabinofuranose-specific antibody with plant polysaccharides and its histochemical application. Phytochemistry **25**: 2041-7.

277 KATAYAMA, Y., MOROHOSHI, N., HARAGUCHI, T., (1980) Formation of lignin—carbohydrate complex in enzymatic dehydrogenation of coniferyl alcohol. II. Isolation of lignin—carbohydrate complex with guaiacylglycerol-β-coniferyl ether or other dimeric compound as lignin moiety. Mokuzai Gakkaishi **26**: 414-20.

278 KATO, Y., MATSUDA, K. (1976) Presence of a xyloglucan in the cell-walls of Phaseolus aureus hypocotyls. Plant Cell Physiol. **17**:1185-98.

279 KATO, Y., MATSUDA, K. (1980) Structure of oligosaccharides obtained by controlled degradation of mung bean xyloglucan with acid and Aspergillus oryzae enzyme preparation. Agric. Biol. Chem. **44**: 1751-8.

280 KATO, Y., NEVINS, D.J. (1984) Enzymic dissociation of Zea shoot cell wall polysaccharides.   II. Dissociation of (1→3),(1→4)-β-D-glucan by purified (1→3),(1→4)-β-D-glucan 4-glucanohydrolase from Bacillus subtilis. Plant Physiol. **75**: 745-52.

281 KATO, Y., NEVINS, D.J. (1984) Enzymic dissociation of Zea shoot cell wall polysaccharides.   IV. Dissociation of xylan by purified endo-(1→4)-β-xylanase from Bacillus subtilis. Plant Physiol. **75**: 759-65.

282 KATO, Y., NEVINS, D.J. (1985) Isolation and identification of 0-(5-0-feruloyl-α-L-arabinofuranosyl)-(1→3)-0-β-D-xylopyranosyl-(1→4)-D-xylopyranose as a component of Zea shoot cell walls. Carbohydr. Res. **137**: 139-50.

283 KATSUMI, M., KAZAMA, H. (1978) Gibberellin control of cell elongation in cucumber hypocotyl sections. Bot. Mag. Tokyo Spe. Iss. **1**: 141-58.

284 KAUFMAN, P.B., LABAVITCH, J., ANDERSON-PROUTY, A., GHOSHEH, N.S. (1975) Laboratory Experiments in Plant Physiology. Macmillan, New York, pp. 47-9.

285 KAUR-SAWHNEY, R., FLORES, H.E., GALSTON, A.W. (1981) Polyamine oxidase in oat leaves: a cell wall-localized enzyme. Plant Physiol. **68**: 494-98.

286 KAUSS, H., SWANSON, A.L., HASSID, W.Z. (1967) Biosynthesis of the methyl ester groups of pectin by transmethylation from S-adenosyl-L-methionine. Biochem. biophys. Res. Comm. **26**: 234-40.

287 KAWASAKI, S. (1987) Synthesis of arabinose-containing cell wall precursors in suspension-cultured tobacco cells. III. Purification and some properties of the major component. Plant Cell Physiol. **28**: 187-97.

288 KAY, E., SHANNON, L.M., LEW, J.Y. (1967) Peroxidase isozymes from horseradish roots.   II. Catalytic properties. J. biol. Chem. **242**: 2470-3.

289 KEEGSTRA, K., TALMADGE, K.W., BAUER, W.D., ALBERSHEIM, P. (1973) The structure of plant cell walls.   III A model of the walls of suspension-cultured sycamore cells based on the interconnections of the macromolecular components. Plant Physiol. **51**: 188-96.

290 KINDEL, P.K., WATSON, R.R. (1973) Synthesis, characterization and properties of uridine 5'-(α-D-apio-D-furanosyl pyrophosphate). Biochem. J. **133**: 227-41.

291 KING, P.J. (1977) Studies on the growth in culture of plant cells. XXII Growth limitation by nitrate and glucose in chemostat cultures of Acer pseudoplatanus L. J. exp. Bot. **28**: 142-55.

292 KIVIRIKKO, K.I., LIESMAA, M. (1959) A colorimetric method for determination of hydroxyproline in tissue hydrolysates. Scand. J. Clin. Lab. Investigation **11**: 128-33.

293 KNEE, M. (1978) Metabolism of polymethylgalacturonate in apple fruit cortical tissue during ripening. Phytochemistry **17**: 1261-4.

294 KOBATA, A., YAMASHITA, K., TAKASAKI, S. (1987) BioGel P-4 column chromatography of oligosaccharides: effective size of oligosaccharides expressed in glucose units. Meth. Enzymol. **138:** 84-94.

295 KOCH, A.L. (1985) Bacterial wall growth and division or life without actin. Trends Biochem. Sci. **Jan 1985:** 11-14.

296 KOLATTUKUDY, P.E. (1981) Structure, biosynthesis and biodegradation of cutin and suberin. Ann. Rev. Plant Physiol. **32:** 539-67.

297 KOLATTUKUDY, P.E., ESPELIE, K.E., SOLIDAY, C.L. (1981) Hydrophobic layers attached to cell walls. Cutin, suberin and associated waxes. In: Encyclopaedia of Plant Physiology, Vol. 13B, eds. W. Tanner and F.A. Loewus. Springer, Berlin, pp. 225-54.

298 KOMALAVILAS, P., MORT, A. (1986) Identification of acetate groups on the backbone of rhamnogalacturonan-I, a pectic polysaccharide of primary cell walls, obtained from suspension-cultured cotton cells. Plant Physiol. **80** (suppl): 30.

299 KOYAMA, T., HAYASHI, T., KATO, Y., MATSUDA, K. (1983) Degradation of xyloglucan by wall-bound enzymes from soybean tissue. II Degradation of the fragment heptasaccharide from xyloglucan and the characteristic action pattern of the α-D-xylosidase in the enzyme system. Plant Cell Physiol. **24:** 155-62.

300 KUTSCHERA, U., BRIGGS, W.R. (1987) Differential effect of auxin on in vivo extensibility of cortical cylinder and epidermis in pea internodes. Plant Physiol. **84:** 1361-6.

301 KUTSCHERA, U., BERGFELD, R., SCHOPFER, P. (1987) Cooperation of epidermis and inner tissues in auxin-mediated growth of maize coleoptiles. Planta **170:** 168-80.

302 KUTSCHERA, U., SCHOPFER, P. (1986) Effect of auxin and abscisic acid on cell wall extensibility in maize coleoptiles. Planta **167:** 527-35.

303 LABAVITCH, J.M. (1981) Cell wall turnover in plant development. Ann. Rev. Plant Physiol. **32:** 385-406.

304 LABAVITCH, J.M., RAY, P.M. (1974) Turnover of cell wall polysaccharides in elongating pea stem segments. Plant Physiol. **53:** 669-73.

305 LAEMMLI, U.K. (1970) Cleavage of structural proteins during the assembly of the head of bacteriophage T4. Nature **227:** 680-5.

306 LAMPORT, D.T.A. (1970) Cell wall metabolism. Ann. Rev. Plant Physiol. **21:** 235-70.

307 LAMPORT, D.T.A. (1977) Structure, biosynthesis and significance of cell wall glycoproteins. In: The Structure, Biosynthesis and Degradation of Wood [Recent Advances in Phytochemistry, Vol. 11], eds. F.A. Loewus and V.C. Runeckles. Plenum, pp. 79-115.

308 LAMPORT, D.T.A. (1980) Structure and function of plant glycoproteins. In: The Biochemistry of Plants, a Comprehensive Treatise, Vol. 3, ed. J. Preiss. Academic Press, New York, pp. 501-41.

309 LAMPORT, D.T.A. (1984) Hydroxyproline glycosides in the plant kingdom. Meth. Enzymol. **106:** 523-8.

310 LAMPORT, D.T.A., CATT, J.W. (1981) Glycoproteins and enzymes of the

cell wall. In: Plant Carbohydrates II Intracellular Carbohydrates [Encyclopedia of Plant Physiology, N.S., vol **13B**], eds. W. Tanner and F.A. Loewus. Springer, Berlin, pp. 133-65.

311 LAMPORT, D.T.A., MULDOON, E.P., KIELISZEWSKI, M. WILLARD, J.J., TERHUNE, B. (1986) Molecular design of a molecular fabric. In: Cell Walls '86, eds. B. Vian, D. Reis and R. Goldberg. Université P. & M. Curie — École Normale Supérieure, Paris, pp. 8-11.

312 LAPIERRE, C., MONTIES, B., ROLANDO, C. (1986) Thioacidolysis of poplar lignins: identification of monomeric syringyl products and characterization of guaiacyl-syringyl lignin fractions. Holzforschung **40:** 113-18.

313 LASKEY, R.A. (1980) The use of intensifying screens or organic scintillators for visualizing radioactive molecules resolved by gel electrophoresis. Meth. Enzymol. **65:** 363-71.

314 LAU, J.M., McNEIL, M., DARVILL, A.G., ALBERSHEIM, P. (1983) Structural characterization of the pectic polysaccharide rhamnogalacturonan I. Plant Physiol. **72:** S60.

315 LAYNE, E., (1957) Spectrophotometric and turbidometric methods for measuring proteins. Meth. Enzymol. **3:** 447-54.

316 LEARY, G.J. (1980) Quinone methides and the structure of lignin. Wood Sci. Technol. **14:** 21-34.

317 LEE, Y.C., SCOCCA, J.R. (1972) A common structural unit in asparagine-oligosaccharides of several glycoproteins from different sources. J. biol. Chem. **247:** 5753-8.

318 LENDZIAN, K.J., SCHÖNHERR, J. (1983) In-vivo study of cutin synthesis in leaves of Clivia miniata Reg. Planta **158:** 70-5.

319 LEUNG, D.W.M., BEWLEY, J.D. (1983) A role for α-galactosidase in the degradation of the endosperm cell walls of lettuce seeds, cv. Grand Rapids. Planta **157:** 274-7.

320 LEVER, M. (1972) A new reaction for colorimetric determination of carbohydrates. Anal. Biochem. **47:** 273-9.

321 LEWIS, D.H., SMITH, D.C. (1967) Sugar alcohols in fungi and green plants. II Methods of detection and estimation. New Phytol. **66:** 185-204.

322 LIENART, Y., BARNOUD, F. (1985) β-D-Glucanase activities in pure cell-wall-enriched fractions from Valerianella olitoria cells. Planta **165:** 68-75.

323 LINDBERG, B., LÖNNGREN, J., SVENSSON, S. (1975) Specific degradation of polysaccharides. Adv. Carbohydr. Chem. Biochem. **31:** 185-241.

324 LLOYD, A. (1984) French push ahead with biomass conversion. New Scientist **26 Apr 1984:** 21.

325 LOEWUS, F.A., LOEWUS, M.W. (1980) myo-Inositol: biosynthesis and metabolism. In: The Biochemistry of Plants, a Comprehensive Treatise, Vol 3, ed. J. Preiss. Academic Press, New York, pp. 43-76.

326 LOEWY, A.G. (1984) The N^{ϵ}-(γ-glutamic)lysine cross-link: method of analysis, occurrence in extracellular and cellular proteins. Meth. Enzymol. **107:** 241-57.

327 LONGLAND, J.M. (1986) The molecular mode of action of abscisic acid in the induction of dormancy. PhD thesis, University of Edinburgh.

328 LUNDQUIST, K., OHLSSON, B., SIMONSON, R. (1977) Isolation of lignin by means of liquid-liquid extraction. Svensk Paperstidn. **80:** 143-51.

329 MACHEIX, J-J., FLEURIET, A. (1986) Les derivés hydroxycinnamiques des fruits. In: Journées Internationales d'Étude, Groupe Polyphénols. Université des Sciences et Techniques du Languedoc, Montpellier, pp. 337-51.

330 MACLACHLAN, G. (1985) Are lipid-linked glycosides required for plant polysaccharide biosynthesis? In: Biochemistry of Plant Cell Walls, eds. C.T. Brett and J.R. Hillman. Cambridge University Press, pp. 199-220.

331 MÄDER, M., MEYER, Y., BOPP, M. (1975) Lokalisation der Peroxidase-Isoenzyme in Protoplasten und Zellwänden von Nicotiana tabacum L. Planta **122:** 259-68.

332 MÄDER, M., SCHLOSS, P. (1979) Isolation of malate dehydrogenase from cell walls of Nicotiana tabacum. Plant Sci. Lett. **17:** 75-80.

333 MARES, D.J., STONE, B.A. (1973) Studies on wheat endosperm. II. Properties of the wall components and studies on their organization in the wall. Aust. J. biol. Sci. **26:** 813-30.

334 MARGNA, U. (1977) Control at the level of substrate supply — an alternative in the regulation of phenylpropanoid accumulation in plant cells. Phytochemistry **16:** 419-26.

335 MARKWALDER, H.-U., NEUKOM, H. (1976) Diferulic acid as a possible crosslink in hemicelluloses from wheat germ. Phytochemistry **15:** 836-7.

336 MARX-FIGINI, M., SCHULZ, G.V. (1966) Über die Kinetik und den Mechanismus der Biosynthese der Cellulose in den höheren Pflanzen (nach Versuchen an den Samenhaaren der Baumwolle). Biochim. Biophys. Acta **112:** 81-101.

337 MASUDA, Y., YAMAMOTO, R. (1970) Effect of auxin on β-1,3-glucanase activity in Avena coleoptile. Dev. Growth Differentn. **11:** 287-96.

338 MAZZA, G., WELINDER, K.J. (1980) Covalent structure or turnip peroxidase 7. Cyanogen bromide fragments, complete structure and comparison to horseradish peroxidase C. Eur. J. Biochem. **108:** 481-9.

339 McCLEARY, B.V. (1983) Enzymic interactions in the hydrolysis of galactomannan in germinating guar: the role of exo-β-mannanase. Phytochemistry **22:** 649-58.

340 McDOUGALL, G.J., FRY, S.C. (1988) Anti-auxin activity of a naturally-occurring xyloglucan oligosaccharide. Food Hydrocolloids (in press).

341 McNEIL, M., ALBERSHEIM, P., TAIZ, L., JONES, R.L. (1975) The structure of plant cell walls. VII. Barley aleurone cells. Plant Physiol. **55:** 64-8.

342 McNEIL, M., DARVILL, A.G., ALBERSHEIM, P. (1979) The structural polymers of the primary cell walls of Dicots. Prog. Chem. Org. Nat. Prods. **37:** 191-249.

343 McNEIL, M., DARVILL, A.G., ALBERSHEIM, P. (1980) Structure of plant cell walls. X. Rhamnogalacturonan I, a structurally complex pectic polysaccharide in the walls of suspension-cultured sycamore cells. Plant Physiol. **66:** 1128-34.

344 McNEIL, M., DARVILL, A.G., ÅMAN, P., FRANZÉN, L.-E., ALBERSHEIM, P. (1982) Structural analysis of complex carbohydrates using high-performance liquid chromatography, gas chromatography, and mass spectrometry. Meth. Enzymol. **83:** 3-45.

345 McNEIL, M., DARVILL, A.G., FRY, S.C., ALBERSHEIM, P. (1984) Structure and function of the primary cell walls of plants. Ann. Rev. Biochem. **53:** 625-63.

346 MEHLTRETTER, C.L. (1963) D-Glucuronic acid. α-D-Glucofuranurono-6,3-lactone by catalytic air oxidation of 1,2-O-isopropylidene-α-D-glucofuranose. In: Methods in Carbohydrate Chemistry, Vol. 2, eds. R.L. Whistler and M.L. Wolfrom. Academic Press, New York, pp. 29-31.

347 MEIDNER, H. (1984) Class Experiments in Plant Physiology. Allen & Unwin, London, pp. 50-52.

348 MEIER, H. (1965) Fractionation by precipitation with barium hydroxide. In: Methods in Carbohydrate Chemistry, Vol. 5, ed. R.L. Whistler. Academic Press, New York, pp. 45-6.

349 MEIER, H., BUCHS, L., BUCHALA, A.J., HOMEWOOD, T. (1981) (1→3)-β-D-Glucan (callose) is a probable intermediate in biosynthesis of cellulose of cotton fibres. Nature **289:** 821-22.

350 MELLOR, R.B., LORD, J.M. (1979) Involvement of a lipid-linked inter-mediate in the transfer of galactose from UDP[^{14}C]galactose to exogenous protein in castor bean endosperm homogenates. Planta **147:** 89-96.

351 MELLOR, R.B., LORD, J.M. (1979) Formation of lipid-linked mono- and oligosaccharides from GDP-mannose by castor bean endosperm homogenates. Planta **146:** 91-99.

352 MELTON, L.D., McNEIL, M., DARVILL, A.G., ALBERSHEIM, P., DELL, A. (1986) Structural characterization of oligosaccharides isolated from the pectic polysaccharide rhamnogalacturonan II. Carbohydr. Res. **146:** 279-305.

353 MENZIES, I.S., MOUNT, J.N., WHEELER, M.J. (1978) Quantitative estimation of clinically important monosaccharides in plasma by rapid thin layer chromatography. Ann. clin. Biochem. **15:** 65-76.

354 MÉTRAUX, J.-P., TAIZ, L. (1978) Transverse viscoelastic extension in Nitella. I Relationship to growth rate. Plant Physiol. **61:** 135-8.

355 MOFFATT, J.G. (1966) Sugar nucleotide synthesis by the phosphoro-morpholidate procedure. Meth. Enzymol. **8:** 136-42.

356 MOLL, C., JONES, R.L. (1981) Calcium and gibberellin-induced elongation of lettuce hypocotyl sections. Planta **152:** 450-56.

357 MØLLER, I.M., LIN, W. (1986) Membrane-bound NAD(P)H dehydrogenases in higher plant cells. Ann. Rev. Plant Physiol. **37:** 309-34.

358 MONTIES, B., LAPIERRE, C. (1981) Donées récentes sur l'hétérogénéité de la lignine. Physiol. Vég. **19:** 327-48.

308

359 MONTREUIL, J., BOUQUELET, S., DEBRAY, H., FOURNET, B., SPIT, G., STRECKER, G. (1986) Glycoproteins. In: Carbohydrate Analysis, a Practical Approach, eds. M.F. Chaplin and J.F. Kennedy. IRL Press, Oxford, pp. 143-204.

360 MOORE, P.J., DARVILL, A.G., ALBERSHEIM, P., STAEHELIN, A.L. (1986) Immunogold localization of xyloglucan and rhamnogalacturonan I in the cell walls of suspension-cultured sycamore cells. Plant Physiol. **82:** 787-94.

361 MOORE, T.S. (1973) An extracellular macromolecular complex from the surface of soybean suspension cultures. Plant Physiol. **51:** 529-36.

362 MORRIS, M.R., NORTHCOTE, D.H. (1977) Influence of cations at the plasma membrane in controlling polysaccharide secretion from sycamore suspension cells. Biochem. J. **281:** 603-18.

363 MORRISON, I.M. (1977) Extraction of hemicelluloses from plant cell-walls with water after preliminary treatment with methanolic sodium methoxide. Carbohydr. Res. **57:** C4-C6.

364 MORT, A.J., BAUER, W.D. (1982) Application of two new methods for cleavage of polysaccharides into specific oligosaccharide fragments: structure of the capsular and extracellular polysaccharides of Rhizobium japonicum that bind soybean lectin. J. biol. Chem. **257:** 1870-75.

365 MORT, A.J., LAMPORT, D.T.A. (1977) Anhydrous hydrogen fluoride deglycosylates glycoproteins. Anal. Biochem. **82:** 289-309.

366 MORVAN, H. (1982) Libération de polymères pectiques acides au cours de la croissance de suspensions cellulaires de Silène. Physiol. Vég. **20:** 671-8.

367 MURASHIGE, T., SKOOG, F. (1962) A revised medium for rapid growth and bio assays with tobacco tissue cultures. Physiol. Plant. **15:** 473-97.

368 NAKAJIMA, R., YAMAZAKI, I. (1979) The mechanism of indole-3-acetic acid oxidation by horseradish peroxidase. J. biol. Chem. **254:** 872-8.

369 NARASIMHAM, S., HARPAZ, N., LONGMORE, G., CARVER, J.P., GREY, A.A., SCHACHTER, H. (1980) Control of glycoprotein synthesis: the purification by preparative paper electrophoresis in borate of glyco-peptides containing high mannose and complex oligosaccharide chains linked to asparagine. J. biol. Chem. **255:** 4876-84.

370 NARI, J., NOAT, G., RICARD, J., FRANCHINI, E., MOUSTACAS, A-M. (1983) Catalytic properties and tentative function of a cell wall β-gluco-syltransferase from soybean cells cultured in vitro. Plant Sci. Lett. **28:** 313-20.

371 NEGREL, J., SMITH, T.A. (1984) The phosphohydrolysis of hydroxycinna-moyl-coenzyme A thioesters in plant extracts. Phytochemistry **23:** 31-4.

372 NEUFELD, E.F., FEINGOLD, D.S., HASSID, W.Z. (1960) Phosphorylation of D-galactose and L-arabinose by extracts from Phaseolus aureus seedlings. J. biol. Chem. **235:** 906-9.

373 NEUFELD, E.F., FEINGOLD, D.S., ILVES, S.M., KESSLER, G., HASSID, W.Z. (1961) Phosphorylation of D-galacturonic acid by extracts from germinating seeds of Phaseolus aureus. J. biol. Chem. **236:** 3102-5.

374 NEUFELD, E.F., GINSBURG, V., PUTMAN, E.W., FANSHIER, D., HASSID, W.Z. (1957) Formation and interconversion of sugar nucleotides by plant extracts. Arch. Biochem. Biophys. **69:** 602-16.

375 NICKELL, L.G., MARETZKI, A. (1970) The utilization of sugars and starch as carbon sources by sugarcane cell suspension cultures. Plant Cell Physiol. **11:** 183-5.

376 NIEDERWIESER, A. (1975) Chromatography of amino acids and oligopeptides. In: Chromatography, a Laboratory Handbook of Chromatographic and Electrophoretic Methods, 3rd edn, ed. E. Heftmann. Van Nostrand Reinhold, New York, pp. 393-465.

377 NISHITANI, K., MASUDA, Y. (1981) Auxin-induced changes in the cell wall structure: changes in the sugar compositions, intrinsic viscosity, and molecular weight distributions of matrix polysaccharides of the epicotyl cell wall of Vigna angularis. Physiol. Plant. **52:** 482-94.

378 NISHITANI, K., MASUDA, Y. (1982/3) Acid pH-induced structural changes in cell wall xyloglucans in Vigna angularis epicotyl segments. Plant Sci. Lett. **28:** 87-94.

379 NISHITANI, K., MASUDA, Y. (1983) Auxin-induced changes in the cell wall xyloglucans. Effects of auxin on the two different subfractions of xyloglucans in the epicotyl cell wall of Vigna angularis. Plant Cell Physiol. **24:** 345-55.

380 NOCK, L.P., SMITH, C.J. (1987) Identification of polysaccharide hydrolases involved in autolytic degradation of Zea cell walls. Plant Physiol. **84:** 1044-50.

381 NORTHCOTE, D.H. (1972) Chemistry of the plant cell wall. Ann. Rev. Plant Physiol. **23:** 113-32.

382 NORTHCOTE, D.H. (1982) The synthesis and transport of some plant glycoproteins. Phil Trans. Roy, Soc. London **B300:** 195-206.

383 NORTHCOTE, D.H. (1985) Control of cell wall formation during growth. In: Biochemistry of Plant Cell Walls, ed. C.T. Brett and J.R. Hillman. Cambridge University Press, pp. 177-97.

384 NORTHCOTE, D.H., PICKETT-HEAPS, J.D. (1966) A function of the Golgi apparatus in polysaccharide synthesis and transport in the root-cap cells of wheat. Biochem. J. **98:** 159-67.

385 NOTHNAGEL, E.A., McNEIL, M., ALBERSHEIM, P., DELL, A. (1983) Host—pathogen relations. XXII. A galacturonic acid oligosaccharide from plant cell walls elicits phytoalexins. Plant Physiol. **71:** 916-26.

386 ODZUCK, W., KAUSS, H. (1972) Biosynthesis of pure araban and xylan. Phytochemistry **11:** 2489-94.

387 OFFORD, R.E. (1966) Electrophoretic mobilities of peptides on paper and their use in the determination of amide groups. Nature **211:** 591-3.

388 OLAITAN, S.A., NORTHCOTE, D.H. (1962) Polysaccharides of Chlorella pyrenoidosa. Biochem. J. **82:** 509-19.

389 O'NEILL, M.A., SELVENDRAN, R.R. (1980) Glycoproteins from the cell wall of Phaseolus coccineus. Biochem. J. **187:** 53-63.

390 OWENS, R.J., NORTHCOTE, D.H. (1980) The purification of potato lectin by affinity chromatography on a fetuin-Sepharose matrix. Phyto-chemistry **19**: 1861-2.

391 OWENS, R.J., NORTHCOTE, D.H. (1981) The location of arabinosyl : hydroxyproline transferase in the membrane system of potato tissue culture cells. Biochem. J. **195**: 661-7.

392 PAN, Y.T., KINDEL, P.K. (1977) Characterization of a particulate D-apiosyl- and D-xylosyltransferase from Lemna minor. Arch. Biochem. Biophys. **183**: 131-8.

393 PANAYOTATOS, N., VILLEMEZ, C.L. (1973) The formation of a β-(1$\rightarrow$4)-D-galactan chain catalysed by a Phaseolus aureus enzyme. Biochem. J. **133**: 263-71.

394 PATAKI, G. (1965) Einfluss der Schichtqualität auf die dünnschicht-chromatographische Trennung von Aminosäuren. J. Chromatogr. **17**: 580-4.

395 PAZUR, J.H. (1986) Neutral polysaccharides. In: Carbohydrate Analysis, a Practical Approach, eds. M.F. Chaplin and J.F. Kennedy. IRL Press, Oxford, pp. 55-96.

396 PECINA, R., BONN, G., BURTSCHER, E., BOBLETER. O. (1984) High-performance liquid chromatographic elution behaviour of alcohols, aldehydes, ketones, organic acids and carbohydrates on a strong cation-exchange stationary phase. J. Chromatogr. **287**: 245-58.

397 PEGG, G.F., VESSEY, J.C. (1973) Chitinase activity in Lycopersicon esculentum and its relationship to the in vivo lysis of Verticillium albo-atrum mycelium. Physiol. Plant Pathol. **3**: 207-22.

398 PENEL, C., GREPPIN, H. (1979) Effect of calcium on subcellular distribution of peroxidases. Phytochemistry **18**: 29-33.

399 PERRY, C.C., WILLIAMS, R.J.P., FRY, S.C. (1986) Cell wall biosynthe-sis during silicification of grass hairs. J. Plant Physiol. **126**: 437-48.

400 PHARMACIA FINE CHEMICALS (undated) Gel Filtration Theory and Practice. (Sephadex manufacturer's literature.)

401 PIERPOINT, W.S. (1985) Phenolics in food and feedstuffs: the pleasures and perils of vegetarianism. In: The Biochemistry of Plant Phenolics, eds. C.F. van Sumere and P.J. Lea. Oxford University Press, pp. 427-51.

402 PIERROT, H., VAN SCHADEWIJK, T.R., KLIS, F.M. (1982) Wall-bound invertase and other cell wall hydrolases are not correlated with elongation rate in bean hypocotyls (Phaseolus vulgaris L.). Z. Pflanzenphysiol. **106**: 367-70.

403 PIERROT, H., VAN WIELINK, J.E. (1977) Localization of glycosidases in the wall of living cells from cultured Convolvulus arvensis tissue. Planta **137**: 235-42.

404 PLASTOW, G.S., BORDER, P.M., HINTON, J.C.D., SALMOND, G.P.C. (1986) Molecular cloning of pectinase genes from Erwinia carotovera ssp. carotovera (Strain SCRI193). Symbiosis **2**: 115-22.

405 POSTLE, A.D., BLOXHAM, D.P. (1980) The use of tritiated water to

measure absolute rates of hepatic glycogen synthesis. Biochem. J. **192:** 65-73.

406 POWELL, D.A., MORRIS, E.R., GIDLEY, M.J., REES, D.A. (1982) Conformations and interactions of pectins. II. Influence of residue sequence on chain association in calcium pectate gels. J. mol. Biol. **155:** 517-31.

407 PRESTON, R.D. (1979) Polysaccharide conformation and cell wall function. Ann. Rev. Plant Physiol. **30:** 55-78.

408 RAGSTER, L., CHRISPEELS, M.J. (1979) Azocoll digesting proteases in soybean leaves. Characteristics and changes during leaf maturation and senescence. Plant Physiol. **64:** 857-62.

409 RANDERATH, K. (1970) An evaluation of film detection methods for weak β-emitters, particularly tritium. Anal. Biochem. **34:** 188-205.

410 RAY, P.M. (1962) Cell wall synthesis and cell elongation in oat coleoptile tissues. Am. J. Bot. **49:** 928-39.

411 RAY, P.M. (1980) Cooperative action of β-glucan synthetase and UDP-xylose xylosyltransferase of Golgi membranes in the synthesis of xyloglucan-like polysaccharide. Biochim. biophys. Acta **629:** 431-44.

412 READ, S.M., NORTHCOTE, D.H. (1981) Minimization of variation in the response to different proteins of the Coomassie Blue G dye-binding assay for protein. Anal. Biochem. **116:** 53-64.

413 REBERS, P.A., WESSMAN, G.E., ROBYT, J.F. (1986) A thin-layer chromatographic method for analysis of amino sugars in polysaccharide hydrolyzates. Carbohydr. Res. **153:** 132-5.

414 REDGWELL, R.J., SELVENDRAN, R.R. (1986) Structural features of cell-wall polysaccharides of onion Allium cepa. Carbohydr. Res. **157:** 183-99.

415 REES, D.A. (1977) Polysaccharide Shapes. Chapman & Hall, London.

416 REES, D.A, WRIGHT, N.J. (1969) Molecular cohesion in plant cell walls. Methylation analysis of pectic polysaccharides from the cotyledons of white mustard. Biochem. J. **115:** 431-9.

417 REID, J.S.G. (1984) Cell wall storage carbohydrates in seeds — biochemistry of the seed "gums" and "hemicelluloses". Adv. bot. Res. **11:** 125-55.

418 REID, J.S.G. (1985) Structure and function in legume-seed polysaccharides. In: Biochemistry of Plant Cell Walls, eds. C.T. Brett and J.R. Hillman. Cambridge University Press, pp. 259-68.

419 REXOVÁ-BENKOVÁ, Ľ, OMELKOVÁ, J., FILKA, K., KOCOUREK, J. (1983) Selective isolation of endo-D-galacturonanase of Aspergillus niger based on interaction with tri(D-galactosiduronic acid) covalently bound to poly(hydroxyalkyl methacrylate). Carbohydr. Res. **122:** 269-81.

420 RICHMOND, P.A. (1984) Cellulose synthesis inhibition and patterns of cell wall deposition. In: Abstracts of the 3rd Cell Wall Meeting, organisers H. Meier, A.J. Buchala, U. Ryser and J. Wattendorf. Université de Fribourg, p. 77.

421 RIDE, J.P. (1975) Lignification in wounded wheat leaves in response to fungi and its possible rôle in resistance. Physiol. Plant Pathol.

312

5: 125-34.

422 RIDGE, I., OSBORNE, D.J. (1971) Role of peroxidase when hydroxy-proline-rich protein in plant cell walls is increased by ethylene. Nature new Biol. **229:** 205-8.

423 ROBERTS, R.M., LOEWUS, F.A. (1973) The conversion of D-glucose-6-^{14}C to cell wall polysaccharide material in Zea mays in presence of high endogenous levels of myoinositol. Plant Physiol. **52:** 646-50.

424 ROBINSON, D.G. (1981) The assembly of polysaccharide fibrils. In: Plant Carbohydrates II [Encyclopedia of Plant Physiology, N.S., vol. **13B**], eds. W. Tanner and F.A. Loewus. Springer, Berlin, pp. 25-8.

425 ROBINSON, D.G., ANDREAE, M., SAUER, A. (1985) Hydroxyproline-rich glycoprotein biosynthesis: a comparison with that of collagen. In: Biochemistry of Plant Cell Walls, eds. C.T. Brett and J.R. Hillman. Cambridge University Press, pp. 155-76.

426 ROBINSON, D.G., EISINGER, W.R., RAY, P.M. (1976) Dynamics of Golgi system in wall matrix polysaccharide synthesis and secretion by pea cells. Ber. deutsch. Bot. Ges. **89:** 147-61.

427 ROLAND, J.-C., VIAN, B. (1981) Use of purified endopolygalacturonase for a topological study of elongating cell walls at the ultra-structural level. J. Cell Sci. **48:** 333-43.

428 ROMBOUTS, F.M., THIBAULT, J.F. (1986) Feruloylated pectin substances from sugar-beet pulp. Carbohydr. Res. **154:** 177-87.

429 ROSEN, H. (1957) A modified ninhydrin colorimetric analysis for amino acids. Arch. Biochem. Biophys. **67:** 10-15.

430 RUBERY, P.H. (1972) Studies on IAA oxidation by liquid medium from crown gall tissue culture cells: the role of malic acid and related compounds. Biochim. Biophys. Acta **261:** 21-34.

431 RUFFINI, G. (1965) Thin-layer chromatography of 2,4-dinitrophenyl-hydrazones of aromatic aldehydes and ketones. J. Chromatogr. **17:** 483-7.

432 RYAN, C.A., BISHOP, P., PEARCE, G., DARVILL, A.G., McNEIL, M., ALBERSHEIM, P. (1981) A sycamore cell wall polysaccharide and a chemically related tomato leaf polysaccharide possess similar proteinase inhibitor-inducing activities. Plant Physiol. **68:** 616-18.

433 SADAVA, D., CHRISPEELS, M.J. (1973) Hydroxyproline-rich cell wall protein (extensin): role in the cessation of elongation in excised pea epicotyls. Devel. Biol. **30:** 49-55.

434 SAKURAI, N., NEVINS, D.J., MASUDA, Y. (1977) Auxin and hydrogen ion-induced wall loosening and cell extension in Avena coleoptile segments. Plant Cell Physiol. **18:** 371-80.

435 SANGER, M.P., LAMPORT, D.T.A. (1983) A microapparatus for liquid hydrogen fluoride solvolysis: sugar and amino sugar composition of Erysiphe graminis and Triticum aestivum cell walls. Anal. Biochem. **128:** 66-70.

436 SARKANEN, K.V., LUDWIG, C.H. (eds.) (1971) Lignins. Occurrence, Formation, Structure and Reactions. Wiley, New York.

437 SCALBERT, A., MONTIES, B., LALLEMAND, J.-Y., ROLANDO, C. (1985) Ether

linkage between phenolic acids and lignin fractions from wheat straw. Phytochemistry **24**: 1359-62.

438 SCHWARZ, K. (1973) A bound form of silicon in glycosaminoglycans and polyuronides. Proc. natl Acad. Sci., U.S. **70**: 1608-12.

439 SCOPES, R.K. (1982) Protein Purification: Principles and Practice. Springer, Berlin.

440 SCOTT, J.E. (1960) Aliphatic ammonium salts in the assay of acidic polysaccharides from tissues. Meth. Biochem. Anal. **8**: 145-97.

441 SCOTT, J.E. (1965) Fractionation by precipitation with quaternary ammonium salts. In: Methods in Carbohydrate Chemistry, Vol. 5, ed. R.L. Whistler. Academic Press, New York, pp. 38-44.

442 SELVENDRAN, R.R. (1985) Developments in the chemistry and bio-chemistry of pectic and hemicellulosic polymers. J. Cell Sci. **Suppl** 2: 51-88.

442a SELVENDRAN, R.R., O'NEILL, M.A. (1987) Isolation and analysis of cell walls from plant material. In: Methods of Biochemical Analysis, Vol. 32, ed. D. Glick. Wiley, pp. 25-153.

443 SELVENDRAN, R.R., STEVENS, B.J.H. (1987) Applications of mass spectrometry for the examination of pectic polysaccharides. In: Gas Chromatography / Mass Spectrometry, eds. H.F. Linskens and J.F. Jackson. [Modern Methods in Plant Analysis, New Series, Vol. 3.] Springer, Berlin, pp. 23-46.

444 SELVENDRAN, R.R., STEVENS, B.J.H., O'NEILL, M.A. (1985) Developments in the isolation and analysis of cell walls from edible plants. In: Biochemistry of Plant Cell Walls, eds. C.T. Brett and J.R. Hillman. Cambridge University Press, pp. 39-78.

445 SEXTON, R., DURBIN, M.L., LEWIS, L.N., THOMSON, W.W. (1980) Use of cellulase antibodies to study leaf abscission. Nature **283**: 873-4.

446 SHANNON, L.M., KAY, E., LEW, J.Y. (1966) Peroxidase isoenzymes from horseradish roots. I: Isolation and physical properties. J. biol. Chem. **241**: 2166-72.

447 SHARP, J.K., McNEIL, M., ALBERSHEIM, P. (1984) The primary structures of one elicitor-active and seven elicitor-inactive hexa(β-D-gluco-pyranosyl)-D-glucitols isolated from the mycelial walls of Phytoph-thora megasperma f.sp. glycinea. J. biol. Chem. **259**: 11 321-36.

448 SHARP, J.K., ALBERSHEIM, P., OSSOWSKI, P., PILOTTI, Å., GAREGG, P., LINDBERG, B. (1984) Comparison of the structures and elicitor activities of a synthetic and a mycelial-wall-derived hexa(β-D-glucopyranosyl)-D-glucitol. J. biol. Chem. **259**: 11 341-5.

449 SHIBUYA, N. (1984) Phenolic acids and their carbohydrate esters in rice endosperm cell walls. Phytochemistry **23**: 2233-7.

450 SHINKLE, J.R., JONES, R.L. (1986) Cell wall peroxidases and stem elongation in Cucumis seedling hypocotyl: a rapid effect of blue light. In: Cell Walls '86 [Proceedings of the 4th Cell Wall Meeting, Paris, 10-12 Sept, 1986], eds. B. Vian, D. Reis and R. Goldberg. Université P. & M. Curie, École Normale Supérieure, Paris, pp. 190-3.

451 SHOEMAKER, H.E., HARVEY, P.J., BOWEN, R.M., PAINTER, J.M. (1985) On the mechanism of enzymatic lignin breakdown. FEBS Lett. **183**: 7-12.

314

452 SIEGEL, S.M. (1969) Evidence for the presence of lignin in moss gametophytes. Am. J. Bot. **56:** 175-9.

453 SINGH, B.D., THOMAS, E.T., HARVEY, B.L. (1974) Effects of gibberellic acid on cell suspension cultures of higher plants. Ind. J. exp. Biol. **12:** 213-15.

454 SMART, C.C., TREWAVAS, A. (1983) Abscisic-acid-induced turion formation in Spirodela polyrrhiza L. I: Production and development of the turion. Plant Cell Env. **6:** 507-14.

455 SMIDSRØD, O., HAUG, A., LARSON, B. (1966) The influence of pH on the rate of hydrolysis rate of hydrolysis of acidic polysaccharides. Acta Chem. Scand. **20:** 1026-34.

456 SMITH, I. (ed.) (1960) Chromatographic and Electrophoretic Techniques. Vol. I, chromatography, 2nd edn. Heinemann, London.

457 SMITH, I. (ed.) (1968) Chromatographic and Electrophoretic Techniques. Vol. II, zone electrophoresis. Heinemann, London.

458 SMITH, J.J., MULDOON, E.P., LAMPORT, D.T.A. (1984) Isolation of extensin precursors by direct elution of intact tomato cell suspension cultures. Phytochemistry **23:** 1233-40.

459 SMITH, J.J., MULDOON, E.P., WILLARD, J.J., LAMPORT, D.T.A. (1986) Tomato extensin precursors P1 and P2 are highly periodic structures. Phytochemistry. **25:** 1021-30.

460 SMITH, M.M., AXELOS, M., PÉAUD-LENOËL, C. (1976) Biosynthesis of mannan and mannolipids from GDP-Man by membrane fractions of sycamore cell cultures. Biochimie **58:** 1195-211.

461 SMITH, M.M., O'BRIEN, T.P. (1979) Distribution of autofluorescence and esterase and peroxidase activities in the epidermis of wheat roots. Aust. J. Plant Physiol. **6:** 201-19.

462 SOJAR, H.T., BAHL, O.P. (1987) Chemical deglycosylation of glycoproteins. Meth. Enzymol. **138:** 341-50.

463 SOMOGYI, M. (1952) Notes on sugar determination. J. biol. Chem. **195:** 19-23.

464 SOWDEN, J.C. (1952) 6-C^{14}-D-Glucose and 6-C^{14}-D-glucuronolactone. J. Am. Chem. Soc. **74:** 4377-9.

465 SPELLMAN, M.W., McNEIL, M., DARVILL, A.G., ALBERSHEIM, P. (1983) Characterization of a structurally complex heptasaccharide isolated from the pectic polysaccharide rhamnogalacturonan II. Carbohydr. Res. **122:** 131-53.

466 SPIK, G., SIX, P., MONTREUIL, J. (1979) Chemical and enzymic degradations of nucleoside mono- and diphosphate sugars. I: Determination of the degradation rate during the glycosyltransferase assays. Biochim. Biophys. Acta **584:** 203-15.

467 SPIRO, R.G. (1976) Isolation of glycopeptides from glycoproteins by proteolytic digestion. In: Methods in Carbohydrate Chemistry, Vol. 7, eds. R.L. Whistler and J.N. BeMiller. Academic Press, Orlando, Florida, pp. 185-90.

468 STAFSTROM, J.P., STAEHELIN, L.A. (1986) Cross-linking patterns in

salt-extractable extensin from carrot cell walls. Plant Physiol. **81**: 234-41.

469 STAFSTROM, J.P., STAEHELIN, L.A. (1986) The role of carbohydrate in maintaining extensin in an extended conformation. Plant Physiol. **81**: 242-6.

470 STAFSTROM, J.P., STAEHELIN, L.A. (1987) A second extensin-like hydroxyproline-rich glycoprotein from carrot cell walls. Plant Physiol. **84**: 820-5.

471 STAHL, E. (ed.) (1969) Thin-Layer Chromatography: a Laboratory Handbook, 2nd edn. Springer, New York.

472 STEPHAN, D., VAN HUYSTEE, R.B. (1981) Some aspects of peroxidase synthesis by cultured peanut cells. Z. Pflanzenphysiol. **101**: 313-21.

473 STEPHEN, A.M., CUMMINGS, J.H. (1980) Mechanism of action of dietary fibre in the human colon. Nature **284**: 283-4.

474 STEPHENS, G.J., WOOD, R.K.S. (1974) Release of enzymes from cell walls by an endopectate-trans-eliminase. Nature **251**: 358.

475 STEVENS, B.J.H., SELVENDRAN, R.R. (1984) Pectic polysaccharides of cabbage (Brassica oleracea). Phytochemistry **23**: 107-15.

476 STEVENSON, T.T., McNEIL, M., DARVILL, A.G., ALBERSHEIM, P. (1986) Structure of plant cell walls. XVIII An analysis of the extracellular polysaccharides of suspension-cultured sycamore cells. Plant Physiol. **80**: 1012-19.

477 STICHER, L., PENEL, C., GREPPIN, H. (1981) Calcium requirement for the secretion of peroxidases by plant cell suspensions. J. Cell Sci. **48**: 345-53.

478 STÖCKIGT, J., ZENK, M.H. (1975) Chemical synthesis and properties of hydroxycinnamoyl coenzyme A derivatives. Z. Naturforsch. **30c**: 352-8.

479 STODDART, J.L., WILLIAMS, P.D. (1980) Interaction of [^{3}H]gibberellin A_1 with a sub-cellular fraction from lettuce (Lactuca sativa L.) hypocotyls. Planta **148**: 485-90.

480 STODDART, R.W, BARRETT, A.J., NORTHCOTE, D.H. (1967) Pectic polysaccharides of growing plant tissues. Biochem. J. **102**: 194-205.

481 STODDART, R.W., NORTHCOTE, D.H. (1967) Separation and measurement of microgram amounts of radioactive polysaccharides in metabolic experiments. Biochem. J. **105**: 61-3.

482 STONE, J., BLUNDELL, M. (1951) Rapid micromethod for alkaline nitrobenzene oxidation of lignin and determination of aldehydes. Anal. Chem. **23**: 771-8.

483 STRAHM, A., AMADÒ, R., NEUKOM, H. (1981) Hydroxyproline-galactoside as a protein—polysaccharide linkage in a water soluble arabinogalactan-peptide from wheat endosperm. Phytochemistry **20**: 1061-3.

484 STREET, H.E. (1977) Cell (suspension) cultures — techniques. In: Plant Tissue and Cell Culture, ed. H.E. Street. Blackwell, Oxford, pp. 61-102.

485 SZU, S.C., ZON, G., SCHNEERSON, R., ROBBINS, J.B. (1986) Ultrasonic irradiation of bacterial polysaccharides. Characterization of the

depolymerized products and some applications of the process. Carbo-hydr. Res. **152**: 7-20.

486 TAIZ, L. (1984) Plant cell expansion: regulation of cell wall mechanical properties. Annu. Rev. Plant Physiol. **35**: 585-657.

487 TAKASAKI, S., MIZUOCHI, T., KOBATA, A., (1982) Hydrazinolysis of asparagine-linked sugar chains to produce free oligosaccharides. Meth. Enzymol. **83**: 263-77.

488 TAKEUCHI, Y., AMINO, S. (1984) Analysis of UDP-sugar content in cucumber cotyledons in relation to growth rate. Plant Cell Physiol. **25**: 1589-93.

489 TAKEUCHI, Y., KOMAMINE, A. (1980) Turnover of cell wall polysaccha-rides of a Vinca rosea suspension culture. I Synthesis and degradation of cell wall components. Physiol. Plant. **48**: 271-7.

490 TALMADGE, K.W., KEEGSTRA, K., BAUER, W.D., ALBERSHEIM, P. (1973) The structure of plant cell walls. I. The macromolecular components of the walls of suspension-cultured sycamore cells with a detailed analysis of the pectic polysaccharides. Plant Physiol. **51**: 158-73.

491 TAN, A.W.H. (1979) A simplified method for the preparation of pure UDP[^{14}C]glucose. Biochim. Biophys. Acta **582**: 543-7.

492 TANAKA, K., NAKATSUBO, F., HIGUCHI, T. (1979) Reactions of guaiacyl-glycerol-β-guaiacyl ether with several sugars. II. Reactions of quinonemethide with pyranohexoses. Mokuzai Gakkaishi **25**: 653-9.

493 TARTAKOFF, A.M. (1983) Perturbation of vesicular traffic with the carboxylic ionophore monensin. Cell **32**: 1026-8.

494 TEPFER, M., CLELAND, R.E. (1979) A comparison of acid-induced cell wall loosening in Valonia ventricosa and in oat coleoptiles. Plant Physiol. **63**: 898-902.

495 TEPFER, M., TAYLOR, I.E.P. (1981) The permeability of plant cell walls as measured by gel filtration chromatography. Science **213**: 761-3.

496 TERRY, M.E., BONNER, B.A. (1980) An examination of centrifugation as a method of extracting an extracellular solution from peas, and its use for the study of indoleacetic acid-induced growth. Plant Physiol. **66**: 321-5.

497 TERRY, M.E., JONES, R.L., BONNER, B.A. (1981) Soluble cell wall poly-saccharides released from pea stems by centrifugation. I Effect of auxin. Plant Physiol. **68**: 531-7.

498 THELEN, M.P., DELMER, D.P. (1986) Gel-electrophoretic separation, detection, and characterization of plant and bacterial UDP-glucose glucosyltransferases. Plant Physiol. **81**: 913-18.

499 THIMANN, K.V., SCHNEIDER, C.L. (1938) Differential growth in plant tissues. Am. J. Bot. **25**: 627-41.

500 THOMAS, R.J., BEHRINGER, F.J., LOMBARD, C.S., SPARKOWSKI, J.J. (1984) Effects of auxin on wall polysaccharide composition and enzyme activity during extension-growth of Pellia (Bryophyta). Physiol. Plant. **60**: 502-6.

501 THORNBER, J.P., NORTHCOTE, D.H. (1962) Changes in the chemical

composition of a cambial cell during its differentiation into xylem and phloem tissues in trees. 3. Xylan, glucomannan and α-cellulose fractions. Biochem J. **82:** 340-6.

502 THOTAKURA, N.T., BAHL, O.P. (1987) Enzymatic deglycosylation of glycoproteins. Meth. Enzymol. **138:** 350-9.

503 TKOTZ, N., STRACK, D. (1980) Enzymatic synthesis of sinapoyl-L-malate from 1-sinapoylglucose and L-malate by a protein preparation from Raphanus sativus cotyledons. Z. Naturforsch. **35c:** 835-7.

504 TOMIMURA, Y., YOKOI, T., TERASHIMA, N. (1979) Heterogeneity in formation of lignin. IV. Various factors which influence the degree of condensation at position 5 of guaiacyl necleus. Mokuzai Gakkaishi **25:** 743-8.

505 TONG, C.B., LABAVITCH, J.M., YANG, S.F. (1986) The induction of ethylene production from pear cell culture by cell wall fragments. Plant Physiol. **81:** 929-30.

506 TOUCHSTONE, J.C., DOBBINS, M.F. (1983) Practice of Thin Layer Chromatography, 2nd edn. Wiley, New York.

507 TRAN THANH VAN, K., TOUBART, P., COUSSON, A., DARVILL, A.G., GOLLIN, D.J., CHELF, P., ALBERSHEIM, P. (1985) Manipulation of the morphogenetic pathways of tobacco explants by oligosaccharins. Nature **314:** 615-17.

508 TUERENA, C.E., TAYLOR, A.J., MITCHELL, J.R. (1982) Evaluation of a method for determining the free carboxyl group distribution in pectins. Carbohydr. Polym. **2:** 193-203.

509 UEKI, K. (1978) Control of phosphatase release from cultured tobacco cells. Plant Cell Physiol. **19:** 385-92.

510 UPDEGRAFF, D.M. (1969) Semi-micro determination of cellulose in biological materials. Anal. Biochem. **32:** 420-4.

511 VALLÉE, J.C., VANSUYT, G., NIGREL, J., PEDRIZET, E., PREVOST, J. (1983) Mise en évidence d'amines liées à des structures cellulaires chez Nicotiana tabacum et Lycopersicon esculentum. Physiol. Plant. **57:** 143-8.

512 VALENT, B.S., DARVILL, A.G., McNEIL, M., ROBERTSEN, B.K., ALBERSHEIM, P. (1980) A general and sensitive chemical method for sequencing the glycosyl residues of complex carbohydrates. Carbohydr. Res. **79:** 165-92.

513 VAN DEN BERG, B.M., VAN HUYSTEE, R.B. (1984) Rapid isolation of plant peroxidase. Purification of peroxidase a from Petunia. Physiol. Plant. **60:** 299-304.

514 VAN DER WILDEN, W., CHRISPEELS, M.J. (1983) Characterization of the isózymes of α-mannosidase located in the cell wall, protein bodies, and endoplasmic reticulum of Phaseolus vulgaris cotyledons. Plant Physiol. **71:** 82-7.

515 VAN DER WILDEN, W., SEGERS, J.H.L., CHRISPEELS, M.J. (1983) Cell walls of Phaseolus vulgaris leaves contain the Azocoll-digesting proteinase. Plant Physiol. **73:** 576-8.

516 VAN GIJSEGM, F., TOUSSAINT, A., SCHOONEJANS, E. (1985) In vivo

cloning of the pectin lyase and cellulase genes of <u>Erwinia</u> <u>chrysanthemi</u>. EMBO J. **4**: 787-92.

517 VAN HUYSTEE, R.B. (1976) A study of peroxidase synthesis by means of double labelling and affinity chromatography. Can. J. Bot., **54**: 876-80.

518 VAN HUYSTEE, R.B. (1987) Some molecular aspects of plant peroxidase biosynthetic studies. Ann. Rev. Plant Physiol. **38**: 205-19.

519 VAN OVERBEEK, J. (1935) The growth hormone and the dwarf type of growth in corn. Proc. natl Acad. Sci. (US) **21**: 292-9.

520 VAN VOLKENBURGH, E., CLELAND, R.E. (1980) Proton excretion and cell expansion in bean leaves. Planta **148**: 273-8.

521 VAN VOLKENBURGH, E., HUNT, S., DAVIES, W.J. (1983) A simple instrument for measuring cell-wall extensibility. Ann. Bot. **51**: 669-72.

522 VENVERLOO, C.J. (1969) The lignin of <u>Populus nigra</u> L. cv. 'Italica'. A comparative study of the lignified structures in tissue cultures and the tissues of the tree. Acta Bot. Néerl. **18**: 241-314.

523 VERACHTERT, H., BASS, S.T., WILDER, J.K., HANSEN, R.G. (1966) The separation of nucleoside diphosphate sugars and related nucleotides by ion-exchange paper chromatography. Meth. Enzymol. **8**: 111-15.

524 VERHAAR, L.A.T., KUSTER, B.F.M. (1981) Liquid chromatography of sugars on silica-based stationary phases. J. Chromatogr. **220**: 313-28.

525 VERMA, D.C., DOUGAL, D.K. (1979) Biosynthesis of <u>myo</u>-inositol and its role as a presursor of cell wall polysaccharides in suspension cultures of wild-carrot cells. Planta **146**: 55-62.

526 VERMA, D.P.S., MACLACHLAN, G.A., BYRNE, H., EWINGS, D. (1975) Regulation and <u>in vitro</u> translation of mRNA for cellulase from auxin-treated pea epicotyls. J. biol. Chem. **250**: 1019-26.

527 VIAN, B., ROLAND, J.-C. (1987) The helicoidal cell wall as a time register. New Phytol. **105**: 345-58.

528 VILLEMEZ, C.L., SWANSON, A.L., HASSID, W.Z. (1966) Properties of a polygalacturonic acid-synthesizing enzyme system from <u>Phaseolus</u> <u>aureus</u> seedlings. Arch. Biochem. Biophys. **116**: 446-52.

529 VOLK, R.J., WEINTRAUB, R.L. (1958) Microdetermination of silicon in plants. Anal. Chem. **30**: 1011-14.

530 VON SACHS, J. (1865) Handbuch der Experimentalphysiologie der Pflanzen. Engelmann, Leipzig.

531 VORAGEN, A.G.J., SCHOLS, H.A., DE VRIES, J.A., PILNIK, W. (1982) High performance liquid chromatographic analysis of uronic acids and oligogalacturonic acids. J. Chromatogr. **244**: 327-36.

532 WADA, S., RAY. P.M. (1978) Matrix polysaccharides of oat coleoptile cell walls. Phytochemistry **17**: 923-31.

533 WAEGHE, T.J., DARVILL, A.G., McNEIL, M., ALBERSHEIM, P. (1983) Determination, by methylation analysis, of the glycosyl-linkage compositions of microgram quantities of complex carbohydrates. Carbohydr. Res. **123**: 281-304.

534 WALDRON, K.W., BRETT, C.T. (1983) A glucuronyltransferase involved in glucuronoxylan synthesis in pea (Pisum sativum) epicotyls. Biochem. J. **213:** 115-22.

535 WALDRON, K.W., BRETT, C.T. (1985) Interactions of enzymes involved in cell wall heteropolysaccharide biosynthesis. In: Biochemistry of Plant Cell Walls, eds. C.T. Brett and J.R. Hillman. Cambridge University Press, pp. 79-97.

536 WALKER-SIMMONS, M., HADWIGER, L., RYAN, C.A. (1983) Chitosans and pectic polysaccharides both induce the accumulation of the antifungal phytoalexin pisatin in pea-pods and anti-nutrient proteinase inhibitors in tomato leaves. Biochem. Biophys. Res. Comm. **110:** 194-9.

537 WALKER-SIMMONS, M., JIN, D., WEST, C.A., HADWIGER, L., RYAN, C.A. (1984) Comparison of proteinase inhibitor-inducing activities and phytoalexin elicitor activities of a pure fungal endopolygalact-uronase, pectic fragments and chitosans. Plant Physiol. **76:** 833-6.

538 WALLNER, S.J., NEVINS, D.J. (1974) Changes in cell walls associated with cell separation in suspension cultures of Paul's Scarlet rose. J. exp. Bot. **25:** 1020-9.

539 WALTER, D.J., EASTWOOD, M.A., BRYDON, W.G. (1986) An experimental design to study colonic fibre fermentation in the rat: the duration of feeding. Brit. J. Nut. **55:** 465-79.

540 WATANABE, T., KAIZU, S., KOSHIJIMA, T. (1986) Binding sites of carbo-hydrate moieties toward lignin in "lignin—carbohydrate complex" from Pinus densiflora wood. Chem. Letts. **1986:** 1871-4.

541 WATERKEYN, L., BIENFAIT, A. PEETERS, A. (1982) Callose et silice épidermiques: rapports avec le transpiration cuticulaire. La Cellule **73:**267-87.

542 WEIGEL, H. (1963) Paper electrophoresis of carbohydrates. Adv. Carbohydr. Chem. **18:** 61-96.

543 WEINSTEIN, L.I., ALBERSHEIM, P. (1979) Structure of plant cell walls. IX Purification of a wall-degrading arabanase and an arabinosidase from Bacillus subtilis. Plant Physiol. **63:** 425-32.

544 WELLINDER, K.G. (1976) Covalent structure of the glycoprotein horse-radish peroxidase. (EC 1.11.1.7). FEBS Lett. **72:** 19-23.

545 WELLS, G.B., KONTOYIANNIDOU, V., TURCO, S.J., LESTER, R.L. (1982) Resolution of acetylated oligosaccharides by reverse-phase high-pressure liquid chromatography. Meth. Enzymol. **83:** 132-7.

546 WELLS, G.B., TURCO, S.J., HANSEN, B.A., LESTER, R.L. (1982) Resolution of dolichylpyrophosphoryl oligosaccharides by high-pressure liquid chromatography. Meth. Enzymol. **83:** 137-9.

547 WEST, C.A. (1981) Fungal elicitors of the phytoalexin response in higher plants. Naturwissenschaften **68:** 447-57.

548 WHISTLER, R.L., BeMILLER, J.N. (1958) Alkaline degradation of poly-saccharides. Adv. Carbohydr. Chem. **13:** 289-329.

549 WHISTLER, R.L., FEATHER, M.S. (1965) Hemicellulose. Extraction from

annual plants with alkaline solution. In: Methods in Carbohydrate Chemistry, Vol. 5, ed. R.L. Whistler. Academic Press, New York, pp. 144-5.

550 WILLIAMS, S.E., BENNETT, A.B. (1982) Leaf closure in the Venus fly trap: an acid growth response. Science **218**: 1120-2.

551 WILLIAMS, V.M, PORTER, L.J., HEMINGWAY, R.W. (1983) Molecular weight profiles of proanthocyanidin polymers. Phytochemistry **22**: 569-72.

552 WILSON, L.G., FRY, J.C. (1986) Extensin — a major cell wall glycoprotein. Plant Cell Env. **9**: 239-60.

553 WING, R.E., BeMILLER, J.N. (1972) Qualitative thin-layer chromatography. In: Methods in Carbohydrate Chemistry, Vol. 6, eds. R.L. Whistler and J.N. BeMiller. Academic Press, New York, pp. 42-53.

554 WOLFROM, M.L., FRANKS, N.E. (1965) Partial acid hydrolysis. Preparation of polymer-homologous oligosaccharides; xyloglycoses from xylan. In: Methods in Carbohydrate Chemistry, Vol. 5, ed. R.L. Whistler. Academic Press, New York, pp. 276-80.

555 WONG, J.R., SUSSEX, I.M. (1980) Isolation of abscisic acid-resistant variants from tobacco cell cultures. I Physiological bases for selection. Planta **148**: 97-102.

556 WONG, Y-S., FINCHER, G.B., MACLACHLAN, G.A. (1977) Cellulases can enhance β-glucan synthesis. Science. **195**: 679-81.

557 WOOD, P.J. (1981) The use of dye-polysaccharide interactions in β-D-glucanase assay. Carbohydr. Res. **94**: C19-C23.

558 WOOD, P.J., SIDDIQUI, I.R. (1971) Determination of methanol and its application to the measurement of pectin ester content and pectin methyl esterase activity. Anal. Biochem. **39**: 418-28.

559 WOOD, T.M., McCRAE, S.I. (1972) The purification and properties of the C_1 component of Trichoderma koningii cellulase. Biochem. J. **128**: 1183-92.

560 WOOD, T.M., McCRAE, S.I. (1978) The cellulase of Trichoderma koningii. Purification and properties of some endoglucanase components with special reference to their action on cellulose in synergism with the cellobiohydrolase. Biochem. J. **171**: 61-72.

561 WOOD, T.M., McCRAE, S.I. (1986) Studies of two low-molecular-weight endo-$(1 \rightarrow 4)$-β-D-xylanases constitutively synthesised by the cellulolytic fungus Trichoderma koningii. Carbohydr. Res. **148**: 321-30.

562 WOOD, T.M., McCRAE, S.I. (1986) Purification and properties of a cellobiohydrolase from Penicillium pinophilum. Carbohydr. Res. **148**: 331-44.

563 WOODING, F.B.P. (1968) Radioautographic and chemical studies of incorporation into sycamore vascular tissue walls. J. Cell Sci. **3**: 71-80.

564 WOODWARD, J.R., FINCHER, J.B. (1982) Purification and chemical properties of two 1,3;1,4-β-glucan endohydrolases from germinating barley. Eur. J. Biochem. **121**: 663-69.

565 WOODWARD, J.R., FINCHER, J.B. (1982) Substrate specificities and kinetic properties of two $(1 \rightarrow 3),(1 \rightarrow 4)$-β-D-glucan endo-hydrolases

from germinating barley (<u>Hordeum vulgare</u>). Carbohydr. Res. **106:** 111-22.

566 WRIGHT, K., NORTHCOTE, D.H. (1975) An acidic oligosaccharide from maize slime. Phytochemistry **14:** 1793-8.

567 YAMAKI, S., KAKIUCHI, N. (1979) Changes in hemicellulose-degrading enzymes during development and ripening of Japanese pear fruit. Plant Cell Physiol. **20:** 301-9.

568 YAMAMOTO, E. TOWERS, G.H.N. (1985) Cell wall bound ferulic acid in barley seedlings during development and its photoisomerization. J. Plant Physiol. **117:** 441-9.

569 YAMAMOTO, R., NEVINS, D.J. (1981) Coleoptile growth-inducing capacities of exo-β-(1$\rightarrow$3)-glucanases from fungi. Physiol. Plant. **51:** 118-22.

570 YAMAOKA, T., CHIBA, N. (1983) Changes in the coagulating ability of pectin during the growth of soybean hypocotyls. Plant Cell Physiol. **24:** 1281-90.

571 YAMAOKA, T., TSUKADA, K., TAKAHASHI, H., YAMAUCHI, N. (1983) Purification of a cell-wall bound pectin-gelatinizing factor and examination of its identity with pectin methylesterase. Bot. Mag. (Tokyo) **96:** 139-44.

572 YAMASHITA, K., MIZUOCHI, T., KOBATA, A. (1982) Analysis of oligosaccharides by gel filtration. Meth. Enzymol. **83:** 105-26.

573 YAMAZAKI, N., FRY, S.C., DARVILL, A.G., ALBERSHEIM, P. (1983) Host—pathogen interactions. XXIV: Fragments isolated from suspension-cultured sycamore cell walls inhibit the ability of the cells to incorporate [^{14}C]leucine into proteins. Plant Physiol. **72:** 864-9.

574 YEUNG, E.C. (1984) Autoradiography. In: Cell Culture and Somatic Cell Genetics of Plants, Vol. 1., ed. I.K. Vasil. Academic Press, Orlando, Florida, pp. 778-84.

575 YORK, W.S., DARVILL, A.G., ALBERSHEIM, P. (1984) Inhibition of 2,4-dichlorophenoxyacetic acid-stimulated elongation of pea stem segments by a xyloglucan oligosaccharide. Plant Physiol. **75:** 295-7.

576 YORK, W.S., OATES, J.E., VAN HALBEEK, H., ALBERSHEIM, P., TILLER, P.R., DELL, A. (1988) Location of the O-acetyl substituents on a nonasaccharide repeating unit of sycamore extracellular xyloglucan. Carbohydr. Res. (in press).

577 YOUNG, R.A., ROWELL, R.M. (eds.) (1986) Cellulose: Structure, Modification, and Hydrolysis. Wiley, New York.

578 ZENK, M.H. (1979) Recent work on cinnamoyl CoA derivatives. Recent Adv. Phytochem. **12:** 139-76.

579 ZINC, R.T., CHATTERJEE, A.K. (1985) Cloning and expression in <u>Escherichia</u> coli of pectinase genes from <u>Erwinia</u> <u>carotovora</u> ssp. <u>carotovora</u>. Appl. Env. Microbiol. **49:** 714-17.

Index

323

324

333